OXFORD READINGS IN SOCIO-LEGAL STUDIES

A Reader on Administrative Law

OXFORD READINGS IN SOCIO-LEGAL STUDIES

Editorial Board
Convenor: MAVIS MACLEAN
DENIS GALLIGAN, HAZEL GENN, MIKE ADLER, ALAN PATERSON, TONY PROSSER, SALLY WHEELER

OTHER TITLES IN THIS SERIES

Criminal Justice
Edited by Nicola Lacey

Family Law
Edited by John Eekelaar and Mavis Maclean

The Law of the Business Enterprise
Edited by Sally Wheeler

Punishment
Edited by R. A. Duff and David Garland

A READER ON

Administrative Law

EDITED BY
D. J. Galligan

OXFORD UNIVERSITY PRESS
1996

Oxford University Press, Walton Street, Oxford OX2 6DP
Oxford New York
Athens Auckland Bangkok Bombay
Calcutta Cape Town Dar es Salaam Delhi
Florence Hong Kong Istanbul Karachi
Kuala Lumpur Madras Madrid Melbourne
Mexico City Nairobi Paris Singapore
Taipei Tokyo Toronto
and associated companies in
Berlin Ibadan

Oxford is a trade mark of Oxford University Press

Published in the United States
by Oxford University Press Inc., New York

© Oxford University Press 1996

All rights reserved. No part of this publication may be reproduced,
stored in a retrieval system, or transmitted, in any form or by any means,
without the prior permission in writing of Oxford University Press.
Within the UK, exceptions are allowed in respect of any fair dealing for the
purpose of research or private study, or criticism or review, as permitted
under the Copyright, Designs and Patents Act, 1988, or in the case of
reprographic reproduction in accordance with the terms of the licences
issued by the Copyright Licensing Agency. Enquiries concerning
reproduction outside these terms and in other countries should be
sent to the Rights Department, Oxford University Press,
at the address above

This book is sold subject to the condition that it shall not, by way
of trade or otherwise, be lent, re-sold, hired out or otherwise circulated
without the publisher's prior consent in any form of binding or cover
other than that in which it is published and without a similar condition
including this condition being imposed on the subsequent purchaser

British Library Cataloguing in Publication Data
Data available

Library of Congress Cataloging in Publication Data
Data available
ISBN 0–19–876408–1
ISBN 0–19–876409–X (Pbk)

1 3 5 7 9 10 8 6 4 2

Typeset by Hope Services (Abingdon) Ltd.
Printed in Great Britain
on acid-free paper by
Bookcraft Ltd., Midsomer Norton, Avon

Acknowledgements

Grateful acknowledgement is made to all the authors and publishers of extract material which appears in this book, and in particular to the following for permission to reprint material from the sources indicated:

Blackwell Publishers: Greer, P. 'The Next Steps Initiative: An Examination of the Agency Framework Documents' 70 Public Administration (1992) pp 89–98; Lloyd-Bostock, S. and Mulcahy, L. 16 Law and Policy (1994) pp 123–146.

Butterworth & Co.: Richardson, G. Law, Process and Custody (1993) pp 186–191, 241–247; Baldwin, R., and McCrudden, C. Regulation and Public Law (1987) pp 4–12, 325–331.

Genn, H., and Genn, Y. The Effectiveness of Representation Before Tribunals (1989) pp 63–4, 107–110.

Open University Press: Graham, C., and Prosser, T. Waiving the Rules (1988) pp 73–94; Harden, I., (ed) The Contracting State (1992) 14–28

Oxford University Press: Baldwin, R. Regulating the Airlines (1985) 186–204; Hawkins, K. (ed) The Uses of Discretion (1992) pp 19–44, 188–204, 213–230; Hawkins, K. Environment and Enforcement (1984) 23–36, 105–128; Craig, P. Public Law and Democracy (1990) pp 162–182; Hanna, J. 3 Medical Law Review (1995) pp 177–188; Peay, J. Tribunals on Trial (1989) pp 41–49, 64–71; Baldwin, J., Wikeley, N., and Young, R. Judging Social Security (1992) pp 31–64.

Public Law Project: Sunkin, M., Bridges, L., Mazeros, G. Judicial Review in Perspective (1993) 1–16, 60, 95–7.

Sweet and Maxwell Ltd.: Wikeley, N., and Young, R. Public Law (Summer 1992) 244–262; Black, J. Public Law (1995) 94–118; Cavadino, M. Public Law (1993) 323–345.

*Every effort has been made to contact copyright holders

Contents

Acknowledgements   v

D. J. GALLIGAN **Introduction: Socio-Legal Readings in Administrative Law**   1

THE STRUCTURE AND COMPOSITION OF ADMINISTRATIVE GOVERNMENT

C. GRAHAM and T. PROSSER **'"Rolling Back the Frontiers"? The Privatisation of State Enterprises'**   63
    From Graham and Prosser (eds.) *Waiving the Rules* (Open University Press, 1988), 73–94

J. HANNA **'Internal Resolution of N.H.S. Complaints'**   90
    (1995) 3 *Medical Law Review* 177–188

I. HARDEN **'The Contractual Approach to Public Services: Three Examples'**   103
    From Harden *The Contracting State* (Open University Press, 1992), 14–28

P. GREER **'The Next Steps Initiative: An Examination of the Agency Framework Documents'**   118
    (1992) 70 *Public Administration* 89–98

THE POLICY-MAKING PROCESS

R. BALDWIN **'Discretionary Justice and the Development of Policy'**   133
    From Baldwin *Regulating the Airlines* (Clarendon, Oxford, 1985), 186–204

R. BALDWIN and C. MCCRUDDEN **'Regulatory Agencies: An Introduction'; 'Conclusions: Regulation and Public Law'**   151
    From Baldwin and McCrudden (eds.) *Regulation and Public Law* (London, 1987), 4–12, 325–331

J. M. BLACK **'"Which Arrow?": Rule Type and Regulatory Policy'**   165
    (1995) *Public Law* 94–118

K. Hawkins **'Setting Standards'** 194
From Hawkins *Environment and Enforcement* (Oxford
Socio-Legal Studies, 1984), 23–36

M. Cavadino **'Commissions and Codes: A Case Study in Law
and Public Administration'** 210
(1992) *Public Law* 333–345

P. P. Craig **'Pluralism: UK'** 225
From Craig *Public Law and Democracy* (Clarendon, Oxford, 1990),
162–182

INDIVIDUALIZED DECISIONS AND PROCESSES

K. Hawkins **'Using Legal Discretion'** 247
From Hawkins (ed.) *The Uses of Discretion* (Oxford
Socio-Legal Studies, 1992), 19–44

D. J. Galligan **'Discretionary Powers in the Legal Order';
'The Exercise of Discretionary Power'** 274
From Galligan *Discretionary Powers*
(Clarendon, Oxford, 1986), 72–84, 128–140

K. Hawkins **'Compliance Strategy'** 299
From Hawkins *Environment and Enforcement* (Oxford
Socio-Legal Studies, 1984), 105–128

J. Baldwin, N. Wikeley, and R. Young **'Adjudication in Local Offices'** 326
From Baldwin, Wikeley and Young *Judging Social
Security* (Clarendon, Oxford, 1992), 31–64

G. Richardson **'Release From Prison'; 'Decision-Making
within Special Hospitals'** 357
From Richardson *Law, Process and Custody: Prisoners and Patients*
(London, 1993), 186–191, 241–247

R. Lempert **'Discretion in a Behavioral Perspective'** 371
From Hawkins (ed.) *The Uses of Discretion* (Oxford
Socio-Legal Studies, 1992), 108–204, 213–230

ACCOUNTABILITY, RECOURSE, AND LEGAL CONTROL

N. Wikeley and R. Young **'The Administration of Benefits
in Britain'** 409
(1993) *Public Law* 250–262

J. PEAY **'Praying Patience: The Patients' Perspectives';**
'The Responsible Medical Officers' 423
From Peay *Tribunals on Trial* (Oxford Socio-Legal Studies, 1992), 41–49, 64–71

S. LLOYD-BOSTOCK and L. MULCAHY **'The Social Psychology of Making and Responding to Hospital Complaints: An Account Model of Complaint Processes'** 438
(1994) 16 *Law and Policy* 123–147

H. GENN and Y. GENN **'Expectations and Experiences of Tribunal Hearings'** 466
From Genn and Genn *The Effectiveness of Representation Before Tribunals* (London, 1989), 63–64, 107–110, 219–226

M. SUNKIN, L. BRIDGES, and G. MAZEROS **'Changing Patterns in Use of Judicial Review'** 477
From Sunkin, Bridges and Mazeros *Judicial Review in Perspective* (London, 1993), 1–16, 60, 95–97

Introduction: Socio-Legal Readings in Administrative Law

D. J. GALLIGAN*

1. The socio-legal contribution

Administrative law lends itself easily and naturally to socio-legal analysis. Being at its most basic a body of rules about the way governmental powers should be exercised, administrative law invites three main questions: what kinds of authorities and institutions make up the administration and what sorts of powers do they have; where do the rules which they apply come from; and to what extent do those rules govern what really happens in every-day practice. These three basic questions mark out an enormous programme of research of a socio-legal kind; but to stop there would be to miss a fourth question which is equally fundamental. For while the first three can broadly be linked to the effective and efficient exercise of powers in order to achieve certain social ends, there is another dimension which concerns the fair treatment of those who are in some way caught up in administrative processes. Effectiveness and efficiency for the one part and fair treatment for the other are the twin components of good and legitimate administrative government. On the basis of this further dimension with which much of administrative law is concerned, a fourth question may be posed: to what extent do the rules and procedures of administrative law achieve the fair treatment of individuals. Here the emphasis is on ensuring that decisions are fairly made and on the remedies and forms of recourse open to an aggrieved party.

One difficulty which a book, avowedly prepared for use by students, faces is that most administrative law courses do not include a socio-legal component. Since the leading textbooks still concentrate on the doctrine of judicial review with minor chapters on such matters as rule-making, tribunals, and ombudsmen, it is probably safe to assume that the great majority of courses follow the same pattern. If this is so, then it is a challenge to persuade the teachers of the subject that a socio-legal approach

* Professor of Socio-Legal Studies in the Faculty of Law and Director of the Centre for Socio-Legal Studies, University of Oxford. The author wishes to acknowledge the helpful comments and suggestions made by N. Lacey and K. Lauer.

has something to offer. That might properly take an essay of its own, but for present purposes it will have to be enough to offer a few passing remarks on two matters: what it means to take a socio-legal approach and what such an approach has to contribute.

Whenever someone asks the 'what is socio-legal studies' question you can be sure you are in for a long journey at the end of which no tangible results are likely to be achieved. While a recent review by the Economic and Social Research Council offered criteria wide enough to include all but the narrowest analysis of legal doctrine,[1] some prefer to abandon the expression socio-legal in favour of the study of law and society, but whether that illuminates the debate is itself debatable. My predecessor at the Oxford Centre for Socio-Legal Studies, Donald Harris, liked to refer to the study of law from the perspective of the social sciences, and that I suggest is a significant step forward. It recognises that law and legal institutions are themselves part of the social context and, like any other part, can be studied from the point of view of sociology, psychology, anthropology and economics. This multi-disciplinary approach, which may include elements of an inter-disciplinary kind, but without that being its main aim, marks the genuinely pioneering work of the Oxford Centre in its studies of compensation, environmental law, and family policy. Studies like these help in our understanding of the administrative world and provide the basis for theoretical generalization about that world. Whether it is possible to move beyond particular disciplines and localized fields of enquiry to create a unified theory of social action is a matter of interesting debate.[2]

Now is not the time to enter into that wider issue; rather, I wish to suggest that there are two quite straightforward levels of socio-legal analysis and that both have a great deal to offer in our understanding of administrative law. At the first level, the object is to discover what really happens in the world; it is concerned to find the facts and to demonstrate that what occurs in practice is not always as simple or predictable as we might assume. At their simplest, studies at this level provide information about the law: that most victims of accidents receive no compensation;[3] that a tiny percentage of first level administrative decisions are subject to any form of review or appeal;[4] that compliance with standards controlling

---

[1] *Review of Socio-Legal Studies* (ESRC, 1994).

[2] For discussion of this issue, see R. Cooter, 'Law and Unified Social Theory' in D. J. Galligan (ed.), *Socio-Legal Studies in Context: The Oxford Centre Past and Future* (Blackwell, 1995).

[3] See D. Harris and others *Compensation and Support for Illness and Injury* (Clarendon, Oxford Socio-Legal Studies, 1984).

[4] See J. Baldwin, N. Wikeley and R. Young, *Judging Social Security* (Clarendon, Oxford 1993).

pollution is achieved less by the threat of prosecution than by bargaining and negotiation between inspectors and polluters.[5] Through pioneering studies like these, we now have available elementary, but not easily acquired, information about a range of legal phenomena.

To know what happens in practice can be both complicated and hard to ascertain. For example, the way in which decisions are made within an administrative organization, the standards being applied, the variables in operation, and the influences upon them, can be enormously complex. The object is still to determine facts about the world, but the facts themselves are no longer so clear cut and hard-edged: they now depend on understanding human practices where the quicksands of intention and the nuances of motive become critical data. Determining such facts also involves capturing a sense of order and coherence where none is obvious, and interpreting social situations which appear varied and changeable. It is here that the disciplines of social science have their parts to play; for as we move from hard and simple facts to soft and complex ones, the dash of common sense and diligence which is enough to discover the former is no longer adequate for the latter. The precise qualities which the disciplines of social science bring are the means to penetrate and comprehend social reality at levels which are beyond simple common sense. Each provides an additional way of looking at human actions and social situations, and of unlocking facts about them which would otherwise be concealed. Therefore, to study law from the social sciences is not fundamentally different from the untutored observations of the layman; drawing on traditions of observation and generalization, it is just to delve more deeply into facts and realities. Knowledge in such contexts is inherently imperfect and our understanding of the facts accordingly contingent. But whatever the difficulties of observation, the limitations of comprehension, and the arcane exclusivity of disciplinary traditions, the quest is still for knowledge of what happens in practice, for comprehension of the way legal rules and institutions function in their social context. My suggestion is that socio-legal research is first and foremost about ascertaining the facts which constitute legal phenomena, whether they be simple facts or facts of the kind that are revealed only through the combination of deep empirical research, the insights of social science, and careful interpretation and reconstruction.

Important as this first level is, there is, however, a second level of socio-legal research. Here the object is to move beyond local facts about this or

---

[5] G. Richardson and others, *Policing Pollution* (Oxford Socio-Legal Studies, 1984) and K. Hawkins, *Environment and Enforcement* (Oxford Socio-Legal Studies, 1984).

that area of law to a more general understanding of recurring phenomena. Generalization can be conducted at various levels of complexity. A study of people appealing to a number of tribunals has shown that they have a better chance of winning if legally represented than if not.[6] On the grounds that the reasons for this apply not just to the few tribunals studied but to tribunals generally, we may justifiably adopt the general hypothesis that legal representation before any tribunal is likely to improve the appellant's chances of success. This is a general hypothesis, for which there is abundant evidence, about tribunals which, although to a degree contingent on particular circumstances, adds to our knowledge about how tribunals in general work. Another example can be drawn from studies of the control of environmental pollution. Here the process of generalization is more complex, but it may be possible, first, to draw general conclusions about the inevitable reliance on bargaining in securing compliance with anti-pollution standards, and, secondly, to extrapolate beyond that context to the enforcement of standards in other areas. Certain conditions could be identified under which elements of negotiation are bound to creep in; these can be complex and their identification might take a lot of careful gathering of data and its painstaking analysis in specific contexts. But a point may come where generalizations can be postulated: to make such postulations is simply to assert certain facts about enforcement processes. Now of course different methods can be used in order to arrive at generalizations. Recurring patterns might be detected through statistical analysis, for example, or alternatively by close study of a few instances from which certain features can be identified as of special significance. However, I shall not here enter into a discussion of these and other methodologies except to note that the study of such matters and the understanding of the strengths and weaknesses of different methods is an important part of socio-legal analysis.

To describe the process of generalization is to describe the process of theorizing. Theory consists in being able to move from the facts particular in one situation to a general claim that the same facts will hold true in other situations. An extensive study of judicial review might enable us to theorize about the conditions under which it will be seen by aggrieved parties as a desirable and viable remedy. A different kind of study might provide a basis for theorizing about the impact of judicial review on administrative bodies. There is nothing mysterious here, nothing to be frightened of, just an attempt to draw general conclusions about aspects

[6] H. & Y. Genn, *The Effectiveness of Representation Before Tribunals* (Lord Chancellor's Department, 1989).

of law and practice from our knowledge of particular instances of them. The gathering and interpreting of information, the conduct of empirical research, is necessarily the basis for generalization. This may seem a low level view of theory, a view which hardly does justice to the efforts of the great theorists Marx and Weber for an older generation, Habermas and Luhmann for a younger. In truth there is no conflict; theory, like generalization, occurs at different levels of abstraction and according to different degrees of complexity. But any theory, no matter how abstract or complex, no matter how removed from the pedestrian events of daily life, is still a claim to knowledge about daily life. Beneath the camouflage is an assertion about the world which is true or false. This might not always be obvious of high theory; indeed some theory could be criticized as no more than speculation without any grounding in social reality. Such criticism is not a decisive argument against high theory since speculation is still a claim to knowledge about the world which might be true or false. The real difficulty is that such theories are neither grounded in nor appear to be concerned about empirical data as to what really does happen in the world. Theories based on speculation are perfectly legitimate and sometimes interesting; but the best theory, whether high or low, is firmly rooted in the knowledge and understanding of what happens in the world.

The process of theorizing is, of course, rather more complex than such a cursory account suggests. Space does not here permit anything like a full analysis of the complexities, but two brief points should be noted. One is that any descriptive theory about the social world is likely to reflect and be shaped by certain understandings and assumptions about the social world in general and even about the very specific part of the social world being considered. The need constantly to examine and re-examine our pre-theoretical assumptions is a necessary step towards objectivity, but just how far we can ever succeed in being objective is itself of theoretical interest. The other point to note concerns the relationship between descriptive theory and normative theory. While descriptive theory purports to tell us what actually happens in the social world, normative theory is about the ends and purposes to which we believe the social world should be directed. Laws generally represent the conclusion of some normative theory, while the extent to which those laws are applied in practice is an element of descriptive theory. But again the relationship between the two is more intricate and problematical that such a simple example suggests. The analysis of that relationship requires an essay of its own, but the point should be made that the relationship is vital to the development of socio-legal studies and should not be ignored in the way that it tends to be.

The discussion may appear to have wandered away from the statement of what socio-legal research has to offer administrative law. Let us, therefore, draw from this discussion precisely what a socio-legal analysis of a subject seeks to do. First, the object is to provide knowledge about administrative bodies and processes: their structure and organization, how they work in practice, how decisions are made and standards formulated, the effect of legal rules and doctrines on them, and the nature and effectiveness of methods of regulation, control, and recourse. Secondly, the methodology for achieving that purpose is twofold: to gather empirical data on aspects of the administrative process, then on the basis of that data to theorize in the sense of formulating general hypotheses about administrative bodies and processes. If this is what the socio-legal study of administrative law has to offer, then the next question to consider is whether its offering is worth the trouble. It is one thing to show what a socio-legal approach consists of, but another thing to persuade those engaged in studying and writing about the doctrines of administrative law why these issues should interest them. Interest them, indeed, to the extent that they would want to include such an approach in the teaching of the subject. It is worth noting a number of reasons why the socio-legal approach has an important contribution to make. I wish to suggest three reasons: first, the socio-legal approach contributes to knowledge about administrative law and process; secondly it may provide guidance in changes and reforms; and thirdly, it can be linked to the protection of rights.

The first needs little explanation. The claim is that knowing how laws, processes, and institutions work in practice is worthwhile for its own sake. Knowledge and the better understanding that comes from it about the social context of law can be valued without its being instrumental to something further. One reason for valuing socio-legal research in this context is that it increases our knowledge. Facts about how often applications for judicial review are made, what sorts of people make them, and how often they succeed, are facts which simply increase our knowledge of judicial review. But such facts can be more than just additions to our stock of information about the issue: they might also deepen our knowledge. Knowledge about the impact judicial review has on administrative bodies, a matter very much one of socio-legal enquiry, deepens our understanding of an issue already closely studied.

The second reason for valuing socio-legal work in administrative law is explicitly instrumental. A better understanding of how law works in its social setting may enable improvements to be made at the institutional and legal levels. For example, research into the psychology of people in lodging complaints about the treatment they have received from an administrative body may have direct implications for the way complaints are dealt with.[7] An elaborate system of tribunals, procedures, and remedies would be wasted if all that is sought by most complainants is a letter of acknowledgment and perhaps an apology. Similarly, from the discovery that Mental Health Review Tribunals invariably accept the recommendation of the doctor responsible for the patient's treatment as to whether the patient should be released, we might conclude that the tribunals are not exercising the independent review required under statute.[8] The ground would then be laid for re-considering the composition of the tribunal and its procedures. Examples like these could be multiplied, the simple point being that while much socio-legal research is not driven by instrumental, policy-based concerns, much of it can be of use for that purpose.

The third reason for taking account of socio-legal research in this area is related but adds an important dimension to the second. Here we may take up the earlier distinction between the public interest in achieving the objects of administrative law and the fair treatment of people affected by it. There is a clear public interest in administrative institutions achieving their objectives effectively and economically, in decisions being made accurately, and in forms of recourse working properly. Socio-legal research can be usefully employed in advancing those ends. But administrative processes also affect individual persons and their rights. A mistaken decision denying that person a benefit or advantage to which he or she is entitled is more than a margin of detriment to the public interest; it is also unfair treatment of the person and deprivation of their rights. So, the high level of mistakes in social welfare decisions, for which there is now convincing evidence, means that many applicants are wrongly being denied their rights. This dimension of fairness and rights adds a moral imperative for improving procedures; socio-legal research can sometimes provide guidance in doing so.

Against this background, we can now consider the four areas of special interest for socio-legal research. Those four areas of questions provide a

---

[7] For studies of psychology of complaining see S. Lloyd-Bostock and L. Mulcahy, 'The Social Psychology of Responding to Hospital Complaints: an Account Model of Complaints Procedures' (1994) 16 *Law and Policy* 123.

[8] For the most helpful study in this area, see J. Peay, *Tribunals on Trial* (Oxford, 1989).

framework for the socio-legal analysis of administrative law which is followed in the rest of this introduction and the readings that are included here. The object is to bring together in an accessible form a selection of research and writing on these issues. Some emphasis should be put on the word selection, for in the limited number of pages available, it is possible to present only examples of socio-legal work without attempting to be comprehensive. Since the selection of examples is a highly subjective matter, it may be wise to mention the main factors guiding my choice. As this is a collection of readings for students of administrative law, one of my aims has been to include materials which are interesting, not too difficult, and easily linked to mainstream administrative law. Another aim has been to emphasize contemporary materials, so that everything included has been published within the last 10 years. This has meant excluding some of the classics, but since the subject is only getting into its stride, there are not many in that class anyhow. The final consideration is that in the absence of a comprehensive bibliography of socio-legal work in the field, there is a risk that something important will have been overlooked.

2. The changing face of administrative government

(i) *The general nature of administrative government*

Anyone attempting to describe the system of administrative government in the UK is confronted with a bewildering array of departments, agencies, public corporations, regulatory bodies, and inspectorates—to name but some of them. The idea that ministers and their advisors develop policies, that those policies wind their way through the legislative process to emerge eventually as statutes, and that those statutes then go back to ministers and the departments for implementation, for which the minister is then answerable to Parliament, is often taken as expressing the heart of British government and administration. Such an idea is of course to some degree accurate, but it has never been completely accurate and just how accurate it is at any time is likely to fluctuate. English administration has always been a mixture of central departments and special boards, offices, and commissions, some created by statute, others operating under prerogative. Government may historically have been small and restrained, but many of the matters we regulate today have been the subject of rudimentary regulation for centuries. The nineteenth century itself was notable initially for the proliferation of statutory bodies, inspectorates, and commissions, sometimes with scandalously wide and unregulated powers, often with neither accountability to Parliament nor ministerial supervision. By the end of the century, however, central departments with ministerial accountability to Parliament were in the ascendancy, to become in the twentieth century the paradigm of good government. With a staff of well-educated, professional civil servants, a minister in charge and through him a direct line of accountability to Parliament, the perfect system of government and administration seemed to have emerged.

Even in its heyday, however, such arrangements constituted only part of the overall system with special bodies, boards, and authorities occupying a substantial part. The truth is that the system of administrative government in the UK has long been varied, unpatterned, and changeable. It should not then be too surprising that, as the end of the century approaches, further shifts and changes are under way. Whether or not those are more mere shifts and changes rather than something fundamental and permanent, they certainly reshape important parts of the administrative landscape. In this section, the object will be to sketch the main features of the new landscape, to identify the ideas behind them, and to draw out the implications for administrative law.

The recent changes in administrative government are broadly fourfold: a fragmentation of central government, a contraction of the administration, the infusion of private sector notions into the public sector, and a proliferation of supervisory bodies and forms of recourse.[9] The first of these, the fragmentation of central government, consists in breaking up departments into special agencies, to carry out administrative tasks previously done within the departments.[10] Each Next Steps Agency, as they are called, has a framework document which defines its objectives, its relations with the department, the minister, and Parliament, and a range of other matters relating to finance, personnel, and performance.[11] The agencies vary greatly across such matters as the distribution of

---

[9] Generally on the shape of modern administrative government see I. Harden and N. Lewis, *The Noble Lie* (Hutchinson, 1986).

[10] On the fragmentation of central government, see further Efficiency Unit, *Making the most of Next Steps* (HMSO, 1991); N. Lewis, 'Change in Government: New Public Management and Next Steps' (1994) *Public Law* 105; G. Fry, A. Flynn, A. Genn, W. Jenkins and B. Rutherford, 'Symposium on Improving Management in Government' (1980) 66 *Public Administration* 429.

[11] The article in this volume by P. Greer contains a helpful analysis of the variety of agencies and their framework documents: P. Greer, 'The Next Steps Initiative: An Examination of The Agency Framework Document' (1990) *Public Administration* 89; see also P. Kemp, *Beyond Next Steps* (1993). On privatization: C. Graham and T. Prosser, 'Rolling Back the Frontiers: Privatization of State Enterprises' in C. Graham and T. Prosser (eds.), *Waiving the Rules* (Open University Press, 1988); see also T. Prosser, *Nationalized Industries*.

welfare, the provision of public services, and the collection of payments for child support. The basic idea is that each agency has well-defined aims which it must achieve and for which typically it is responsible to the minister. Those aims should be confined to implementing policy not making it; policy matters are for ministers, putting them into operation for the agencies. The ethos of the agency is good management, and efficiency and effectiveness in realizing its aims. The stated aim of the government is that the greater part of administration, together with most civil servants, will be hived-off to agencies, leaving only the barest departmental shells. At the time of writing more than sixty agencies have been formed with many more to follow.

The second major change, the contraction of the public sector and the consequential reduction of administrative government, has several parts: the privatization of public industries, the conversion of public agencies into private bodies, and the transfer of public services to the private sector. The privatization of public industries consists of converting the old public corporations relating to such matters as power, communications, and transport, into private corporations, with shares to be held privately or to be bought and sold on the stock market. As public bodies working within a statutory framework, such industries were subject to ministerial control and accountability and often had to tailor their essentially commercial activities to bureaucratic constraints. With privatization, such controls and constraints are replaced by the canons of commerce and the power of shareholders. However, the public interest in the activities of these industries remains strong. Since the industries often have a monopoly, they are not constrained by the forces of the market. In addition, there is a continuing need to ensure that essential industries and services are conducted according to the public interest. For these reasons political and administrative checks are not relinquished entirely, and one fairly standard form of control is to create by statute special regulatory bodies with certain supervisory powers, particularly in relation to the prices charged for goods and services.[12]

The other two aspects of the reduction of the public sector can be linked to the process just described. The transfer of certain public services to private bodies is well under way. Here the idea is that an administrative body, whether a department, local authority, or other agency, enters into a contract with a private firm or company under which the latter will perform a certain service. The administrative body decides any policy

[12] For other forms of continuing governmental control, see Graham and Prosser, 'Rolling Back the Frontiers' (extracts included in this work).

issues in relation to the service, but its implementation is contracted-out.[13] In one sense this process is not new since public authorities have regularly entered into contracts with private firms for the performance of one task or another.[14] What is new is the drive to contract-out tasks which have long been regarded as properly performed by public bodies. The running of prisons is a good example; traditionally considered a public service, it is now to be contracted to private bodies. Other examples are aspects of police work, services provided by local authorities, and parts of the court system. The contracting-out process is still in its early stages, but once the general policy is adopted that much of what is now done by public bodies has no specifically public character and could be done equally well by private bodies, then the scope for contracting-out is vast. It would then be only a short step to replace many administrative agencies with private organizations. The fragmentation of central government, the creation of large numbers of Next Steps Agencies, and the strong ideological commitment to privatizing the public sector have together created a logic and a momentum which could prove hard to resist. Take as an example the Child Support Agency which is responsible for collecting money from fathers for the support of children from an earlier marriage or relationship. A government committed to reducing the public sector might well consider that such a task could well be done by a private firm, a debt collection agency for example. Once that commitment is made, as it appears to have been by the present government, it is difficult to see where the process will stop.

An alternative to contracting-out public services to the private sector is to introduce contractual ideas into the public sector itself.[15] This process which we may call contracting-in, is the third major shift in the nature of administrative government. The basic idea is to create contractual relations between different parts of an administrative authority for the delivery of a public service. One example relates to Next Steps Agencies; the chief executive of each enters into a contractual arrangement with the Minister and department which sets out the duties of the agency in providing a service. Another, perhaps more notable example, is the National Health Service. Here various bodies within the service enter into contracts with each other for the provision of health care. Some parts of the

---

[13] A good analysis of the several forms of contracting-out is I. Harden, *The Contracting State* (Open University Press, 1992) (extracts included here).

[14] On the use of governmental contracting power, see T. Daintith, 'The Executive Poser Today; Bargaining and Economic Control' in J. Jowell and D. Oliver (eds.), *The Changing Constitution* (2nd ed., Clarendon Press, 1989).

[15] See I. Harden, *The Contracting State*, (n. 13 above) ch. 3.

service, such as hospitals, provide health care, other parts, such as district health authorities or general practitioners, buy health care. The idea is that the purchasers of health care will be able to buy from different providers and that the former will naturally seek services at the best price available. This will create a sense of competition amongst providers, a kind of internal market. The hope is that this will lead to the more efficient and cost-effective provision of medical services. This process of contracting-in is of course very different from contracting-out, although the commitment of each to a contractual foundation creates a natural affinity between them. The driving force is also similar in both cases. It is to entrust the provision of a distinct public service to a distinct body, whether public or private, where the terms and conditions, the levels of performance, and the criteria of success, are spelt out in a contract-like document. Contracting-in, like contracting-out, is potentially of enormous scope. The main experiment so far has been the NHS followed by Next Steps Agencies, but consideration has been given to applying contracting-in to other areas of administration, such as the court system, legal aid, and the prisons.

The fourth and final feature of significance in administrative government is the proliferation of supervisory bodies. By this is meant the creation of one range of authorities the specific functions of which are to scrutinize and review the activities of other bodies. This development is not as new or dramatic as the other three, but the suggestion is that in recent years a great emphasis has been put on creating supervisory bodies and providing the substantial resources needed for their operation. They are of various kinds. Some are directed to the complaints or appeals of individual persons affected by an administrative process. The two notable cases are tribunals and ombudsmen. Since the Franks Committee Report in 1948, the idea has spread that a person subject to an administrative process should have the opportunity to appeal on the merits of an independent tribunal. There are still many exceptions, not least in the areas of deportation and homelessness, but the general idea is gradually finding its way into the most fusty corners of administration. The ombudsman idea, that a person should be able to have a complaint about his or her treatment by an administrative body investigated by an independent authority, had developed in a similar way. Since the Parliamentary Commission for Administration was set up in 1967 to investigate complaints about central government, we have seen the creation of ombudsmen for local government, the health service, the police, the inland revenue, and most recently the prisons.[16]

[16] It is interesting to note that the ombudsman idea has also swept areas of the private sector including banking, financial services, and building societies.

A different type of supervisory body is concerned not with the complaints of individuals but with the effective and efficient working of the administrative authority. Here again several types can be seen. One is the inspectorate: the authority whose duty it is to examine the activities of administrative bodies to ensure that they are properly discharging their responsibilities. Inspectorates are an old form of supervisory body reaching back to the early nineteenth century. Amongst the many areas now covered by inspectorates are schools, prisons, and mental hospitals. It is common to find new specialized inspectorates being created to supervise very specific matters. Another type of supervisory body is specially concerned with the financial management of an administrative body, and here the outstanding example is the Audit Commission.

(ii) Underlying ideas

The changes and trends described above are the results of a number of ideas which have become prominent in the UK, but which possibly tap deeper currents in Western societies than their local expression suggests. Amongst the many ideas that could be shown to have some connection, three are especially prominent: new managerialism, privatization, and the protection of citizens.

New managerialism has been much discussed and written about in recent years. It is a loose term for a set of related ideas.[17] First and perhaps most important is the notion that administrative authorities should be committed to active and efficient management. That means having clear objectives and then achieving them in the most cost-effective and economical way possible. Secondly, a distinction should be drawn between making policy and implementing it. Policy issues properly belong to ministers who are accountable through the political process, while the implementation of policy should be governed by principles of good management. Thirdly, good management is generally best achieved by small administrative units rather than larger central departments. Fourthly, administrative bodies charged with the implementation of policy should have not only a clear statement of objectives, but also precise targets or performance indicators. Fifthly, the basis of accountability for such authorities is partly to Parliament or a minister in the traditional

---

[17] From a vast literature, the following are especially recommended: C. Wood, 'A Public Management for All Seasons' (1991) 69 *Public Administration* 3; C. Pollit, *Managerialism and the Public Services: The Anglo-American Experience* (Blackwell, 1990) and S. Zifcak, *The New Managerialism* (Open University Press, 1994).

way, but is also partly internal in that it should be judged according to principles of good management.

The influence of these ideas can be seen at many points in contemporary British government.[18] In common with other countries, the public sector in Britain has come to be seen as unwieldy, inefficient, and a significant consumer of resources. The Financial Management Initiative taken by the Prime Minister in the early 1980s promoted the notion of good management within the administration as a policy objective in its own right. The idea of accountability through good management became a guiding principle for the reform of administration and had a major impact on the civil service. The recommendations of the Efficiency Unit in 1988 to create Next Steps Agencies were similarly based on the commitment to new managerialism. The idea lies behind so many other acts and reforms of government in recent years that it may not be an exaggeration to consider it the most influential principle in contemporary administrative government.

If new managerialism is the dominant idea, privatization is close behind. The earlier discussion of changes in administrative government shows how much they are driven by a commitment to privatization.[19] The idea is simple: if an aspect of the public sector can be performed better or even as well by the private sector, then it should be transferred to the private sector. The grounds might be that the private sector, governed by commercial considerations and by market forces (where there is a market), will be more economical and cost-effective in providing the service. The merits of these arguments are arguable; but whether convincing or not, the commitment to privatization is not entirely dependent on its merits but also draws on an ideology which regards privatization as a good thing. We saw above the different ways in which privatization can manifest itself in administrative government. In its clearest and strongest sense it means transferring public services to private bodies to be conducted on essentially commercial lines, subject to residual controls from ministers and regulatory agencies. In a secondary and less direct sense, it involves introducing commercial norms and private law notions of contract into public authorities. The lack of a real market and a commercial

---

[18] For an interesting analysis of the influence of new managerialism on the Royal Commission in Criminal Justice, see N. Lacey, 'Missing the Wood . . . Pragmatism versus Theory in the Royal Commission' in M. McConville and L. Bridges (eds.), *Criminal Justice in Crisis* (E. Elgar, 1994) and N. Lacey, 'Government as Manager, Citizen as Consumer: The Case of the Criminal Justice Act 1991' (1994) *Modern L. Rev.* 534.

[19] For a useful review of existing literature, see D. Marsh, 'Privatization Under Mrs Thatcher: A Review of the Literature' (1991) 69 *Public Administration* 459.

environment can be partly overcome by creating them artificially and notionally, the underlying policy being that the insertion of a private sector ethos will have beneficial effects on the performance of public bodies; a public sector infused with private sector ideas, so its proponents believe, will necessarily be to the good.

The third idea behind the transformation of administrative government is the protection of citizens. Some might prefer the more evocative expression the empowerment of citizens but, however put, the idea is that since people are often the recipients, subjects, or even victims of administrative decisions and processes, they ought to have open to them the means for making sure they are properly treated, and, if not, then the opportunity to seek recourse. This fits the distinction made earlier between good administration as part of the common good and as the fair and proper treatment of the individual person. The distinction is of great importance, although it would be more accurate to say that the common good in sound administration includes the fair and proper treatment of individuals.

The concern to protect citizens in these ways lies behind many of the changes and developments of recent years, including: the extension of a right of appeal to tribunals and the work of the Council on Tribunals; the proliferation of ombudsmen; the placing of great emphasis on complaints procedures; and even the increased vitality of judicial review. One direct manifestation of the concern for greater protection of citizens is the *Citizens' Charter*[20]. Proclaimed by the government in 1991, the Charter has two main themes: one that central departments and other public authorities should publish the standards of performance that the members of the public may expect; the other that every authority should establish procedures for dealing with citizens' complaints when the standards are not met. The Charter does not link the new complaints procedures to the existing framework of recourse, but presumably the intention is for the former to augment the latter. As a result of the Charter, departments and administrative authorities at both central and local levels have begun to set and publicize standards of performance and to put in place procedures for complaints.

Another example of the protection of citizens principle is the work of the Council on Tribunals in monitoring the work of tribunals and tribunal-like bodies. Created in the aftermath of the Franks Committee,

[20] *Citizens Charter* (HMSO, 1991). For commentary see Select Committee Report on the PCA, *The Implications of the Citizens' Charter for the Work of the PCA* (HMSO, 1992) and A. Barron and C. Scott, 'The Citizens' Charter Programme' (1992) 55 *Modern L. Rev.* 526.

the Council's main task has been to scrutinize the performance of tribunals and make recommendations in relation to their procedures. Its approach of quietly chipping away at procedural issues has been criticized for a lack of dynamism, but considering the severe limit to its powers and resources, it has had a notable influence in improving the standards of procedural fairness in tribunals. Indeed, the recent formulation of a model of procedures for tribunals might become a milestone in the development of administrative due process with an impact extending beyond tribunals. To these two examples of the vitality of the protection of citizens principle, others could be added, particularly the ombudsman which has become an increasingly common form of recourse. We may conclude, nevertheless, that despite the vitality of the principle, reforms expressing it are neither co-ordinated nor systematic; they respond as much to political motives as to a real concern to protect citizens in the face of government. However, ideas launched and institutions created acquire a life of their own and may develop in ways which confound their origins. Whether this will mean that the protection principle really is promoted, remains uncertain; but one thing that is certain is that the scope for socio-legal research into such innovations is extensive and the need for it pressing.

Having chartered some of the features of the new administrative landscape, I shall conclude this section by drawing out the implications for administrative law. This requires an essay in itself, but at least some of the more obvious points can be noted. The first is the effect on ministerial responsibility, on the idea that the primary basis of accountability for administrative action is through the minister to Parliament. While the shortcomings of that principle have long been recognized, the recent changes leave no doubt that it is to be regarded as one basis of accountability amongst several. The proliferation of agencies, the separation of policy and its implementation, the setting of performance targets, and the absorption of an ethos drawn from accounting and commerce, all bear witness to a new administrative autonomy and rationality. Ministerial responsibility for high policy remains, and in practice ministers will continue to have general responsibility for administration; as the predominant form of accountability, however, its light has faded.[21]

This leads onto a second consequence, namely, the new bases of accountability and their consequences. Performance indicators and contractual terms now provide the tests of good administration. Those tests will be applied, moreover, only partly by Parliament and ministers, with

[21] For further analysis see D. Oliver, 'Law, Politics, and Public Accountability' (1994) *Public Law* 238.

other boards, inspectorates and commissions, themselves more-or-less independent statutory bodies, now having a major part in scrutiny and supervision. All this takes place within a statutory framework and Parliament can always have the final say in stipulating aims and objectives and in fixing standards. In practice, the statutory framework is likely to consist in fixing the parameters, leaving what happens within them for settlement by lesser administrative agencies. The direct line between statute and administrative action is fractured, for in between are inserted targets and conditions, partly at the discretion of intermediate administrative bodies and partly the products of managerialist notions. One victim of the new dispensation may be the idea that good administration consists in the informed and balanced judgement of professional civil servants. Professional judgement raises its own problems, but its considerable influence as the cornerstone of the civil service no longer holds sway in the face of objectives based more on quantity, speed, and economy, than quality and accuracy. These are just some of the issues in the new accountability which in the future will be a fertile ground for socio-legal research.

A third point of interest is how administrative law will deal with notions and practices which occur within the new administrative landscape. I shall here do little more than list some of the issues arising. One relates to the Next Steps Agencies: what is their legal status; to what extent will their decisions be subject to judicial review; will there be problems about the division between policy and operation; what effect will the performance targets and quasi-contracts within the agency have on questions of judicial review. Another set of issues relates to the prevalence of contracts and quasi-contracts within the administration generally. We have seen some of the contexts; contracts between administrative authorities and private firms for the performance of services; contracts between the chief executives of Next Steps Agencies and ministers regarding objectives and targets; contracts between departments or ministers and other public agencies for the delivery of services; and contracts or notional contracts between bodies and agencies within a defined area, as in the case of the NHS. Again, several issues arise about the status and enforceability of such contracts, whether real or factitious. Issues also arise about whether the contracting process and the terms of the contracts are subject to legal scrutiny, judicial review in particular. In dealing with these and related issues, administrative law principles will have to be created anew or extrapolated from the rather lean offerings now available.[22] These will be

[22] For an analysis of some of these issues see M. R. Freedland, 'Government by Contract and Public Law' (1994) *Public Law* 1986.

interesting and challenging matters which would benefit from the deeper understanding that socio-legal research can encourage.

3. The policy-making process

The making of policy occurs at many points in an administrative system. At the highest levels it is made by Cabinet and becomes legislation; at lower levels it is made by ministers and departments, by statutory agencies and regulatory bodies. The object at these lower levels may be to settle a specific issue, such as where to allow a major development or the siting of an airport; but also and more often its object is to make subordinate legislation or statutory rules, or to settle on standards and guidelines of the more informal and sometimes unofficial kind. Elements of policy-making are pervasive throughout the administrative system, whatever the nature of the authority or no matter how lowly the official. Attempts are sometimes made to distinguish policy-making from policy implementation, as in the case of the Next Steps Agencies, but it is well known that bodies applying even the clearest rules often build around them a web of discretion and the policy element which inevitably comes with it. The idea that because of their political accountability, only ministers should make policy has long been abandoned. There are often good reasons, political, organizational, and as to effectiveness, for delegating aspects of policy to statutory bodies and officials. A certain distance from the political process can itself be an advantage which, together with the need for expert or specialist knowledge, often lies behind the vesting of agencies and authorities with policy-making powers.

What then are the specifically socio-legal issues in relation to policy-making? Policy-making is so clearly at the crossroads of political science, economic analysis, and administrative theory, that it lends itself naturally to analysis from different stand-points, any one of which could have socio-legal interest. Here I shall mention the four issues which appear specially significant: the different types of policy-making and policy-making bodies; the procedures for policy-making; the basis for participation in policy-making by groups, interests, and individuals; and the influences on policy. The aim here will be to single out some of the issues arising under each of these headings and to note the contribution socio-legal research has made to them.

(i) *Policies, rules, and policy-making bodies*
An initial division can be drawn between making a policy decision in order to settle a specific issue and doing so in order to create general stan-

dards for future application. In the case of the first, the context might be a general enquiry into an issue, including the gathering and testing of evidence in order to resolve an issue of high policy. Examples include where to site a new road or airport or whether to develop a nuclear energy policy. Issues of general policy might also occur in the course of an individualized decision in that it is about how a person should be treated. An example would be a licensing decision in areas such as civil aviation or broadcasting; in making the decision whether to grant a licence, matters of high policy may have to be determined. It is also the case that many individualized processes involve low level policy, but here I shall concentrate on what we might call high policy, the suggestion being that the subject matter has wide ranging significance for the community collectively.

The second category noted above involves policy-making in order to create general standards for future application. This might range from formal and binding delegated legislation to the most informal and advisory guidelines. This branch of policy-making is often referred to as rule-making, as if the idea were always to formulate strict rules. The terminology is well established, but it must be kept in mind that it refers, somewhat misleadingly, to the whole range of general standards, whether rules or something lesser. Another term which is perhaps more familiar in the UK is quasi-legislation, indicating that the creation of standards is a legislative function, although not legislative in the fullest sense, considering that many such standards lack the force of binding rules. This legislative, rule-making process may be expressly authorized by Parliament in the grant of statutory powers or it may be implicit in the exercise by the agency of its discretionary authority. The Parole Board, for example, has no express power to make rules, but in the exercise of its wide discretion in recommending the release of prisoners, it formulates general standards. A statutory grant of express rule-making power is likely to provide for the procedures to be followed, while the procedures for setting standards in the exercise of discretion are typically left for the authority itself.

The rule-making process has attracted much attention in recent years. One issue of interest has been the different kinds of standards which may be set. K. C. Davis was one of the early writers to advocate the structuring of discretion through the setting of general standards: guidelines, policy statements, directives, etc.[23] Other writers, stimulated by Davis' ideas, have attempted to give a fuller analysis of the types of standards; amongst

---

[23] K. C. Davies, *Discretionary Justice* (Louisiana, 1964).

them R. Baldwin and J. Houghton offer the most comprehensive,[24] while the work of C. Driver has also been important in trying to match different kinds of standards to different tasks and contexts.[25] Writers like these have shown that the rules in rule-making are of many different kinds and shades. Their work has also been significant in showing that the different kinds of standards can have distinct and specific functions according to their context; sometimes fairly strict rules are appropriate, while at other times a set of guidelines would be more suitable. The essay by Julia Black included in this volume is an original and up-to-date analysis of the different forms of rules and standards.[26]

As for the different kinds of policy-making bodies, little need be said here beyond drawing attention to their great variety. Ministers of State are of course at the centre of the policy process, and much rule-making is carried out by them and their departments; but just as policy-making occurs at all points and levels of administration, so does rule-making. Of special interest are the statutory agencies and authorities, such as the Gaming Board, The Health and Safety Executive, and the Office of Fair Trading, which have clear policy-making functions, sometimes with ministerial supervision and control, at other times with substantial autonomy. Case studies of some of these agencies are now available, but there is still ample scope for further work in analysing their policy-making role. Amongst the most recent and interesting studies are R. Baldwin's detailed analysis of the Civil Aviation Authority,[27] the case studies of several agencies brought together in R. Baldwin's and C. McCrudden's *Regulation and Public Law*,[28] and J. Black's analysis of agencies in the financial services industry.[29] These and other studies throw light on a number of issues: the different kinds of policy-making bodies; the relationship between the nature of the authority and the nature of its particular policy tasks; the influence exerted in shaping the agency by such factors as whether policy is to be made by negotiation and agreement or by enforcing general standards; and the relationship between the agency and government.

---

[24] See R. Baldwin and R. Houghton, 'Circular Arguments: the Status and Legitimacy of Administrative Rules' (1986) *Public Law* 239. See also R. Baldwin, *Rules and Government* (Clarendon Press, 1995).

[25] C. Driver, 'Policy-Making Paradigms in Administrative Law' (1981) 95 *Harvard L. Rev* 393.

[26] J. Black, 'Which Arrow?: Rule Type and Regulatory Policy' (extracts included in this work).

[27] R. Baldwin, *Regulating the Airlines* (Clarendon, Oxford 1985) (extracts included in this work).

[28] Weidenfeld and Nicholson, London 1987.

[29] J. Black, *Regulators as Rule-Makers* (doctoral thesis, Oxford University, 1994).

(ii) Means and procedures for making policy

The means and procedures for making policy, whether in settling a specific issue or creating standards, are naturally of special interest to administrative lawyers and are the subject of an extensive literature. Here we are not concerned so much with the contents of policies and standards, but rather the mechanics for making them. To American lawyers, the choice has been between rule-making and trial-type adjudication. That means a choice between two kinds of procedures: those directed at making rules by notice and comment procedures, and those directed at settling policy issues in the course of a trial-type procedure about how to decide a particular case.[30] Notice and comment procedures are based around giving notice of the terms of the proposed rule, including information about the material and evidence being relied on, and the method of analysis employed. The public should be invited to make representations, which the agency must in turn deal with in its statement of reasons for the final form of the rule.[31] Trial-type procedures, on the other hand, mix together policy-making and adjudication. The procedures are those of the adversarial trial, with parties, evidence, examination, and argument, where the object is to settle both the case in question and a wider policy issue. The policy made in one case will then apply in later cases. Both procedural forms are used and their respective merits have been much debated.

That debate can be linked to deeper currents in the literature and to wider theoretical issues.[32] The image of the policy-making authority collecting information, analysing the problem, considering possible solutions and then arriving at a conclusion, as J. Black has nicely put it, reflects the idea of comprehensive planning.[33] The idea is that the social world can be fully comprehended and general standards to deal with problems formulated. Social realities are rather more unruly than this idea allows, so that grander versions of comprehensive planning have to be modified to take account of inadequate knowledge, overbearing complexity, and imperfect judgement. The idea of being able rationally and effectively to deal with an issue through general standards nevertheless remains firmly at the hub of policy-making. An alternative, which also runs deep in administrative

[30] See the Administrative Procedure Act 1946, s. 4.
[31] See M. Asimow, 'Delegated Legislation: United States and United Kingdom' (1984) 5 *Oxford J.L.S.* 253.
[32] For a fuller account, see D. J. Galligan, *Discretionary Powers* (Clarendon, Oxford, 1986) p. 117 ff (extracts included in this work).
[33] J. Black, *Regulators as Rule-Makers* (n. 29 above).

theory and practice is incrementalism, that is, the step-by-step approach dealing with specific issues as they arise, deciding only those issues, and in this way, over time, developing a policy stream. Its cautiousness, responsiveness to concrete issues, and the flexibility it leaves for the future are said to be its strengths. Its natural link to the adjudicative approach, the trial-type procedure, is clear to see.

Important as both the notice and comment and the trial-type procedures are, together with the deeper models which ground them, both in the USA and the UK, it would be a mistake to imagine that they provide an adequate account of the procedures followed in policy-making. Things are never quite so neat or so simple. The notice and comment procedures are often employed in the UK in one form or another, but whether they are depends entirely on the relevant statute rather than any general legal requirement.[34] There is no set pattern to follow, so that when they are relied on, various procedural forms are followed. Those variations can be seen in the notice provisions, the imparting of information, the manner and extent of consultation with interested parties, and the nature of the reasons given.[35] It is also common enough, particularly in statutory agencies, to find policy being made in the course of adjudicative procedures; but again the procedures, while broadly trial-type, vary from one context to another.[36] Finally, it ought to be noted that the distinction between notice and comment, and the trial-type procedures has never been as important in the UK as in the USA. But even if much policy-making broadly follows the two forms discussed, we should not imagine that they, no matter how loosely construed, cover the field. One of the major contributions of socio-legal studies has been to show how often elements of negotiation and bargaining, followed by agreement, enter into the policy process. Studies of agencies such as the Monopolies and Megers Commission,[37] the Health and Safety Executive[38] and the Office of Fair Trading,[39] all attest to the importance of negotiation and agreement,

[34] Compare the American position where a general statute applies: Administrative Procedure Act 1946, s. 4.

[35] For discussion of the procedural variations, see G. Ganz, *Administrative Procedures* (Sweet & Maxwell, London 1980); Baldwin and McCrudden, *Regulation and Public Law* (see n. 28 above).

[36] R. Baldwin's study of the Civil Aviation Authority's licensing decisions is a good example of trial-type procedures: R. Baldwin, *Regulating the Airlines* (n. 27 above) p. 143 following.

[37] P. Craig, 'The Monopolies and Mergers Commission: competition and administrative rationality' in Baldwin and McCrudden, *Regulation and Public Law* (n. 28 above).

[38] R. Baldwin, 'Health & Safety at Work: Consensus and Self-Regulation' in Baldwin and McCrudden, *Regulation and Public Law* (n. 28 above).

[39] I. Ramsay, 'The Office of Fair Trading: Policing the Consumer Market Place' in Baldwin and McCrudden, *Regulation and Public Law* (n. 28 above).

whether officially provided for in the statute or simply relied upon as a practical necessity in policy-making. The same has been found in areas of policy-making which are themselves of rather lower visibility and of more uncertain legal status, an outstanding example being enforcement procedures. Studies of the regulation of water pollution have shown that negotiation and agreement are essential elements in setting the standards and in securing compliance with them.[40]

The discussion of negotiation and agreement as elements of the policy process leads on to the wider discussion of the impact of interest groups on policy-making. But first one or two remarks to conclude this section. From this briefest of surveys we can gain some idea of how the unpatterned and unruly nature of the policy process is reflected in procedures. It is not helped by the fact that in the UK there is no general legal framework governing the making of policy and the setting of standards. Common law is mainly silent and, in the absence of a general statutory framework in the nature of the American Administrative Procedure Act, everything is left to the particular statute. Many would agree that the existing approach to policy-making is both unduly diverse and informal, and that the case for a general procedural framework is convincing. However, those who crave such order and certainty should not overlook the lesson socio-legal research teaches, namely, that policy-making is a messy and complex business, that what happens at the formal procedural levels is only ever a part of the story, with so much else happening informally and out of sight. Legal reforms which fail to take account of that nether world are doomed to be largely ineffective.

(iii) Interest groups and policy communities

Empirical studies showing that policy-making bodies are apt to negotiate with interest groups have given us the evidence for what is not so surprising, considering the close and complex interrelationship between administrative bodies making policy and the groups and interests directly affected by them.[41] There may be areas where policy is formally made and handed down with the expectation that those subject to it will comply. But generally the more accurate image is one of a policy community, that is to say, a complex network of groups and interests clustered around the agency,

---

[40] See K. Hawkins, *Environment and Enforcement* (Oxford, 1984) ch. 2 (extracts included in this work) and G. Richardson, *Policing Pollution* (Oxford 1982).

[41] The article by P. Cavadino included in this book gives an interesting snapshot of one corner of interest-group activities: P. Cavadino, 'Commissions and Codes: A Case Study in Law and Public Administration' 1992 *Public Law* 332.

where the lines of contact and intercourse are many and varied. Policy emerges out of that community, more the result of informal communications, discussions, negotiations, and understandings, than a unilateral determination by the agency.[42] This idea of a policy community opens up a fascinating study in itself of how interest groups operate, of the different kinds of groups, and of the distribution between insider and outsider groups. Indeed the notion of a policy community sometimes gives way to the idea that the interests affected are so strong that they capture the agency itself.[43] A substantial literature has grown around this phenomenon, colourfully known as 'agency capture' or 'capture theory'.[44] The underlying idea is that the policy made by an agency will serve the interests of the very groups and organizations meant to be regulated. Evidence of agency capture has been found in the experience of the American regulatory agencies, where interest in the subject originated; in the UK, however, there is little hard data to support agency capture.[45] For the British experience at least, the sense of a policy community appears to be a more useful explanatory model.

While the realities of policy-making at its various levels remain at the centre of socio-legal research, it is also important to understand the normative and theoretical base upon which policy-making should turn. This is one of the points of intersection between socio-legal work and normative theory. Political theorists for their part have written at great length about the nature of the policy process, the basis of participation of groups and interests within it, and the procedures and institutions best suited to ensure better representation. The extracts from P. Craig's book *Public Law and Democracy* included in this work are important in setting the normative framework in which those processes occur.[46]

---

[42] See W. Grant, *Pressure Groups, Politics and Democracy* (1989); J. J. Richardson and G. Jordan, *Governing under Pressure* (1979).

[43] For recent case studies of the workings of a policy community see M. J. Smith, 'Changing Agencies and Policy Communities: Agricultural Issues in the 1930s and the 1980s' (1989) 67 *Public Administration* 149 and M. Wright, 'City Rules OK? Policy Community, Policy Network, and Takeover Bids' (1988) 66 *Public Administration* 389.

[44] Examples of an extensive literature are R. E. Cushman, *The Independent Regulatory Commissions* (Oxford, 1942); W. L. Carey, *Politics and the Regulatory Agencies* (1967); A. Downes, *Inside Bureaucracy* (1967); Baldwin and McCrudden, *Regulation and Public Law* (n. 28 above) pp. 9–11.

[45] This seems to be the conclusion drawn in several of the case studies contained in Baldwin and McCrudden, *Regulation and Public Law* (n. 28 above).

[46] Clarendon Press, Oxford 1990.

(iv) Influences on policy-making

Let us now move from means and procedures to content and substance. It is hardly necessary to drive home the point that the policy process is subject to many influences and pressures. In this section, I shall make reference to some of them, drawing attention at the same time to socio-legal work which may help us in understanding the policy process. But first a word of caution; we should be careful not to draw a tight line between procedures and content, since here as elsewhere the distinction between the two is watertight neither in theory nor in practice. Each is shaped by the other.[47] With that reservation in the background, let us note three of the main factors in influencing the policy-making process: the statutory regime, factors explicitly or implicitly authorized by statute, and considerations deriving from the social context. Nothing too rigorous should be expected of these divisions, since their purpose is only to help in locating different kinds of influences.

The first sets out the parameters within which policy is to be made, the objects and goals to be achieved, and the procedures to be followed. Good legislation, we might expect, should make clear and full provision for each of these matters. Such expectations are in practice likely to be severely disappointed. How little help the statute gives is a point often made in studies of policy-making. It may do nothing more than create the policy-making authority, present it with a series of issues, and let it get on with the task as it thinks best. Baldwin and McCrudden conclude from studies of several agencies that 'one theme which emerges is the vagueness of the mandate under which most agencies operate'.[48] That, moreover, is the norm rather than the exception. Parliamentary guidance will often consist in little more than a mandate to deal with the substantive issue—pollution, financial services, monopolistic practices—buttressed by only the vaguest indications of the guiding principles.[49] We should not be too shocked since the very point of the delegation is that Parliament and the executive themselves have no clear view on the matter but want the agency or authority to fill the gap. This delegation of 'complex, specialist, and flexible functions' can be linked to the wider question of why administrative bodies play such a major role in policy-making. The reasons include the pressures of time, the complexity and variability of the tasks, the need for specialist knowledge, the concern to

---

[47] J. Black brings this interrelationship into clear focus in her study of policy-making in the financial world: J. Black, *Regulators as Rule-makers* (n. 29 above) p. 49 ff.
[48] Baldwin and McCrudden, *Regulation and Public Law* (n. 28 above) p. 327.
[49] See further J. Black, *Regulators as Rule-Makers* (n. 29 above).

accommodate interest groups, and the need for a flexible approach to day-to-day shifts and changes.[50]

The second group of influencing factors refers not to the direct guidance the statute itself gives but to the sources from which guidance might be drawn. One such source is the minister or even the government itself. Where policy-making is conferred on a statutory body, a close link is likely to be maintained with the minister who becomes a source of policy guidance. The minister, as the representative of political authority, might give advice, issue directions, or exercise a veto. The minister will be woven into the policy community, and apart from any formal powers of intervention, will be a valuable asset in informally determining what is politically acceptable. Studies of policy-making refer time and again to the importance of the relationship between policy-maker and government. The second source of guidance comes from the composition of the policy-making body itself. Issues of a technical, specialist, or expert nature need, by definition, to be left to those with the requisite knowledge. The need for the specialist has always been a reason for passing policy matters to administrative agencies. If the constituent members do not themselves have that knowledge they may nevertheless bring in professionals who do. Baldwin's study of civil aviation, particularly the licensing of routes, brings out the heavy reliance on experts and on the gradual accumulation by the authority of banks of technical knowledge.[51] The concept of specialist knowledge and professional judgement raises a host of interesting issues: what counts as specialist knowledge, how can it be tested, to what extent is it open to lay judgement, in what ways should it be relied on in settling complex social issues.[52] These are matters which researchers have not so far fully explored.

The third source of guidance often sanctioned by statute is interest groups. We have noted how pervasive their influence can be, how complex consultative networks develop, and how those networks can be more significant in influencing policy than formal consultative procedures. Again the study of interest groups and their role in policy-making is a sub-discipline in itself which can be only mentioned here rather than explored in any detail. Julia Black's study of regulation of the financial services industry demonstrates the roles of different interest groups, and

---

[50] These factors are analysed more fully in relation to the allocation of discretionary powers generally in D. J. Galligan, *Discretionary Powers* (n. 32 above).
[51] R. Baldwin, *Regulating the Airlines* (n. 27 above) 143 ff.
[52] See further M. Stacey, *Regulating British Medicine: The General Medical Council* (Wiley, 1992) and R. Klein, *Complaints Against G.P.s* (C. Knight, 1973).

their capacity not only to influence policy but in effect to veto proposals which are unacceptable to them.[53] The scope for more research in this area is clearly open and the work of political scientists should be better known to administrative lawyers.[54]

These are examples of what might be regarded as authorized sources of guidance; we must now consider the unauthorized but nevertheless enormously influential factors in shaping policy. The policy process occurs in a social context and is subject to the tides which ebb and flow within it. Each context is made up of competing groups and interests, ideas and values, and pressures and influences. The agency itself will have a certain ethos, an organizational culture, a way of viewing itself and its tasks.[55] It must relate to the social and political environment, respond to the pressures upon it from various quarters, and satisfy at least some of the expectations held with respect to it. The professional outlook of the officials will itself be a factor in the way they approach their functions and interpret their roles. The agency, as we have noted, must develop its network for maintaining contact and discussion with interest groups, other administrative bodies, and government. Indeed, it is not only organized interest groups that can influence the agency, since the public itself may be a constituency from which powerful currents arise.[56] The authority is likely to develop a set of norms which reflect and accommodate this social mêlée, norms which will not necessarily correspond, except at certain points, with those of the law. The legal norms might indeed be as much a resource to be used by the agency as a set of standards to apply. This sociology of decision-making, and in particular of the policy process, is well known and much written about. P. Manning's early study of police culture,[57] K. Hawkins' research into securing compliance with anti-pollution standards,[58] and J. Black's detailed analysis of regulatory bodies in the financial world, are just a few examples of a rich and extensive literature. From studies like these we now know that any administrative process, whether directed to policy-making or otherwise, is a complex one which can only partly be understood by looking at the legal framework. The legal framework is itself part of that social context and is often in

---

[53] J. Black, *Regulators as Rule-Makers* (n. 29 above).
[54] See the references in nn. 43 and 44.
[55] An illuminating discussion of the organizational dimension is M. Feldman, 'Social Limits to Discretion: an Organizational Perspective' in K. Hawkins (ed.), *The Uses of Discretion* (Oxford Socio-Legal Studies, 1992).
[56] See K. Hawkins, *Environment and Enforcement* (n. 40 above) p. 9.
[57] P. Manning, *Police Work: The Social Organization of Policing* (1977).
[58] K. Hawkins, *Environment and Enforcement* (n. 40 above).

competition with other parts. The better we understand what is happening within the context, the more complete our knowledge will be of how decisions are made and policies shaped.

4. Individualized decisions and processes

Once policies have been made and standards formulated, the task of administrative bodies is to determine individualized cases according to them. In this section I shall concentrate on individualized decision-making but we must avoid the temptation of treating these as neatly separate from standard-setting. The convenience of dividing the two tasks should not obscure the fact that standards are often set or at least extended and developed in the course of individualized decisions. With that note of caution in the background, it is nevertheless the case that consideration of individualized decisions, or the determination of specific cases, takes us to the heart of administrative government and administrative law. We noted earlier how individualized processes have to be understood and evaluated at two levels. At one level they are about the effective achievement of social goals, at the other they raise issues about the fair treatment of the individuals and groups subject to them. Taken together they constitute the common good or the public interest, of which both are essential elements. A certain degree of tension naturally holds between the two: effectiveness tends to ally itself with considerations of efficiency, economy and cost, where what matters is overall aggregates rather than specific cases; while fair treatment for its part is founded on a proper and accurate determination in each case. Fair treatment can be expressed as rights, rights that is to have one's case dealt with according to the authoritative standards prevailing within an administrative context.[59]

(i) The public interest: effectiveness and fair treatment

This brings us to the first major aspect of individualized processes; the primary decision. What happens in administrative contexts, how authorities make their decisions and use their powers, and the treatment meted-out to those affected, are matters of great moment. Administrative law and the interests of administrative lawyers tend to concentrate on the legal remedies available when things go wrong. The great expansion of forms of recourse—complaints procedures, reviews, appeals, and investiga-

---

[59] A fuller analysis of rights and fair treatment is contained in my book *Due Process and Fair Procedures* (Clarendon, Oxford 1996).

tions—was identified earlier as a feature of contemporary government. The availability of recourse is of course important, but one of the lessons socio-legal research has taught is that a tiny fraction of people seek any form of recourse from the primary administrative decision.[60] The reason for this will be examined in the next section, but the point for our present purposes is that, unless primary decisions are made properly and accurately, the chance of their being corrected is small. Where they are not corrected, the result is administrative injustice and ineffectiveness; we should not imagine, moreover, that the standards of primary decision-making are high. The research here is not extensive, but a recent study of decision-making in social welfare, where a great mass of matters is dealt with each year by officials working under pressure, gives good reasons for concluding that the levels of accuracy are low.[61] Baldwin, Wikeley and Young show the tension between the organizational constraints under which officials work, and the concern for decisions of good quality. They show that the sheer volume of cases, limited resources, and the need for quick decisions, are just some of the factors which militate against the cause of accurate outcomes.

This is only one area of primary decision-making and others may be different. The need for further research into decision-making in other areas is clear, although the results can be anticipated: it is likely that in any area of administration, organizational and contextual factors will greatly influence the way decisions are made and will often compete against considerations of accuracy. Besides showing exactly what happens within different administrative contexts, such research should attempt to show how an area could be restructured to overcome the obstacles to accuracy. This might well become one of the main goals of socio-legal research in this area. Forms of recourse with respect to particular decisions will never be more than marginal; few cases will be challenged, and those that are, even when successfully, are likely to have only a minor impact on the future practices of the administrative body. This point is considered more fully in the next section, but we may take it that, in general, appeals and reviews do not have a major influence on primary decision processes. If processes are to be improved it must be directly by understanding the

---

[60] A study of social welfare decisions shows that less than 1% seek any form of recourse: J. Baldwin, N. Wikeley, and R. Young, *Judging Social Security* (Clarendon, Oxford 1993) (extracts included in this work). Further evidence comes from the small number of complaints made to ombudsmen: see the annual reports of the Parliamentary Commission for Administration and the Commission for Local Authorities.

[61] Baldwin, Wikeley and Young, *Judging Social Security* (n. 60 above). See also I. Loveland, *Housing Homeless Persons* (Clarendon, Socio Legal Studies, 1995).

context and restructuring it in order to ease the way to better quality decisions. Changes and reforms would of course be policy matters, but here is a clear opening for socio-legal research which would have an impact on the policy arena. What is meant by restructuring here is that methods and techniques can be developed for improving quality: they might take the form of rigorous internal quality control through checking, scrutinizing, and reviewing; they might be aimed at structuring the decision process through clearer instructions to low level officials and by emphasizing the need for clear standards rather than leaving matters to discretion; they might emphasize the training of officials; or they might try to achieve better quality by requiring that each step be recorded, explained and justified.

A good example of the last is administrative processes in the police station. The investigation, detention and questioning of suspects are administrative processes carried out by police officers as public officials. The police station is, however, an environment which is notoriously difficult to control; for that reason the approach under the Codes of Practice is especially interesting: a distribution of tasks among different officers, extensive duties to document, record, and give reasons for each step in the process, and close scrutiny of one officer by another.[62] The empirical evidence to show the effects of this new regime on the quality of pre-trial processes is now beginning to be collected, and we have good reason to think that the improvements have been significant. The rationale is to ensure both effective policing and fair treatment of suspects, and, while the procedural framework created is not perfect, it is an imaginative and progressive attempt to achieve those two aims to a high level. I shall not here attempt to analyse other areas of administration, but any one of them would benefit, first, from close research into how decisions are made, and secondly, from bold and imaginative thinking about how better quality in the sense of effectiveness and fair treatment could be achieved.

(ii) Models of individualized process

Administrative processes usually take the form of decisions about how to deal with a person's case according to authoritative standards. That is the basic process in areas of mass decision-making like social security, but it also characterizes a great range of other areas, such as licensing in its many forms, immigration, and the granting of parole. This standard case of an individualized process I have elsewhere referred to as adjudication

---

[62] Police and Criminal Evidence Act 1984 and Codes of Practice.

and modified adjudication: adjudication just in the sense that standards are applied to facts; modified adjudication in the sense that the standards may allow for discretion, and the more they do, the more the process moves away from a normal and simple sense of adjudication.[63] This is not adjudication in the sophisticated sense that we identify with a court, nor with the distinctive sense of adversarial adjudication familiar to Anglo-American lawyers. It is nevertheless adjudication in the elementary sense that a deliberate judgement has to be made by an official about a person on the basis of facts and standards. It is then a short step to postulating a procedural model suited to administrative adjudication. The components of such a model would be: obtaining evidence, finding facts, ascertaining the standards, and judging the facts according to the standards. More detailed procedures would then be drawn up to ensure that the evidence is reliable, that conclusions of fact are accurately made, and that final judgements carefully and impartially apply the standards.

One of the first things a socio-legal approach can tell us is that, while that procedural model of individualized processes appears to be the result of common sense and logical analysis, it is also the product of certain implicit values and assumptions about the administrative process. The socio-legal approach helps us to recognize how those values and assumptions create pressures and tensions which influence the way decisions are made and distort any simple procedural model. It is possible then to sketch a number of procedural models, each emanating from a specific value or set of values, each in a state of tension with the others. I shall comment briefly on the five following models: the adjudicative model; the routine administration model; the discretionary model; the professional model; and the negotiation model.

Adjudicative model. This model is the one that perhaps fits best the outline given above of individualized processes.[64] Its basic value is the fair treatment of persons which is likely to be achieved by clear and certain standards, reliable evidence and information, careful evaluation of the evidence and the drawing of factual conclusions, followed by an impartial and deliberative judgment applying the standards to the facts. The object is to reach a proper and accurate decision according to the standards; that in turn means that the person's legal rights are respected and that he or she is treated fairly. There is some leeway as to the precise procedures

---

[63] See D. J. Galligan, *Discretionary Powers* (n. 32 above).
[64] See J. Mashaw, *Due Process in the Administration State* (Yale, 1945).

required in each context, but since the governing principle is to reach a proper and accurate decision, the procedures are likely to be exacting.

Routine administration. This suggests a concern to discharge administrative tasks as effectively, efficiently, and cheaply as possible.[65] Its general object is to carry out certain statutory goals, whether to distribute welfare, house the homeless, or grant driving licences. But because that aspect of the public interest tends to predominate, the interest in how any particular person is treated is a secondary matter. However, once the overall public interest is divided in this way, it is easy for an administrative body to become overly concerned with aggregates and general patterns rather than individual cases. The way is open for a kind of bureaucratic rationality to take over; its effects will be uniformity and routinization, an emphasis on quantity and turnover, and an overriding commitment to achieving targets within strictly limited resources. This sense of routine administration is inherent in any organization; it draws on a certain psychology and responds to constant and persistent demands for ever more economy and efficiency.

Discretionary model. How much discretion there is in any context is a matter of degree and where it occurs a matter of great diversity. Nevertheless, a discretionary model can be identified according to which each case is to be decided as appears best to the official where what appears best will be determined according to factors which go beyond the circumstances of the case and draw on a wider sense of the public interest. One example is whether to grant a broadcasting licence and what conditions to impose on it, where the public interest is the main consideration; another is the minister's discretion to grant parole where his primary object is to protect certain aspects of the public interest. This is not to suggest that the individual person and his situation counts for nothing; there is a right to have one's case properly considered, but the dominant mode of reasoning and logic is the public interest in its wider sense, with the fair treatment of the person of lesser concern.

Professional judgement model. It was pointed out earlier that often the reasons for delegating authority to administrative bodies is to enable matters to be settled by professional, expert, or specialist judgement. This may range from highly technical scientific matters to the clinical judgement of

---

[65] This mode of administration has a direct line to Max Weber's notion of formal legal rationality and bureaucratic organization: M. Weber, *Economy and Society* (California, 1978).

medical practitioners. The idea of professional judgement (using the term broadly) is that the standards being applied require a specialist knowledge and training which is the preserve of a few.[66] The standards depend on specialists for this application and scrutiny is outside the competence of laymen. The result is a closed system, impatient of the usual methods and procedures of adjudication and impervious to supervision by other non-professionals. It has its own internal logic and points to a particular procedural framework.

Informal negotiation model. It has long been observed that elements of negotiation and bargaining often creep into individualized processes. This might be an express requirement as in many areas of family law,[67] or it might be relied on unofficially and informally as a way of handling certain issues, notable examples being the exercise of prosecutoral powers and enforcement processes more generally.[68] Although this area has not been a major interest for socio-legal research, informal negotiation has a significant role in many areas of decision-making.[69] The basic idea is that the official and the party negotiate and finally reach agreement about how the power or some part of it should be used or what decision should be made. Provided that each party can see some advantage to itself, the conditions are right for exploring the options and possibly agreeing on an outcome. This model again introduces a characteristic reason and logic, and points to a procedural approach different from each of the other models.

All five models have a distinct place in individualized administration. Sometimes an area of decision-making will fit neatly into one, to the virtual exclusion of the others; clear examples of routine administration, individualized justice, and strong discretion can easily be found. However, in many areas the position will be more complex with different models in competition with each other, often resulting in confusion and complexity. Scratch the surface of the most simple and straightforward cases of adjudication and we may find the strong pull of routine administration, a noticeable element of informal negotiation, and a dash of

[66] J. Mashaw was one of the first to notice the significance of this mode of process: see J. Mashaw, *Due Process in the Administrative State* (Yale, 1985).

[67] See R. Dingwall and J. Eekelaar, *Divorce Mediation and the Legal Process* (Clarendon, Oxford 1985).

[68] On enforcement processes, the leading studies are K. Hawkins, *Environment and Enforcement*, G. Richardson, *Policing Pollution*, and B. Hutter, *The Reasonable Arm of the Law* (Clarendon, Oxford Socio-Legal Studies, 1988).

[69] I. Loveland's study of administrative processes relating to housing the homeless has some interesting comments on the bargaining element in that area: I. Loveland, *Housing Homeless Persons* (Oxford Socio-Legal Studies, 1995).

professional judgement for good measure. This is in the nature of administrative processes; each is a complex matter, reflecting and responding to the competing values and influences within administrative organizations. Some of these values and concerns derive from deep social currents, others from organizational and bureaucratic constraints, while others still owe their origins to the ethos and outlook of officials and professional groups. Studies of decision processes which neglect the influence each of these models has both on content and procedure will give a false impression of the social processes at work. Socio-legal work in this area is in its early stages with many aspects of the competing models still to be explored.

The discussion of different models of process, each with its distinct value base and rationality, leads onto another point: the relationship between the fair treatment of persons and the social reality which is marked by the different models. On the one hand, lawyers and, dare one say, society at large are determined to improve the levels of individualized justice in administration.[70] The natural vehicle for individualized justice is the adjudicative mode. On the other hand, the social reality is that administrative processes are often not easily channelled into the adjudicative mode. It is not enough to adopt an adjudicative form hoping it will tame the unruly social forces at work. The new property movement was based on assumptions of that kind: if administrative procedures were moulded into an adjudicative form then the rights of persons to proper and accurate decisions would be protected.[71] A connection can be made with K. C. Davies' crusade for the structuring of discretion: the more fully an area is subjected to guidelines the closer it moves to the adjudicative mode, and in turn to the better protection of people affected.[72]

Both movements, new property and structuring discretion, were enormously influential. But each in its own way failed to take adequate notice of these complexities. One of the first to draw attention to the difficulties of imposing an adjudicative mode on administrative process was

---

[70] Again the literature is extensive; examples are J. Mashaw, *Due Process in the Administrative State* (Yale, 1985), R. Sainsbury, *Deciding Social Security Claims: A Study in the Administrative Theory and Practice of Social Security* (Ph.D. Thesis, Edinburgh 1988).

[71] C. Reich was the founder of the new property movement: C. Reich, 'The New Property' (1964) 73 *Yale L.J.* 733; for commentary on the new property see H. D. Monaghan, 'Of Liberty and Property' (1977) 62 *Cornell L. Rev.* 405 and W. Van Alstyen, 'Cracks in the New Property; Adjudicative Due Process in the Administrative State' (1977) 62 *Cornell L. Rev.* 445.

[72] K. C. Davies, *Discretionary Justice* (n. 23 above). For commentary see K. Hawkins and R. Baldwin, 'Discretionary Justice: Davis Reconsidered' 1984 *Public Law* 570.

J. Handler.[73] Welfare administration appears well-suited to adjudication; but Handler's study demonstrates how, even here, the relationship between officials and claimants cannot easily be broken down into discrete and separate decisions. What went before and what is likely to come later may be as important as any specific decision, so that procedures which concentrate on the decision itself may be ineffectual in controlling administrative behaviour. The authority and the parties subject to it will often have a continuing relationship, a factor well brought out in studies of social welfare and housing the homeless; the relationship is likely to be complex, based on past understandings and future expectations, and may be disrupted in an unhelpful way by attempts at judicialization. If to these factors are added the various competing forces considered above, the obstacles to an effective adjudicative mode are formidable.

We should not conclude, however, that the concern for fair treatment and respect for rights has to be abandoned. That concern remains a normative principle at the heart of administrative government; the only question is how it can be achieved, given the difficulties of maintaining an adjudicative mode in many areas of administration. One answer is to recognize that despite the natural affinity of the two, fair treatment and rights can be protected by means other than adjudication. This takes us back to the earlier discussion where it was suggested that other procedures and methods can be used to secure the protection of rights. Without repeating that discussion, it is perhaps worth reiterating that this is an area where a socio-legal approach may be especially helpful in developing such procedures.

(iii) The socio-legal analysis of decisions

In this section we move from procedural models and their underlying values to a more direct analysis of decision-making itself. One way of beginning is to contrast two approaches, the legal and the socio-legal. According to the first, administrative decisions consist largely in officials rationally and impartially applying clear and settled legal rules and principles to cases that come before them. This draws on a Benthamite vision of social goals being set, and law and legal institutions being effective instruments for achieving them.[74] The socio-legal approach does not

[73] J. Handler, 'Controlling Official Behaviour Welfare in Adjudication' (1966) 54 *Calif. Rev.* 479. His ideas are developed further in a book published 20 years later: *The Conditions of Discretion* (1986).

[74] For discussion of the Benthamite approach and its relationship to a socio-legal approach, see D. J. Galligan, 'Introduction' in D. J. Galligan (ed.), *Socio-Legal Studies in Context* (Blackwell, 1995).

contest this vision as an ideal, as something worth aiming at; it merely insists that decision processes are complex social processes, that the apparently straightforward decision is likely to be buffeted by a host of forces and influences, and that the clearest legal rules are shot through with discretion and will often have to compete with other norms originating in deeper social and organizational conditions. The socio-legal approach also replaces the detached and dispassionate official of the law with deeply human and fallible persons whose attitudes and actions are shaped and constrained by their social and administrative environments. This is not to say that all is chaos and anything goes; on the contrary, while the socio-legal approach has good reason for being sceptical of the simplicity of the legal approach, it nevertheless uncovers patterns and regularities in the way officials and organizations behave. But it is only by the fuller analysis of social reality that those patterns and regularities can be detected and explained. The socio-legal approach, moreover, does not drive a wedge between legal standards and social reality; rather, the legal standards are part of that reality interacting with other parts and having a greater or lesser role depending on the nature of the interaction. To neglect the legal materials would be as serious a mistake as to imagine that only those materials matter.[75]

For these reasons, the focus of much socio-legal research has been on decision-making, on understanding how decisions are made and what influences bear on them. The disciplines of the social sciences have much to contribute to that understanding, particularly sociology, administrative and organizational theory, and psychology.[76] Hawkins' recent synthesis of that contribution is included in these readings and need not be repeated here.[77] He notes that while much of the research of the social sciences has not been directed specifically at legal or administrative decisions, it can nevertheless be applied to them. He draws a contrast between perceiving decisions as rational means to achieving given objectives, a view which reflects the legal, Benthamite approach etched above, and perceiving them as naturalistic processes.[78] Naturalism stresses the influences of context, the need for actors to make sense of the conflicting considerations

---

[75] A deeper analysis of this idea is contained in P. Fitzpatrick, 'Being Social in Socio-Legal Studies' in D. J. Galligan (ed.), *Socio-Legal Studies in Context* (see n. 74 above).

[76] An example of the contribution from a social science perspective is D. J. Hickson, R. J. Butler, D. Cray, G. R. Malory and D. C. Wilson, 'Decisions Organization—Process of Strategic Decision-Making and their Explanations' (1989) 67 *Public Administration* 373.

[77] K. Hawkins, 'The Use of Legal Discretion: Perspectives from Law and Social Science' in Hawkins, *The Uses of Discretion* (n. 55 above).

[78] Ibid, p. 25.

and meanings within it, and the relation between that context and other social spheres.

Hawkins' own study of the way inspectors construct an approach to the enforcement of anti-pollution standards well illustrates this theoretical analysis.[79] His study demonstrates how their approach is shaped by such factors as: matters internal to the organization, their relations with external social systems, their own moral attitudes and the attitudes of the wider public to pollution, and the relationship between inspectors and polluters. Inspectors are able to build up from this complex environment a set of attitudes and a coherent strategy towards setting standards and gaining compliance with them. That strategy emphasizes the need for the inspectors to gain the co-operation and to a certain extent the consent of the polluters. Negotiation and agreement between the two are vital to that strategy, not as deviations from good enforcement policy, but the most effective way of gaining compliance.[80] In other words, once we understand the social environment in which the inspectors work, we can understand why they adopt the approach they do and why, far from being aberrant, it is coherent and rational. That strategy, at the same time, is developed within a legal framework where the possibility of resort to legal enforcement is always present.

This is one kind of decision amongst the many that might be cited. A recent study of decisions in the area of social security reinforces the importance of the organizational and social context in understanding what happens. Here the contextual factors are very different. The volume of cases to get through, the need for speed and turnover, and the sense of distributing scarce resources to an inexhaustible source of claimants, are amongst the matters which define the attitudes of officials and the character of their decisions. At the formal, legal level, this is a rights-based area, tightly bound by rules and guidelines, with as little discretion as possible in the adjudicating officials. And yet the reality is that the levels of accuracy are low: the moralistic categorization of claimants as deserving or undeserving, as genuinely needy or unscrupulous exploiters, and the heavy reliance of adjudication officers on junior staff to create the file on each person, point to a system of welfare distribution which at its best is a rough approximation of that intended by Parliament and the minister. These are just two case studies; others included in the readings are the workings of a board with the task of evicting tenants for non-payment of

---

[79] K. Hawkins, *Environment and Enforcement* (n. 5 above).
[80] Ibid, pp. 126–8.

rent;[81] the way decisions are made within closed institutions about patients and prisoners;[82] and the decisions of a tribunal reviewing the recommendations of medical officers.[83]

(iv) The pervasiveness of discretion

One concept that has been of special interest for socio-legal research is discretion. Some of the issues which have attracted attention are: the nature of discretion, its different senses, the many ways it occurs in decision-making, and its relationship with rules. Discretion is also a concept which brings together diverse disciplinary approaches—legal, jurisprudential, and socio-legal—in a way that has added to the range and level of analysis. Some of R. Dworkin's early ideas on judicial discretion[84] could easily be transferred to the administrative process;[85] once it was recognized just how pervasive discretion was within administration and how useful it could be in analysing decision-making, the way was open for some of the most interesting socio-legal analysis. Raising as it does some fundamental questions about the nature of government and administration, discretion opens a direct line to deeper issues of social theory of the kind explored by Habermas, in his idea of communicative action,[86] and Luhman in relation to complexity.[87] Little wonder then that this complex, multi-faceted, and mercurial concept should have such a hold over the socio-legal imagination, as evidenced in the recent volume of studies *The Uses of Discretion*. The essays in that book, written from a range of disciplinary perspectives, show how complex and interesting discretion is; they also show how useful the concept is as a way of venturing deep into the very interstices of administrative processes. The few extracts contained in these readings should give a taste of the range of research and writings available.

These readings, however, are no substitute for a fuller study of the extensive literature on discretion.[88] But it may be worth mentioning several reasons why the study of discretion can be a good way of highlighting

---

[81] R. Lempert, 'Discretion in behavioural Perspectives: the Case of a Public Hearing Conviction Board' in Hawkins, *The Uses of Discretion* (n. 55 above).

[82] G. Richardson, *Law, Process and Custody* (Weidenfeld & Nicholson, 1993).

[83] J. Peay, *Tribunals on Trial* (Clarendon, Oxford, 1989).

[84] R. Dworkin, *Taking Rights Seriously* (Duckworth, 1978).

[85] This was one of my aims in *Discretionary Powers* (n. 32 above).

[86] J. Habermas, *Legitimation Crisis* (London, 1973).

[87] N. Luhmann, *Trust and Power* (London, 1969) and 'Differentiation in Society' (1971) 11 *Can. J. of Sociology* 29.

[88] Both *Discretionary Powers* (n. 32 above) and *The Uses of Discretion* (n. 55 above) contain extensive bibliographies.

important aspects of individualized processes. The first is that discretion conferred on officials appears to challenge certain legal and political values. Notions of the rule of law, of certainty and stability in legal relations, and of the protection of rights, all seem to be threatened by discretion. Discretion is associated with arbitrariness, with a lack of concern for consistency, and with inherent uncertainty. We need not here examine this normative dimension, except to note its importance as the frame of reference which attracts lawyers and political theorists and within which discretion is seen as a problem. But the normative aspect has a direct consequence of interest to socio-legal analysis: it shows why lawyers and judges consider rules and discretion to be opposites. What they have in mind is that discretion can be removed by rules; that is to say, the famous hole in the doughnut can be filled in by adding more doughnut. Discretion, for lawyers, is by definition the absence of legal rules, or at least of rules of sufficient detail and comprehensiveness that they could pass as rules; the more rules, the less discretion. This is why Davis' idea of structuring is so attractive at the legal level: it removes discretion by setting guidelines which, if not exactly rules, have family connections with them.

That conception of discretion, while suitable within a legal paradigm, is thrown into jeopardy when we move to a fuller understanding of decision processes, to what we may call the social paradigm. This brings me to the second reason why discretion is so important, for within the social paradigm, discretion is all pervasive. It no longer connotes the absence of legal rules, but occurs within and in relation to legal rules; it recognizes the discretion in interpreting and applying rules, in ignoring rules, departing from, making exceptions to, or even changing them. In short, discretion is all over the place so that the mutual exclusiveness of legal rules and discretion disappears. The reason is that within the social paradigm all sorts of other normative standards and restraints come into play.[89] Legal standards are then just one set of norms which may have to compete with others. Moreover, what looks like unbridled discretion from the point of view of legal rules, may be closely bound by normative standards arising from other sources. Once we move into the social paradigm to consider what really happens, the purely legal conception has to be replaced by another which looks at how officials decide matters, what constraints are

---

[89] For an analysis of the social paradigm and some of the factors at work within it, see my *Discretionary Powers* (n. 32 above) pp. 72–84 and 128–140 (included in this work).

40   D. J. Galligan

on them, and how they view their own positions.[90] This is another way of saying that a study of discretion takes us to the heart of administrative decision-making. And here there may be some surprises: what looks discretionary from the legal paradigm may be highly structured in social reality; what looks closely rule-bound from the legal point of view may turn out to be heavily discretionary.[91] The recognition of different levels and of different frameworks of analysis is of great importance in understanding discretion. We can then see how at one level discretion is normatively charged, while at another level it is an instrument for descriptive analysis. Both conceptions are important: the normative emphasizing the need to bring administrative officials into line with legal and political values, while the descriptive, which is to say the more socio-legal, provides the knowledge of decision-making and administrative behaviour which is a necessary condition of effective normative reconstruction.

A third reason why discretion has received so much attention in the analysis of individualized processes is that it draws attention to low level policy-making. By that is meant not the big policy matters deliberately conferred by parliament (which may involve a high degree of discretion), but the kind of policy involved when, for example, social security officers distinguish between deserving cases and scroungers, or inspectors decide that one polluter is more blameworthy than another, or the parole board determines that a prisoner is not yet rehabilitated. Low level policy choices of these kinds are nevertheless real policy choices upon which the treatment of people depends. They are policy choices which are often hard to identify, difficult to control, and without proper legal authority; they are the products of the moral and social attitudes of officials, which are in turn to a large degree the results of the social and organizational ethos of a department or agency. At these low levels of policy, stereotyping is rife and the failure to act for good reasons is evident. Socio-legal research has been of great service in isolating low level policy, in showing how such policies are formed, and how important they are in understanding administrative decisions. Studies of policing,[92] social security,[93]

---

[90] What Lempert calls the behavioural and phenomenological conceptions: R. Lempert, 'Discretion in a Behavioural Perspective' in Hawkins, *The Uses of Discretion* (n. 55 above).
[91] On the relationship between the legal and the social, see further the essay by N. Lacey, 'The Jurisprudence of Discretion: Escaping the Legal Paradigm' in Hawkins, *The Uses of Discretion* (n. 55 above).
[92] P. K. Manning, *Police work: The Social Organization of Policing* (Mass, 1977).
[93] Baldwin, Wikeley and Young, *Judging Social Security* (n. 60 above).

housing,[94] pollution control,[95] and mental health,[96] have all contributed substantially to our understanding of discretion and of the pervasiveness of low level policy within it.

5 Accountability, recourse and legal control

In this final section we turn to the legal framework itself, to the doctrines and institutions which make up that framework and are responsible for enforcing it. This is the most specifically legal part of the analysis, for here we are concerned with the supervision of administrative processes, with the forms of accountability of authorities, and the methods of recourse available when things go wrong. Following the suggestion that a new emphasis on supervisory methods and institutions is a noticeable feature of modern government, we can now consider the contribution socio-legal research has made, or might in the future make, to our understanding of how well the supervisory processes work in practice.

It may be wise to begin by reminding ourselves of the dual elements of the public good in relation to administrative processes: the effective and economical realization of statutory objectives and the fair treatment of persons. This dual character is highly pertinent in examining the various forms of accountability and recourse. Accountability appears most appropriate in describing how one authority must account to another for its actions, where the emphasis is on showing that it has carried out its statutory functions properly and efficiently. The minister accounts to Parliament, the Next Steps Agency to the minister, the subordinate official to his superior. The Select Committee may scrutinize a department or agency to see that it is working well, while the Audit Commission will concentrate on its financial side. The Inspectors of Probation investigate and report on whether the probation service is discharging its functions according to the directions of the Secretary of State, while the Magistrates' Courts Service Inspectorate has a similar role in relation to the magistrates' courts.

These are examples of the methods and institutions of accountability which appear to be expanding in many directions. Some are concerned with policy, others with financial management, and others still with overall effectiveness in delivering a public service. The historic line of accountability through a minister to Parliament has been ruptured, and the base

---

[94] I. Loveland, *Housing the Homeless* (n. 69 above).
[95] K. Hawkins, *Environment and Enforcement* (n. 5 above).
[96] J. Peay, *Tribunals on Trial* (n. 83 above).

of accountability broadened, so that the grounds now recognized and the bodies responsible for them are diverse and often separate from ministerial accountability. Here is another area in which socio-legal research might make a major contribution by analysing the different institutions of accountability, the grounds they apply and the procedures they follow, and how effective they are in practice. These are important issues and well suited to socio-legal examination, although so far they have been neglected.

Administrative law itself is more concerned with the way individual persons are dealt with by administrative bodies and the forms of recourse open to them. The terminology here is not all that precise, but this could be viewed as accountability in the sense that the primary decision-maker has to explain and justify its actions in the particular case to an appeal or review body. Whatever the best terminology, the emphasis is on creating forms of recourse the main object of which is to ensure that individual cases are decided properly, and if not, that a complaint can be lodged and dealt with by a different, often superior authority. This is the thread which links the different forms of internal review, appeals to tribunals, complaints to ombudsmen, and even judicial review. Properly decided here means decided in accordance with the prevailing statutory standards and any general common law principles, such as procedural fairness. For a person to have his case determined in this way is to have his rights respected and to be treated fairly. This normative framework stimulates a number of interesting socio-legal issues, among which the following are of special importance: first, the nature of the process of complaining, seeking a review, or lodging an appeal; secondly, the different forms of recourse; and thirdly, how effective they are.

(i) The process of seeking recourse

In an age which is prepared to commit substantial resources to the creation of forms of recourse and the institutions needed to make them effective, it is important to know why people seek recourse and what they hope to achieve by it. At one level the answer is simple: they complain, appeal, or seek review in order to have their rights upheld. This is not only a simple answer but a powerful one since fair treatment and the protection of rights is at the centre of administrative processes: the person seeking recourse has been denied something to which he is entitled and the various forms of recourse are all seen as means of righting that wrong.

Simple and powerful as it is, however, that answer does not give a full explanation of why people seek recourse. Recent research by social psychologists on the nature of complaining is of direct relevance. Here I can make only brief reference to a thriving sub-discipline of its own, but a study by S. Lloyd-Bostock and L. Mulcahy sets out the basic ideas.[97] While complaints procedures (using that term in its wider sense) are often goal-oriented or rights-based in the way suggested above, they may also have a different logic. Complaining may be part of an 'account episode' where the object is to require the offending party to provide an account of its action to the complainant. This approach 'views the complainant as putting forward a negative evaluation of some event or circumstance for which he or she holds the hospital responsible, or calls on the hospital to respond'.[98] The study by Lloyd-Bostock and Mulcahy of complaints by patients within the NHS demonstrates the force of this account model. Many of the complaints about the treatment received were best analysed within that model: the recourse sought was often vague, sometimes no more than an acknowledgment of fault, an apology, or an undertaking that the system would be improved for the future.

This approach potentially has great significance for administrative law and process. It opens the way to a better and fuller understanding of why people seek recourse and how their claims can best be dealt with. It raises the question whether conciliatory and mediatory procedures might sometimes be more suitable than other forms of recourse. The more that is known about the psychology of complaining and seeking recourse, the better the design of procedures and institutions should be.

(ii) The forms and institutions of recourse

Returning now to the forms and institutions of recourse, we have noted the number and variety of internal reviews, external reviews, appeals to tribunals, complaints to ombudsmen, and ultimately judicial review. Each of these warrants a study of its own; here I shall do no more than comment briefly on each.

Internal review. The use of internal reviews does seem to be on the increase. The idea is that the first stage in the recourse process should be a review of the case by an official within the department or agency in which the primary decision was made. Internal review has the obvious advantages of speed, informality, and economy, as well as satisfying a more

[97] S. Lloyd-Bostock and L. Mulcahy, 'The Social Psychology of Responding to Hospital Complaints: an Account Model of Complaints Procedures' (1994) 16 *Law and Policy* 123 (included in this work).
[98] Ibid, p. 133.

general sense that the primary authority should have the main responsibility for ensuring that its decisions are correct and that it has adequate procedures for reviewing and correcting them. There are also disadvantages: a tendency to be secretive, a sense that the same officials are checking each other, and that the reviewing officers will be under the same pressures and constraints as those being reviewed. Just how these merits and de-merits work out in practice is a matter we know little about.

The one notable case study in recent years which looks closely at internal review is by Baldwin, Wikeley and Young on social security.[99] This is a valuable study because it compares two parallel but different forms of internal review, one where the primary decision makers, the adjudication officers, review their own decisions, the other where the primary decision is reviewed by separate appeals officers. The second proved to be much more effective than the first in detecting and correcting errors in primary decisions. The reasons are not hard to discern: the appeals officers had more time to examine cases, were better trained, and were committed to accuracy. They knew that if they did not find for the claimant, they would have to defend their decisions before the appeal tribunal. It is easier, moreover, to overturn someone else's decision that one's own. The system of self-review, on the other hand, is working badly, largely it seems because of the reluctance or perhaps the inability of adjudication officers to think critically about their own decisions.

Much can be learned from case studies of this kind. One lesson is the rather obvious one that a well conducted system of internal review will reduce substantially the number of cases going on to external appeal. The other is that there can be within a department or agency relative independence of one set of officials from another for the purposes of arm's length review. But while there are valuable lessons to be learnt from a valuable study, the reality is that we know very little about the workings of many of the systems of internal review.[100] What we do know is that many such systems are in operation, some of a formal and official basis, others created informally and unofficially. The signs are that internal, administrative review of one kind or another is the method favoured by modern

---

[99] *Judging Social Security* (n. 60 above) (extracts included in this work).
[100] This was the conclusion reached nearly 10 years ago by R. Rawlings in his review of socio-legal work; apart from a very few studies, the same conclusion is true today: see R. Rawlings, *The Complaints Industry: A Review of Socio-Legal Research on Aspects of Administrative Justice* (ESRC, 1986). A recent addition to the literature is R. Sainsbury, 'Internal Reviews and the Weakening of Social Security Claimant's Rights of Appeal' in S. Richardson and H. Genn (eds.), *Administrative Law and Government Action* (Clarendon Press, 1994).

governments for dealing with complaints.[101] This will in turn reduce the level of recourse to external bodies, especially tribunals and courts.

Tribunals. Since the report of the Franks Committee in 1948, the specialist tribunal has been the basic forum for appeals from primary decisions. Not all primary decisions are subject to appeal, but as the idea that there ought to be an appeal grows, so does the number, range, and variety of tribunals. This is added to by the valuable work of the Council on Tribunals in monitoring areas of decision-making and recommending the creation of tribunals where none exists.[102] This is also an area which has long attracted socio-legal research with major studies of tribunals in the areas of social security, industrial disputes, mental health review, and immigration to name but a few.[103] The range and quality of such studies is impressive and this is one area of administrative law where socio-legal work has flourished.

This is not surprising. Tribunals handle a vast number of appeals each year and for most people the tribunal is the one chance to have their treatment by administrative officials examined by an outside body; it is likely to be their only encounter with law and the possibility of justice. If there is a slight criticism to be made of the socio-legal contribution, it is only that some of the work on tribunals might have been usefully directed to primary decision-making. Although the volume of cases decided by tribunals is considerable, it is only a fraction of the numbers of primary decisions; the importance of the primary processes and the need for more research into them is all too obvious. However, the benefit is that we know a lot about tribunals. What then are the main issues about tribunals for socio-legal research? A good starting point is the model of tribunals put forward by the Franks Report: tribunals should be cheap, easily accessible, unburdened by formal or technical procedures, able to decide quickly, drawing on the professional expertise of their members. These are noble ideals behind which lies a vision of tribunals as the keys to administrative justice and respect for the rights of all. That is the normative basis which guides and justifies the tribunal system; socio-legal

[101] This is reinforced by the Citizens Charter and the requirement that administrative agencies create complaints procedures.
[102] The Annual Report of The Council on Tribunals gives a good indication of its work and its impact on tribunals.
[103] A full bibliography of research up to 1986 is contained in R. Rawlings, *The Complaints Industry.* Amongst the major studies since then are: Peay, *Tribunals on Trial* (n. 83 above) and H. Genn and Y. Genn, *The Effectiveness of Representation Before Tribunals* (London, 1989) (extracts included in this work).

research has made a major contribution in showing how difficult it is to achieve that ideal and how far short of it the reality falls. If tribunals are the great fortresses of rights and fair treatment, then socio-legal research has shown that those within are highly vulnerable.

The reasons for this are well documented. The first is the low rate of appeal. From primary decisions in social security, which is an area of high volume, it is less than 1%; and while other areas may be higher, the percentage of cases going on appeal is still small and out of proportion to the likely rates of error contrary to the applicant's interests at the primary level.[104] The reasons for such low rates lead onto the second point. The difficulties of mounting an appeal and carrying it through to a hearing are greater than expected. Lack of knowledge on the part of the claimants, their low expectations of the system, the complexity of the law, the scarcity of proper advice and representation, capped by an endemic sense of civic torpor, all contribute to the low appeal rate.[105] While each of these is important, the factor which stands out most prominently is the scarcity of professional help, both for advice in lodging an appeal and assistance in presenting a case at the hearing. The Genn study has supplied the hard and convincing evidence for what other studies had foreshadowed and many suspected, namely, that legal or other professional representation makes a significant difference both in mounting a case and in the result.[106] Across the four tribunals studied—social security, mental health, immigration, and labour relations—the level of professional advice varied, but the conclusion common to all tribunals is that having it makes a big difference. As well as being invaluable in dealing with the complexities of law, professional assistance helps in many other ways. Again this finding is not surprising; lawyers are trained to understand the law, to be able to marshal the facts and evidence needed to present a case in its best light, and to reveal the weaknesses in the opposing case.

There are, however, other matters which affect the proper functioning of tribunals. Jill Peay's study of Mental Health Review Tribunals is a classic example of socio-legal research revealing several important features: the problems for a tribunal reviewing the professional and clinical judgement of medical practitioners; the difficulties for a tribunal of being inde-

---

[104] Even allowing for the increased effectiveness of internal review, the rate of mistaken decisions which are not remedied is still high.

[105] See the study of H. Genn and Y. Genn, *The Effectiveness of Representation Before Tribunals* (n. 103 above); also Baldwin, Wikeley and Young, *Judging Social Security* (n. 60 above).

[106] Extracts from the Genn study are included in this work.

pendent of those whose decisions are under consideration; the attitudes of patients and doctors towards the tribunal; the way tribunals can easily become vehicles for legitimating first instance decisions; and finally the difficulties even at that level of having the law properly and accurately applied, no matter how clear and unambiguous it is.[107] In their study of social security, Baldwin, Wikeley, and Young found a real tension between the inquisitorial and adversarial mode of tribunal hearings, with some tribunals committed to getting to the truth of the matter, others content to be more like umpires deciding the cases put before them.[108] Another interesting aspect which calls for further research is the way in which elements of negotiation and agreement enter into the pre-hearing stage, sometimes with the tribunal privy to the discussion, at other times without reference to it. These are just some of the areas which have attracted fruitful investigation, with many others still to be considered. Since tribunals are likely to retain, even extend, their importance as the basic forum of appeal, it is important for socio-legal research to show their strengths and weaknesses. The implications for policy-making are also clear to see, should policy-makers bother to take notice.

Ombudsmen. The impressive socio-legal contribution to tribunals has not been matched for ombudsmen.[109] For while an enormous amount has been written about ombudsmen, much of it of great interest, few major empirical studies have been made.[110] Why this should be the case is not clear; perhaps they are still seen as marginal forms of recourse, even though the concept has spread in both the public and private sectors. Certainly the numbers of complaints made seem small and those investigated tiny, although the point has been made that the important thing is the quality and range of the investigations rather than their quantity. It may also be that some ombudsmen have not been that keen on opening their files and practices to researchers. Whatever the full explanation, the result is that ample opportunities exist for empirical studies in the future, not only of the much discussed Parliamentary Commissioner for

---

[107] J. Peay, *Tribunals on Trial* (n. 83 above).

[108] Baldwin, Wikeley, and Young *Judging Social Security* (n. 60 above).

[109] For a review up to 1986, see R. Rawlings, *The Complaints Industry* (ESRC, 1986) pp. 49–69. There appear to have been few major studies since then: see further details in M. Senevaratne, *Ombudsmen in the Public Sector* (Open University Press, 1994).

[110] A useful and recent general examination of ombudsmen is M. Senevaratne, *Ombudsmen in the Public Sector* (n. 109 above). The leading early study is R. Gregory and M. Hutchesson, *The Parliamentary Ombudsmen* (1978).

Administration (PCA) and Commission for Local Administration (CLA), but also the Police Complaints Authority,[111] the Health Service Commissioner,[112] the Revenue Adjudicator and the Ombudsman for Prisons.[113]

However, let us sketch out a few of the issues which are important to our understanding of ombudsmen and on which socio-legal research could be helpful. The first issue should be to gain a better understanding of just what it is that ombudsmen do and what effect they have on the administrative bodies they investigate. Their statutory duties are primarily to investigate the complaints of individuals who have been the subject of bad administration and who have, as a result, suffered injustice. One very basic issue is then simply to know more about the investigation itself: how is it conducted; what obstacles are encountered; the attitudes of the officials; how much bargaining and negotiation goes on between the investigators and the officials themselves in resolving an issue. At the moment our knowledge of the investigatory process depends almost entirely on brief and rather formal annual reports from each office. The close study of the investigatory process is likely to lead onto a second aspect which is perhaps even more fundamental, namely, what exactly are the functions of ombudsmen. Simply to say they investigate complaints is neither adequate nor all that illuminating. We should distinguish, moreover, between what they should be doing and what they are in practice doing. What they are doing in practice might also turn out to differ from one type of ombudsman to another, with the Prison Ombudsman for example, having a rather different role from that of the PCA.[114] The context, the client group, and the relations with the authority being investigated will all bear heavily on defining exactly what the functions of each are.

The final aspect of this question of what the ombudsmen do is the impact they have on the authorities being investigated. The impact might result directly from investigation, or it might stem from an ombudsman's ranging more widely and taking positive steps to identify and remedy more general administrative practices. How far the ombudsman is authorized to take this wider view of his functions is a matter on which views

[111] See A. J. Goldsmith, *Complaints Against the Police* (1991).
[112] See D. Longley, *Public Law and Health Service Accountability* (1993).
[113] Created in 1994.
[114] Some of the difficulties in investigating complaints in a closed institution are well documented: I. Matheson, 'The Ombudsmen and Prison Complaints' (1982) 12 *Victoria U. of Wellington L. Rev.* 265; P. Barton, 'Ombudsmen in Corrections—The Power of Presence' in G. Caiden (ed.), *International Handbook of the Ombudsmen* (1983).

and practices differ, although none of the ombudsmen takes a strongly positive view towards detecting and remedying general administrative defects. The investigation of specific complaints by individuals is for each the main object with any wider role very much incidental. Nevertheless, a more precise assessment of these attitudes and their practical approaches would be valuable. Also of great interest from a socio-legal point of view is the impact and influence that their actions have on administration. This point is taken up again later on.

Another general issue which has been much discussed is access to this form of recourse and why so few take advantage of it. A perception prevails that the level of complaints, particularly complaints to the PCA, is low. Low one might ask in relation to what, for it is difficult to know what we should be expecting, considering the various other forms of recourse available. It may also be that a few well-conducted investigations are an effective and adequate use of the ombudsman idea. This issue of expectations and the way the ombudsman fits in with other forms of recourse, although fundamental to understanding and evaluation of the office, needs much closer analysis than it so far has received. Quite apart from that basic issue, questions of knowledge and access are crucial to an effective system. Some of the more obvious factors, such as the need to go first to a member of Parliament rather than directly to the PCA, are much debated,[115] but a deeper analysis about knowledge and access is missing. The attitudes of people affected are vital and the work of social psychologists in relation to the complaining process could be extended to this context. Other social and structural factors might well be uncovered in a searching study of this area. Recent research by Lewis and others opened the way to some of these issues, but the scope for further work is considerable.[116]

(iii) Judicial Review

The final form of recourse to comment on is judicial review. Judicial review is first and foremost an inquiry into a specific complaint to determine whether certain legal standards have been complied with by the administrative decision-maker. As with other forms of recourse, it may also have a more general effect on the administrative processes and organizations subject to review. In this section I shall concentrate on judicial

---

[115] On this see G. Drewry and C. Harlow, 'A Cutting Edge? The Parliamentary Commissioner and MPs' (1990) 55 *Modern Law Review* 753.
[116] N. Lewis, M. Senevaratne and S. Cracknell, *Complaints Procedures in Local Government* (Centre for Socio-Legal Studies, Sheffield 1990).

review as a method for resolving grievances, leaving questions about the wider impact for the final section.

As a method of resolving grievances and investigating complaints and upholding rights, judicial review is often proclaimed a success story. The late Lord Diplock, whose own contribution as a senior judge was outstanding, stated that the development of administrative law (meaning judicial review) had been 'the greatest achievement of the English courts in my judicial lifetime'.[117] Many lawyers, judges and academics share that view. Judicial review is certainly much written about; the grounds of review, the limits of judicial intervention, and its legitimacy, all continue to be the subject of intense debate. There has not, however, been any systematic attempt to study judicial review in its social context. Important studies have been made of the senior judges, but not with specific reference to their role in judicial review.[118] A strong tradition of critical writings, developed mainly this century by such figures as H. Laski,[119] W. Robson,[120] and J. A. G. Griffith[121] has been directed mainly at exposing the values and premises upon which judicial review is based.[122] These have been important in showing that there is more to judicial review than appears in legal textbooks and that the judges are unavoidably part of a wider social and political process. However, these writings, interesting as they are, are not themselves based on close empirical study nor have they succeeded in stimulating socio-legal research.

It was only with the formation of the Public Law Project in very recent years that research into the most basic issues of judicial review began. Within the general question of how effective and useful judicial review is as a form of recourse, researchers began to make some very interesting discoveries. First, they found that while there has been a marked increase in the number of cases of judicial review, the majority of these are attributable to two areas of central government, homelessness and immigration, where inadequate statutory forms of recourse exist; it also found that a high percentage of other cases are against local authorities rather than central government.[123] If these three areas were taken away, the

---

[117] R. v. *Inland Revenue Commissioners ex parte National Federation of Self-Employed and Small Businesses* [1982] 2. A.C. 237 at 285.
[118] See A. A. Paterson, *The Law Lords* (Weidenfeld and Nicholson, 1979).
[119] H. Laski, *Authority in the Modern State* (1919) and *A Grammar of Politics* (1938).
[120] W. Robson, *Justice and Administrative Law* (3rd edn, 1951).
[121] J. A. G. Griffith, *The Politics of the Judiciary* (4th edn, 1991).
[122] See further M. Loughlin, *Public Law and Democracy* (Clarendon, Oxford 1993).
[123] M. Sunkin, L. Bridges, and G. Mazeros, *Judicial Review in Perspective* (Public Law Project, London 1993).

power of judicial review as a way of controlling government would be much reduced. Secondly, a study of the procedures for bringing judicial review, particularly the leave requirement, shows how meritorious cases can fail at this early hurdle. In exercising their discretion to grant or refuse leave, the courts take account of policy factors, including the need to protect themselves from being swamped and governmental bodies from litigiousness. Leave appears to be easier to obtain in some areas of administration than others and the courts generally insist that other forms of recourse first be exhausted.[124] Basic research like this is vital to a proper understanding of judicial review, and we now have a better understanding of how judicial review fits into the general framework of forms of recourse, of its use, its availability, and its effectiveness. This, however, is only a beginning and it is hoped that the work of Sunkin, Bridges and Le Sueur will continue and inspire others.

Another aspect of judicial review which has so far largely escaped the attentions of socio-legal research is the grounds and doctrines of review. Notions of relevance and irrelevance, reasonableness and unreasonableness, objects and purposes, fettering and procedural fairness, are the very terms of discourse of judicial review. They have attracted a critical literature of the kind referred to above,[125] but have not been the subject of more social and contextual study. And yet such a study is both absolutely central to the socio-legal enterprise and highly rewarding. An analogy can be drawn with the concept of fault in accident cases. The Oxford Centre's study of compensation showed how interesting and complex the notion of fault is and how illuminating a socio-legal analysis can be.[126] Similarly for the doctrines and concepts of judicial review: where did they come from; what do they mean; how do they relate to similar non-legal concepts; do they vary from one context to another; how are they seen and understood by those outside the courts and the legal profession, particularly by administrators and applicants. The one area in which some work of this kind has been done is due process and procedural fairness. D. McBarnet's study of these concepts in the pre-trial processes of criminal justice shows how fertile this field can be.[127] According to her study

[124] See A. Le Sueur and M. Sunkin, 'Applications for Judicial Review: The Requirement of Leave' 1992 *Public Law* 1021.

[125] See for example M. Loughlin, 'Courts and Governance' in P. B. H. Birks (ed.), *The Frontiers of Liability* (Oxford, 1994).

[126] See D. Harris and others, *Compensation for Support for Illness and Injury* (Oxford Socio-Legal Studies, 1984) especially the essay by S. Lloyd-Bostock, 'Fault and Liability for Accidents; The Accident Victim's Perspective'.

[127] See D. McBarnet, *Conviction, State, Law and Order* (Oxford 1975).

many of the problems of pre-trial procedures stem not from lawlessness or deviance on the part of officials, but from the legal concept of due process itself. A very different kind of approach has been taken by a group of social psychologists who have studied the perception of procedural fairness held by people subject to administrative processes.[128] They show to what extent these perceptions correspond to legal notions of due process and in what ways they differ.[129] This kind of work could be extended with great profit to other doctrines and concepts of judicial review and administrative law more generally.

(iv) The impact of recourse on administration

In considering the forms of recourse, I have concentrated so far on their role in correcting errors in the specific case. Recourse can also affect the administrative body in ways which will lead to better decisions in the future. This second dimension of recourse is often commented on but has not been the subject of close study. It is easy to see how it might have an impact: the ombudsman investigating a complaint finds that it results from a more general fault in the system and makes recommendations about how to correct it; the tribunal's decision on a specific issue in one case might have wider implications for other cases; the court exercising judicial review not only decides the law in that case, but sets more general standards to be followed in other cases. Each form of recourse has the potential to influence administrative processes in this more general, forward-looking manner, but precisely how far it does is hard to measure. In an early study of judicial review, T. Prosser examined the impact of an important case relating to entitlements to social security.[130] He found the impact on the administrators to be both mixed and complicated, dispelling any assumption that a general standard laid down in judicial review would be directly assimilated into administrative practice. In a later study, based on a number of prominent judicial review cases in which ministers and departments suffered serious reverses, Sunkin and Le Sueur distinguished between the formal and the informal consequences.[131] The formal consequences were to encourage departments

---

[128] See T. Tyler, *Why People Obey the Law* (Yale, 1990).

[129] For a critique of this work see my forthcoming *Due Process and Fair Procedures*, ch. 2.

[130] T. Prosser, 'Politics and Judicial Review: The Atkinson Case and its Aftermath' 1983 *Public Law* 59.

[131] M. Sunkin and A. Le Sueur, 'Can Government Control Judicial Review?' 1991 *C.L.P.* 161.

and agencies to be more careful in the drafting of laws and in that way to reduce the risk of challenge in the courts. The informal consequences related to 'the internal, informal and private world of government departments'.[132] The threat of judicial review set off a flurry of activity at both levels. The conclusion reached by Sunkin and Le Sueur is that despite attempts by governments and departments to take positive steps to absorb judicial review rulings, those who make the mass of front line decisions remain largely ignorant of the most elementary legal principles, including principles laid down in judicial review.

The impact of the decisions of tribunals is one area which has been studied. In their study of social security, Baldwin, Wikeley and Young examined the educative effect of the tribunal on primary decision-makers. When the rate of appeal is very low, as is generally the case in areas of mass decision-making, the educative effect of decisions is especially important in improving the quality at first instance. The fewness of appeals, however, leads one to ask how there could be a significant educative effect. A number of tentative answers can be suggested: first, a generalized educative effect flows from the very existence of the appeal tribunal. This consists in primary decision-makers knowing that their decisions might be scrutinized by an independent external body.[133] The strength of this effect is hard to measure and will depend on factors such as the likelihood of an appeal being brought, how soon it will be brought and whether the particular official will be personally responsible. The second aspect of the educative effect is the example set by the appeal body. The adjudicative nature of the appeal procedures, the impartiality of the tribunal's members, and the concern to gather sufficient evidence and to weigh it carefully are all factors which set an example of how things should be done. The third aspect of the educative nature of tribunal decisions concerns the level of involvement of primary officials in the appeal process. An important feature of the study of social security is that the adjudicating officer, who made the initial decision, often had to present and defend the case before the appeal tribunal.[134] The authors conclude that, at least in this area of decision-making, these factors combine to give the appeal a significant educative impact on the accuracy of primary decisions.[135]

[132] Ibid, p. 166.
[133] On this see also I. Loveland, *Housing the Homeless* (n. 69 above).
[134] See N. Wikeley and R. Young, 'The Administration of Benefits' 1993 *Public Law* 238, p. 284 ff (extracts included in this work).
[135] The authors do note, however, that half the adjudication officers claim they were not influenced by the tribunal: ibid, p. 252.

6. Conclusion

The socio-legal approach to administrative law requires a much fuller study than this introduction and the accompanying readings provide. It is hoped, however, that this will be at least a beginning on which others may build. The general object has been to present in a compact form a sample of writings which show how interesting and illuminating a socio-legal approach to administrative law and administrative institutions can be. The range of studies now available is impressive, but clearly there are many gaps. In the course of this introduction, I have indicated numerous areas about which we know very little, and again it is to be hoped that researchers might be stimulated enough to take up the challenges these areas offer. But above all, this volume is aimed at students and teachers of administrative law. If it succeeds in enticing them away even briefly from the endlessly fascinating but ultimately narrow study of what judges do and say in judicial review towards the social context of law and legal authorities, then it will have succeeded in its purposes.

Select bibliography and references for further reading

The Changing Face of Administrative Government

The General Nature of Administrative Government

DAINTITH, T. (1989) 'The Executive Power Today; Bargaining and Economic Control' in Jowell, J. and Oliver, D. (eds), *The Changing Constitution*. Clarendon, Oxford.

EFFICIENCY UNIT (1991) *Making the Most of Next Steps*. HMSO, London.

FRY, G., FLYNN, A., GENN, A., JENKINS, W. and RUTHERFORD, B. (1980) 'Symposium on Improving Management in Government' 66 *Public Administration* 429.

GRAHAM, C. and PROSSER, T. (1988) 'Rolling Back the Frontiers: The Privatization of State Enterprises' in Prosser, T. and Graham, C., *Waiving the Rules*. Open University Press.

GREER, P. (1990) 'The Next Steps Initiative: An Examination of the Agency Framework Document' 68 *Public Administration* 89.

HARDEN, I. (1992) *The Contracting State*. Open University Press.

—— and LEWIS, N. (1986) *The Noble Lie*. Hutchinson.

KEMP, P. (1993) *Beyond Next Steps*.

LEWIS, N. (1994) 'Changes in Government: New Public Management and Next Steps' *Public Law* 105.

PROSSER, T. (1986) *Nationalized Industries and Public Control*. Blackwell.

Underlying Ideas

BARRON, A. and SCOTT, C. (1992) 'The Citizens Charter Programme' 55 *Modern Law Review* 526.
FREEDLAND, M. R. (1994) 'Government by Contract and Public Law' *Public Law* 86.
HOOD, C. (1991) 'A Public Management for All Seasons' 69 *Public Administration* 3.
LACEY, N. (1994a) 'Missing the Wood . . . Pragmatism versus Theory in the Royal Commission' in McConville, M. and Bridges, L. (eds), *Criminal Justice in Crisis*. E. Elgar.
—— (1994b) 'Government as Manager, Citizen as Consumer: The Case of the Criminal Justice Act 1991' 57 *Modern Law Review* 534.
MARSH, D. (1991) 'Privatization Under Mrs Thatcher: A Review of the Literature' 69 *Public Administration* 459.
OLIVER, D. (1994) 'Law, Politics, and Public Accountability' *Public Law* 238.
POLLIT, C. (1990) *Managerialism and the Public Services: The Anglo-American Experience*. Blackwell.
POWER (1994) 'The Audit Society' in Hopwood and Miller (eds), *Accounting as Social and Institutional Practice*. Cambridge.
STEWART, J. and WALSH, K. (1992) 'Change in the Management of Public Services' 70 *Public Administration* 499.
The Citizens' Charter; Raising the Standard (1991). HMSO.
The Implications of the Citizens' Charter for the Work of the PCA (1992) Select Committee Report on the PCA. HMSO.
ZIFCAK, S. (1994) *The New Managerialism*. Open University Press.

The Policy-Making Process

Policies, Rules, and Policy-Making Bodies

BALDWIN, R. (1985) *Regulating the Airlines*. Oxford Socio-Legal Studies.
—— (1995) *Rules and Government*. Oxford Socio-Legal Studies.
—— and McCRUDDEN, C. (1987) *Regulation and Public Law*. London, Weidenfeld and Nicholson.
—— and HOUGHTON, R. (1986) 'Circular Arguments: The Status and Legitimacy of Administrative Rules' *Public Law* 239.
BLACK, J. (1994) *Regulators as Rule-Makers*. Doctoral Thesis, Oxford University.
—— (1995) 'Which Arrow? Rule Type and Regulatory Policy' *Public Law* 94.
DAVIES, K. C. (1964) *Discretionary Justice*. Louisiana.
DRIVER, C. (1981) 'Policy-Making Paradigms in Administrative Law' *Harvard Law Review* 393.

Means and Procedures for Making Policy

Administrative Procedure Act (USA) (1946).
ASIMOW, M. (1984) 'Delegated Legislation: United States and United Kingdom' 5 *Oxford J.L.S.* 235.

BALDWIN, R. (1987) 'Health and Safety at Work: Consensus and Self-Regulation' in Baldwin and McCrudden, *Regulation and Public Law*.
BLACK, J. (1994) *Regulators as Rule-Makers*.
CRAIG, P. 'The Monopolies and Mergers Commission: Competition and Administrative Rationality' in Baldwin and McCrudden, *Regulation and Public Law*.
GALLIGAN, D. J. (1986) *Discretionary Powers*. Clarendon, Oxford.
GANZ, G. (1980) *Administrative Procedures*. Sweet & Maxwell, London.
HAWKINS, K. (1984) *Environment and Enforcement*. Oxford Socio-Legal Studies.
RAMSAY, I. (1987) 'The Office of Fair Trading: Policing the Consumer Market Place' in Baldwin and McCrudden, *Regulation and Public Law*.

Interest Groups and Policy Communities
CAREY, W. L. (1967) *Politics and the Regulatory Agencies*.
CAVADINO, P. (1992) 'Commissions and Codes: A Case Study in Law and Public Administration' *Public Law* 332.
CRAIG, P. (1990) *Public Law and Democracy*. Clarendon, Oxford.
CUSHMAN, R. E. (1942) *The Independent Regulatory Commissions*. Clarendon, Oxford.
DOWNES, A. (1967) *Inside Bureaucracy*.
GRANT, W. (1989) *Pressure Groups, Politics and Democracy*.
JORDAN, G. (1979) *Governing Under Pressure*. M. Robertson.
SMITH, M. J. (1989) 'Changing Agencies and Policy Communities: Agricultural Issues in the 1930's and the 1980's' 67 *Public Administration* 149.
WRIGHT, M. (1988) 'City Rules OK? Policy Community, Policy Network, and Takeover Bids' 66 *Public Administration* 389.

Influences on Policy-Making
BALDWIN, R. (1985) *Regulating the Airlines* pp. 143 ff. Clarendon, Oxford.
—— and McCRUDDEN (1987) *Regulation and Public Law* pp. 327 ff. London, Weidenfield and Nicholson.
BLACK, J. *Regulators as Rule-Makers*. Doctoral Thesis, Oxford University.
FELDMAN, M. (1993) 'Social Limits To Discretion: An Organizational Perspective' in K. Hawkins (ed.), *The Uses Of Discretion*. Oxford.
Klein, R. (1973) *Complaints Against G.P.'s*. C. Knight.
MANNING, P. (1977) *Police Work: The Social Organization of Policy*. MIT Press.
ROCK, P. (1986) *A View From The Shadows*. Oxford Socio-Legal Studies.
SIMPSON, A. W. B. (1991) *In The Highest Degree Odious*. Clarendon, Oxford.
STACEY, M. (1992) *Regulating British Medicine: The General Medical Council*. Wiley.
UNSWORTH, C. (1987) *The Politics of Mental Health Legislation*. Oxford Socio-Legal Studies.

Individualized Decisions and Processes

The Public Interest: Effectiveness and Fair Treatment

BALDWIN, J., WIKELEY, N. and YOUNG, R. (1993) *Judging Social Security*. Clarendon, Oxford.
GALLIGAN, D. J. (1996) *Due Process and Fair Procedures*. Clarendon, Oxford.
LOVELAND, I. (1994) *Housing Homeless Persons*. Oxford Socio-Legal Studies.
MASHAW, J. (1974) 'The Management Side of Due Process' 59 *Cornell Law Review* 777.
MCCONVILLE, M., SANDERS, A. and LENG, R. (1991) *The Case for the Prosecution*. Routledge.
SAINSBURY, R. (1989) 'The Social Security Chief Adjudication Officer: The First Four Years' *Public Law* 323.

Models of Individualized Processes

BALDWIN, R. and HAWKINS, K. (1984) 'Discretionary Justice: Davis Reconsidered' *Public Law* 540.
DAVIES, K. C. (1964) *Discretionary Justice*. Louisiana.
GALLIGAN, D. J. (1996) *Due Process and Fair Procedures*, chs. 8 and 9. Clarendon, Oxford.
GENN, H. (1990) *Hard Bargaining*. Oxford Socio-Legal Studies.
HANDLER, J. (1966) 'Controlling Official Behaviour in Welfare Adjudication' 54 *California Law Review* 479.
HANDLER, J. (1986) *The Conditions of Discretion*. New York.
HUTTER, B. (1988) *The Reasonable Arm of the Law*. Oxford Socio-Legal Studies.
MASHAW, J. (1983) *Bureaucratic Justice*. Yale.
—— (1985) *Due Process in the Administrative State*. Yale.
MONAGHAN, H. D. (1977) 'Of Liberty and Property' 62 *Cornell Law Review* 405.
REICH, C. (1964) 'The New Property' 73 *Yale Law Journal* 733.
RICHARDSON, G., BURROWS, P. and OGUS, A. (1984) *Policing Pollution*. Oxford Socio-Legal Studies.
SAINSBURY, R. (1988) 'Deciding Social Security Claims: A Study in the Administrative Theory and Practice of Social Security'. Ph.D. Thesis, Edinburgh.
VAN ALSTEYN, W. (1977) 'Cracks in the New Property; Adjudicative Due Process in the Administrative State' 62 *Cornell Law Review* 445.
WEBER, M. (1978) *Economy and Society*. G. Roth and C. Wittick (eds.), California.

The Socio-Legal Analysis of Decisions

CAIN, M. (1995) 'Horatio's Mistake' in D. J. Galligan, *Socio-Legal Studies in Context: The Oxford Centre, the Past, and the Future*.
FITZPATRICK, P. (1995) 'Being Social in Socio-Legal Studies' in D. J. Galligan, *Socio-Legal Studies in Context: The Oxford Centre, the Past, and the Future*.
GALLIGAN, D. J. (ed.), (1995) *Socio-Legal Studies in Context: The Oxford Centre, the Past, and The Future*. Journal of Law and Society and Blackwells.

HAWKINS, K. (1993) 'The Use of Legal Discretion: Perspectives from Law and Social Science' in Hawkins, *The Uses of Discretion*.
—— (1993) *The Uses of Discretion*. Clarendon, Oxford Socio-Legal Studies.
JACKSON, D. J., BUTLER, R. J., CRAY, D., MALORY, G. R. and WILSON, D. C. (1989) 'Decisions Organisation—Process of Strategic Decision-Making and their Explanations' 67 *Public Administration* 373.
LEMPERT, R. (1993) 'Discretion in Behavioural Perspective: The Case of a Public Hearing Conviction Board' in K. Hawkins, *The Uses of Discretion*.
PEAY, J. (1989) *Tribunals on Trial*. Clarendon, Oxford.
RICHARDSON, G. (1993) *Law, Process and Custody*. Weidenfeld & Nicholson.

The Pervasiveness of Discretion

DWORKIN, R. (1978) *Taking Rights Seriously*. Duckworth.
GALLIGAN, D. J. (1986) *Discretionary Powers*. Clarendon, Oxford.
HABERMAS, J. (1973) *Legitimation Crisis*. London.
LACEY, N. (1993) 'The Jurisprudence of Discretion: Escaping the Legal Paradigm' in Hawkins, *The Uses of Discretion*.
LEMPERT, R. (1993) 'Discretion in Behavioural Perspectives: The Case of a Public Hearing Conviction Board' in Hawkins, *The Uses of Discretion*.
LUHMANN, N. (1969) *Trust and Power*. London.
—— (1971) 'Differentiation in Society' 11 *Canadian Journal of Sociology* 29.
MANNING, P. K. (1977) *Police Work: The Social Organisation of Policing*. MIT Press.
TEUBNER, G. (1983) 'Substantive and Reflexive Elements in Modern Law' 17 *Law and Society Review* 239.

Accountability, Recourse and Legal Control

The Process of Seeking Recourse

BALDWIN, J., WIKELEY, N. and YOUNG, R. (1993) *Judging Social Security*. Clarendon, Oxford.
LLOYD-BOSTOCK, S. and MULCAHY, L. (1994) 'The Social Psychology of Responding to Hospital Complaints: an Account Model of Complaints Procedure' 16 *Law and Policy* 123.
LOVELAND, I. (1994) *Housing Homeless Persons*. Oxford Socio-Legal Studies.
PEAY, J. (1989) *Tribunals on Trial*. Clarendon, Oxford.

The Forms and Institutions of Recourse

BALDWIN, J., WIKELEY, N. and YOUNG, R. (1993) *Judging Social Security*. Clarendon, Oxford.
BARTON, P. (1983) 'Ombudsmen in Corrections—The Power of Presence' in Caiden (ed.), *International Handbook of the Ombudsmen*.
DREWRY, G. and HARLOW, C. (1990) 'A Cutting Edge? The Parliamentary Commissioner and MP's' 55 *Modern Law Review* 753.

GENN, H. and GENN, Y. (1989) *The Effectiveness of Representation Before Tribunals*. Lord Chancellor's Department, London.
GOLDSMITH, A. J. (1993) *Complaints Against the Police*.
GREGORY, R. and HUTCHESSON, M. (1978) *The Parliamentary Ombudsmen*.
LEWIS, N., SENEVARATNE, M. and CRACKNELL, S. (1990) *Complaints Procedures in Local Government*. Centre for Socio-Legal Studies, Sheffield.
LONGLEY, D. (1993) *Public Law and Health Service Accountability*. Open University Press.
MATHESON, I. (1982) 'The Ombudsmen and Prison Complaints' 12 *Victoria University of Wellington Law Review* 265.
PEAY, J. (1989) *Tribunals on Trial*. Clarendon, Oxford.
RAWLINGS, R. (1986) *The Complaints Industry: A Review of Socio-Legal Research on Aspects of Administrative Justice*. ESRC.
RICHARDSON, G. and GENN, H. (eds) (1994), *Administrative Law and Government Action*. Clarendon, Oxford.
ROBSON, W. (1951) *Justice and Administrative Law* (3rd edn).
SAINSBURY, R., 'Internal Reviews and the Weakening of Social Security Complainants' Rights of Appeal' in Richardson and Genn, *Administrative Law and Government Action*.
SENEVARATNE, M. (1994) *Ombudsmen in the Public Sector*. Open University Press.

Judicial Review
GALLIGAN, D. J. (1996) *Due Process and Fair Procedures*. Clarendon, Oxford.
GRIFFITH, J. A. G. (1991) *The Politics of the Judiciary* (4th edn).
HARRIS, D. et al. (1984) *Compensation for Support for Illness and Injury*. Clarendon, Oxford Socio-Legal Studies.
LASKI, H. (1919) *Authority in the Modern State*. New Haven.
—— (1938) *A Grammar of Politics*. New Haven.
LE SUEUR, A. and SUNKIN, M. (1992) 'Applications for Judicial Review: the Requirement of Leave' *Public Law* 1021.
LLOYD-BOSTOCK, S. (1984) 'Fault and Liability for Accidents; The Accident Victim's Perspective' in Harris, *Compensation for Support for Illness and Injury*.
LOUGHLIN, M. (1993) *Law and Democracy*. Clarendon, Oxford.
—— (1994) 'Courts and Governance' in P. B. H. Birks (ed.), *The Frontiers of Liability*. Oxford.
PATERSON, A. A. (1979) *The Law Lords*. Weidenfeld and Nicholson.
SUNKIN, M., BRIDGES, L. and MAZEROS, G. (1993) *Judicial Review in Perspective*. Public Law Project, London.
TYLER, T. (1990) *Why People Obey the Law*. Yale.
—— (1988) 'What is Procedural Justice?' 22 *Law and Society Review* 103.

The Impact of Recourse on Administration
PROSSER, T. (1983) 'Politics and Judicial Review: The Atkinson Case and its Aftermath' *Public Law* 59.

SUNKIN, M. and LE SUEUR, A. (1991) 'Can Government Control Judicial Review?' *Current Legal Problems* 161.

WIKELEY, N. and YOUNG, R. (1993) 'The Administration of Benefits' 1993 *Public Law* 238.

THE STRUCTURE AND COMPOSITION OF
ADMINISTRATIVE GOVERNMENT

'Rolling Back the Frontiers'? The Privatisation of State Enterprises

C. GRAHAM AND T. PROSSER

The privatisation programme

The programme of privatisation of nationalised industries would at first sight appear to be the clearest example of the implementation of a full-blooded strategy of reducing the role of the state and replacing it with the discipline of market forces. Thus by the end of the second Thatcher Government, the state's involvement in production had been reduced by almost half; moreover, the number of individual shareholders in Britain had doubled or trebled, one survey suggesting that almost a quarter of the adult population now owned shares (*Guardian*, 21 September 1987). Moreover, the scale of the programme had continually increased from fairly modest beginnings to achieve a scope beyond the wildest dreams of most early advocates of privatisation. In autumn 1986 the target for the net proceeds of privatisation was raised to £5 billion per annum in each of the following three financial years. The 1987 Conservative Election Manifesto promised sales of the water authorities and electricity industry, and a bill to pave the way for these was introduced almost immediately Parliament reassembled (Public Utility Transfers and Water Charges Bill). Indeed, it appeared that the target of £5 billion pounds per year was likely to be overshot.

Recently some of the lustre has rubbed off the privatisation programme, in part because of the mounting criticism of British Telecom's performance. Perhaps more importantly, the fall in share prices on the world stock markets raises questions about the feasibility of further sell-offs. The sell-off of the Government's remaining shares in BP was clearly a failure and only went ahead through the Bank of England supporting the share price by a special buy-back scheme for BP shares (*Independent*, 3 November 1987; *Financial Times*, 6 November 1987). However, the Government remains determined to press ahead: indeed, shortly after the stock-market collapse the Financial Secretary to the Treasury pledged that the programme of £5 billion per year would continue for the following three years (*Financial Times*, 18 November 1987). In the following weeks plans were announced for the sale of British Rail Engineering Ltd and of the British Steel Corporation.

If the scale of the privatisation programme is extraordinary, so are the claims made as to its effect on political and industrial life. According to the then Financial Secretary to the Treasury, who had played a major role in preparing the programme:

> In the course of two Parliaments, we ... have nearly halved government involvement in state-owned business and liberated a substantial portion of economic activity from suffocation by the state. I have no doubt that the successful conclusion of this Parliament's programme will produce an irreversible shift in attitudes and achievement which will bring lasting benefits to the United Kingdom.
>
> (Moore 1986a: 97)

As is apparent even from this brief quotation, the privatisation programme has been associated with what are essentially constitutional claims in that they are based on a particular view as to the legitimate role of government. To quote the Financial Secretary once more:

> Less government is good government. This is nowhere truer than in the state industrial sector. Privatisation hands back, to the people of this country, industries that have no place in the public sector.
>
> (Moore 1986b: 93)

Indeed, privatisation's justification as cutting back the state as a matter of principle has grown in importance as the programme has proceeded. In the early days justifications for privatisation concentrated on the disciplines to be provided by exposing the industries privatised to a competitive environment: indeed, most of the enterprises privatised under the first Thatcher Government existed in an environment in which there was some degree of competitive pressure. However, this has now changed; the enterprises sold in the second term do not necessarily face any real competition in important markets, as is notably the case with the British Gas Corporation. Moreover, opportunities to increase competition have in important cases not been taken up on privatisation (see e.g. Beesley and Littlechild 1983; Vickers and Yarrow 1985; Hammond, Helm and Thompson 1985). Indeed, in two cases privatisation has served to reduce competition; the acquisition of the Royal Ordnance Factories by British Aerospace, thereby creating the largest Western defence manufacturing company outside the United States, and the takeover of the airline British Caledonian by British Airways.

Instead of the virtues of competition, two linked themes have come to prominence as the new justifications for the privatisation programme. The first is that privatisation will free the industries from the governmental intervention which had bedevilled the nationalised industries. In one sense, this is used to mean that the industries will be able to raise finance for their extensive investment programmes of the next few years outside the artificial constraints of the Treasury's external financing limits. In another sense it refers to the freeing of the industries from bureaucratic constraints on their commercial judgement: the first reason given for the proposed sale of the water authorities was that 'the authorities will be free of Government intervention in day to day management and protected from fluctuating political pressures' (Cmnd 9374 para. 3).

The second theme in this argument is that ineffective mechanisms of political accountability will be replaced by accountability to shareholders. This forms part of one of the central presentational themes of the programme: the promotion of wider share ownership. This is not, of course, an argument based on a more equal distribution of wealth, but rather that individual responsibility, independence and freedom will be increased through ownership of a stake in a major industry and that a more direct form of accountability can be exercised through the capital market and the company meeting. This will be more effective than diffused and indirect political control (see e.g. the Secretary of State for Trade in relation to the Telecommunications Bill at 48 HC Debs, cols. 26–38, 18 July 1983). The encouragement of wider share-ownership suffered something of a battering with the fall of the stock market and the failure of the sale of the government's final stake in BP, 'in which the only significant new recruits to wider share-ownership turned out to be the Bank of England and the Kuwait Investment Office' (*Financial Times*, 7 December 1987), and it remains to be seen how central this theme will become in future flotations. Our major aim in this contribution will be to assess the claims that privatisation will free the industries from governmental intervention and that democracy through share ownership offers a more direct form of political accountability than does public ownership.

It is worth reminding ourselves, however, that these claims feed on a real and justified dissatisfaction with nationalisation in Britain. At the time of the major examples of nationalisation in the late 1940s it had been assumed that government could remain mainly at 'arm's length' from the industries, which could discover an unproblematic 'public interest' to provide an objective once the profit motive had been removed. This unsurprisingly proved a will-o'-the-wisp and instead governmental intervention on a range of matters (especially pricing) came to dominate industry decision-making. The intervention, however, did not provide coherent objectives over anything but the shortest of terms, and later attempts to rationalise government intervention through the provision of financial targets or, more recently, external financial limits, have not resolved the problems (Prosser 1986, chs. 2–4; for a recent assessment see the Public Accounts Committee in HC 343 1985–6).

Nor did nationalisation provide effective public accountability. Government intervention was not implemented through published directions but through informal and usually secret processes. As a result, accountability was attenuated to vanishing-point: who could be accountable if it was not clear whether responsibility for decisions rested with the industry boards or with government? Parliamentary accountability has been deliberately limited by ministers and industry chairmen (Prosser 1986, ch. 10; for a recent critical assessment see the Public Accounts Committee, HC 26 1986–7). Moreover, the consumer councils established to protect consumer interests were generally weak, unimaginative and hampered by an inability to gain information from the industries (Prosser 1986, chs. 8–9).

As with so many other themes within Thatcherism, then, the dissatisfactions on which privatisation draws are real and the failure to develop effective and accountable structures for nationalisation has been a source for much of the attraction of the programme. However, whether the privatisation programme will solve the problems is a different matter. In a complex and interdependent economy with an inevitable role for government, privatisation does not solve the problems characteristic of nationalisation simply through change in ownership. Rather, it merely provides the opportunity for the design of fresh institutions which may or may not be superior to those of public ownership. We will now assess the extent to which the claims made to justify the privatisation programme can be justified.

Privatised industries and government

It is quite clear that the Government possesses enormous discretion in the actual process of privatisation, for example as regards pricing and timing of disposals. This discretion is largely unscrutinised, at least until after the event (for descriptions by the National Audit Office of the sales of British Telecom, the Trustee Savings Banks, British Gas and British Airways making this clear, see HC 495 1985–6; HC 237 1986–7; HC 22 1987–8 and

HC 37 1987–8). Even as regards the recent BP sale, the major negotiations took place in private between the lead underwriter, the Treasury and BP after which, in the absence of agreement, they were contractually obliged to seek the Bank of England's advice, the final decision resting with the Treasury (see 121 HC Debs col. 169, 27 October 1987). However, pricing of issues has become a major issue of political controversy with allegations of underpricing and, as a result, substantial premiums being available to share purchasers (see e.g. Mayer and Meadowcroft 1985; Buckland 1987; Public Accounts Committee in HC 35 1985–6). In other cases the Government has actually taken added powers to compel the nationalised industries to dispose of assets or subsidiaries (this is something that has also occurred in relation to local government). The most notable example would have been the power proposed in the Treasury Consultative Proposals on Nationalised Industries Legislation of December 1984. This would have given the sponsoring minister power to require the privatisation of any assets and activities and would have had the power to amend any statute applicable to a nationalised industry by statutory instrument to facilitate the exercise of his powers (Treasury 1984). These proposals were shelved, but there are nevertheless such powers in other statutes. (See e.g. the British Telecommunications Act 1981, s. 62(3), providing such powers in relation to the Post Office, the British Shipbuilders Act 1983, ss. 1–2, and the Iron and Steel Act 1982, ss. 2 and 5. For fuller discussion see Graham and Prosser 1987: 24–30.)

However, disposal does not end the Government's role. Whilst a major theme in the justification of privatisation has been that once denationalised the industries will be free from political intervention and able to concentrate on their own commercial interests, the history of nationalisation should give us immediate pause here. It was precisely the objective of keeping the nationalised industries free from extensive governmental interference that led to the adoption of the Morrisonian model in which the industries were to be at 'arm's length' from government. As outlined above, this was a manifest failure: political intervention became extensive but largely *ad hoc*. It could be argued that this was because the nationalised status of the industries made the temptation to intervene irresistible, but it is not hard to think of a variety of situations in which governments of any political complexion might wish to intervene in the affairs of privatised industries These include the threat of bankruptcy (particularly of industries with a major defence role such as British Aerospace or occupying a strategic role in the economy such as British Telecom), unwelcome takeover bids either by the privatised industry or for it, the threatened curtailment of socially desirable services or abandonment of British suppliers, industrial action threatening the rest of the economy, pressure from foreign governments party to contracts with the privatised firm, and other types of political pressure (see Steel 1984: 105–8). Given such temptations, the key question must be whether government possesses the tools by which it can successfully intervene in the affairs of privatised concerns. The answer to this must be a firm 'yes', and indeed it has taken care to provide itself with the necessary legal powers for intervention.

Government and regulation

It must first be stressed that a number of the most important privatised industries will necessarily be surrounded by a pattern of strategic decisions to be taken by government. For example, despite some liberalisation of gas import and export policy, important governmental powers remain, an these are likely to be crucial in the future strategy to be adopted by the British Gas Corporation. Imports will be allowed subject to consent for laying pipelines across the Continental Shelf (paraphrased by one commentator as, 'if we don't like your next gas import, you will only be able to get the stuff into the UK in balloons' (*Financial Times*, 11 March 1987) and in appropriate cases the conclusion of inter-governmental treaties: British Gas has given the Government an assurance that it will be consulted on its import plans. As regards exports, the Government will consider waiving the requirement to land gas in the United Kingdom on a case-by-case basis (83 HC Debs cols. 211–12 [written answer], 6 March 1986). Even more importantly, the Department of Energy retains control of the allocation of licences for the development of North Sea oilfields, and the Department of Transport has an important role in the allocation of route licences to airlines, in addition to the role of government in international negotiations for the designation of air carriers. As we shall see, these powers are central to two key areas of controversy involving recently privatised industries. In other contexts, privatisation has been accompanied by the provision of new regulatory powers directly in the hands of government. Examples are the promulgation of traffic distribution rules at airports and the limitation of aircraft movements after the sale of the British Airports Authority (Airports Act 1986, ss. 31–4). In the case of the water authorities, current plans are for a Director-General of Water Services, whilst the powers of environmental regulation previously held by the authorities themselves will be transferred to a National Rivers

Authority. However, river quality objectives will be set by the Secretary of State and he will have the power to direct the Authority to implement specific policies of an environmental nature (Department of the Environment 1987, paras. 4.6–4.8).

Planning competition

In the telecommunications and gas industries, new regulatory bodies independent of government have been established. This has served to disguise the fact that government has retained important powers and the operation of the regulatory bodies is to a large degree dependent on prior decisions of the Secretary of State. Thus in telecommunications, the previous British Telecom monopoly has been replaced by licensing of telecommunications systems by the Secretary of State (after consultation with the Director-General of Telecommunications, the new regulator). This means that the development of competition is effectively in the hands of government, and there has already been criticism of 'the government's illiberal policies on the licensing of public networks and resale' (Vickers and Yarrow 1985: 45, and see generally ch. 3 and pp. 81–3 therein). This may or may not be justified (and there are in fact forceful arguments in favour of initial restrictions on competition to prevent 'cream-skimming' of the most profitable services and to protect 'infant industries' as potential forces for increased competition). What it makes clear, however, is that paths to greater competition must be *planned*; as the Director-General of Telecommunications has put it:

although I believe that a presumption exists in favour of competition, and careful consideration must be given to the justification for any inhibitions of competition, nevertheless some planning of the path to competition and some limitation of the ultimate scope of competition is likely to be in the public interest.

(HC 457 1984–5)

In Britain, such planning has been retained firmly in the hands of government. As regards the gas industry, similar powers for the licensing of public gas suppliers have been given to the Secretary of State, though here the extent of competition is bound to be far less than in telecommunications.

Apart from such special provisions for the regulation of privatised industries, the industries are subject to the normal provisions of competition law. The point must be firmly made that the operation of British domestic competition law has *not* been privatised. Under domestic law the right to bring a private action arises only in narrowly defined instances, although it is possible to invoke European law in the domestic courts. Nevertheless, the central characteristic of UK competition law remains the extent of the discretion given to the Department of Trade and Industry, the Office of Fair Trading (OFT) and the Monopolies and Mergers Commission (MMC). As would be expected, the enforcement of competition law depends on a complex network of bargaining between these institutions and the affected parties. Two examples will suffice: anti-competitive practices and merger references to the MMC. The former are policed by the OFT which conducts preliminary investigations, such as that on the complaints by industrial users about British Gas's charges, negotiates undertakings after MMC investigations and supervises adherence to undertakings. All these activities involve bargaining between the OFT and the affected firms. O'Brien (1982) has argued that there is a built-in incentive to co-operate with the OFT at an early stage, in order to avoid referral to the MMC, although the pricing policy of British Gas in regard to industrial customers is currently under investigation by both the MMC and the European Commission.

On merger references, the Director-General of Fair Trading advises the Secretary of State, but before doing so the advice of a mergers panel, consisting of representatives of the OFT, the MMC and interested government departments is sought (Fairburn 1985). The MMC has a wide discretion as to the procedures it adopts. For example, it was reported that the Commission indicated that BA's original bid for British Caledonian was likely to be seen as against the public interest; BA therefore produced new proposals to meet the MMC's fears (see Cmnd 247 1987 paras. 5.36–5.38, 8.70–8.71). Other interested parties were not given sight of, nor opportunity to comment on, these proposals, which convinced the MMC the bid was not against the public interest. One participant then unsuccessfully challenged the Commission's decision through an action for judicial review on the grounds of procedural unfairness (*Financial Times*, 12 November 1987; *Guardian*, 21 November 1987). The Secretary of State need not accept the advice of the OFT nor of the MMC that a merger is against the public interest, and this discretion is structured by the most rudimentary of guidelines (see HC Debs vol. 63, cols. 213–14, 5 July 1984 [written answer]; Fair Trading Act 1973, s. 84) and there are no requirements of openness.

The role of the Secretary of State in deciding whether to make a reference to the MMC has already become the central issue in one case involving a privatised concern. When the Royal Ordnance Factories were sold to the recently privatised British Aerospace there were a number of complaints to the OFT from other defence contractors that this would inhibit competition in key areas of military procurement and, in particular, would give British Aerospace a near-monopoly in the United Kingdom in making major missile and munitions systems. The sale had been made conditional on non-referral to the MMC; to no one's surprise the Secretary of State did not make such a reference, and instead assurances were received from the company that the Ministry of Defence would have the right to inspect British Aerospace's books to check the price of items supplied by Royal Ordnance. It was hoped that this would enable the ministry to detect whether British Aerospace was using its Royal Ordnance subsidiary as a preferred source of supplies at below market rates (see *Financial Times*, 23 April 1987).

Although the Secretary of State for Trade and Industry played a less prominent role in the takeover of British Caledonian by British Airways, the troubles of British Caledonian, leading the way to its takeover, seem largely to stem from the earlier decision of the government not to follow the recommendations of the Civil Aviation Authority that the company be awarded some of British Airways' route licences before the latter was privatised (see Cmnd 9366 1984). This decision has been widely attributed to a preference for a successful flotation over increased competition. In any event, it is indeed nonsense to talk of decisions determined simply by market forces in an area in which government has such a major role on competition issues and in the allocation of such basic assets as route licences and designation for international flights (see *Financial Times*, 12 November 1987).

Regulatory controls

The licence issued by the Secretary of State to telecommunications and gas operators will also contain the key regulatory provisions in the form of licence conditions: the most important of these is that controlling price increases (for the complexities of the practical operation of these formulas, see Helm 1987). The formula adopted is to relate certain prices to the retail price index; thus in the case of British Telecom a basket of charges cannot be increased by more than the retail price index minus 3 points (the latter figure representing the desired efficiency gains) each year until 31 July 1989. The negotiations on the actual figure to be set and on the range of prices to be covered were treated as a private matter between the Department and British Telecom, and, for example, the Post Office

Users' National Council, representing consumer interests in telecommunications, was not allowed to participate.

The price formula appears straightforward and near-automatic in its application. In practice, however, its operation has proved highly controversial. Although total tariff increases have remained within the formula, its terms have permitted extensive 'rebalancing' of prices at the expense of rentals and some local calls. This was clearly envisaged when the licence was issued, and was indeed an inevitable result of the form of economic regulation adopted in an industry where prices did not directly reflect costs and where a degree of cross-subsidisation existed. Nevertheless, it does represent in effect a direct redistribution of resources from domestic and small-business users to the larger telecommunication users, and responsibility for this lies strictly with the Government (see Hills 1986, chs. 5 and 7). Criticism of the Director-General of Telecommunications over the scale of the tariff rebalancing is beside the point as his basic responsibility is to enforce the formula as included in the Licence, and he has (quite justifiably decided only to seek a modification of the formula if British Telecom were making an excessive return on capital not attributable to increased efficiency. After 1989 a new formula is likely to be agreed through a licence amendment involving a report from the Monopolies and Mergers Commission. However, by then the rebalancing process is likely to have been completed, so the most controversial result of telecommunications privatisation will have been the result of direct governmental decision-making rather than deriving from principles drawn up by any independent regulator.

In the case of the British Gas Corporation, the pricing formula is more complex. As well as permitting the company to increase its domestic tariffs by a figure 2 per cent below the rise in the retail price increase, it is allowed to take into account in full the average cost of gas acquired during the year. This appears to assume that British Gas is a passive price-taker in gas purchase, but in fact it will remain the dominant buyer: the formula appears to remove the incentive for British Gas to minimise the cost paid for gas, thus creating the familiar problems associated with profit regulation by the regulatory agencies in the United States without the means available to US regulators for the scrutiny of efficiency (see Littlechild 1983, 1986; cf. Henney 1986). It would also be possible in principle for British Gas to purchase expensive gas to satisfy demand in more competitive areas of its markets and then to cross-subsidise it from consumers within the captive tariff market. Because of this possibility the Energy Select Committee of the House of Commons recommended that the Director-General of Gas Supply, the new regulatory authority for gas, should have the power to satisfy himself that contracts for the purchase of gas were prudently incurred to meet the needs of tariff customers and to disallow any costs not so allowed: the Government firmly rejected this proposal (HC 15 1985–6, para. 27: Cmnd 9759, paras. 46–9). The problem has rather disappeared from view because of the fall in oil prices and a resulting decrease in gas tariffs, but it can be expected to reappear in the future, and the question of cross-subsidy will be central to the MMC investigation into the corporation's pricing policies for industrial users.

A related difficulty can be anticipated as regards the electricity industry, because over 75 per cent of the CEGB's fuel input is purchased from British Coal. The purchase costs are the result of negotiations between the CEGB and British Coal which are embodied in an unpublished Joint Understanding. The Understanding has been criticised for limiting imports and forcing the CEGB to pay a higher than necessary price for coal, which is due, in part, to the Understanding being a mechanism for the implementation of government policy towards the coal industry (Electricity Consumers' Council 1987).

It is clear, then, that government remains at the heart of the regulatory process which has been created after privatisation. It is true that once licences have been issued, there is a separation of functions between the Secretary of State and the new regulatory authority, enforcement of licences being for the latter, and modification against the will of the licensee being the task of the regulatory authority after a report from the MMC. However, the initial licence is to run for a considerable period of time (25 years in the case of British Telecom), and the procedures for amendment are cumbersome. Indeed, the procedure for licence amendment has already been bypassed in relation to the merger between British Telecom and Mitel, where the majority of the MMC recommended that anti-competitive practices be restrained by undertakings rather than licence modifications because the 'provisions of the Telecommunications Act which govern amendments are such that in the short term it would be difficult to make the necessary changes to the licence without involving further delay and a risk to the future of Mitel'. This has raised doubts as to the enforceability of the undertakings (Cmnd 9715: paras. 10.79, 10.82; see also Gist and Meadowcroft 1986) and British Telecom has already asked for them to be withdrawn (*Financial Times*, 21 October 1987). The really important aspects of regulation have been retained in government hands, and as the examples above show, this has effectively determined the major directions of strategy of privatised enterprises on such important matters as pricing.

'Golden' shares

The discussion so far has concerned government involvement in the affairs of privatised industries where regulation has been deemed necessary. The most important reason for regulation is that competition in the market-place for the sale of the industry's products is limited, and so intervention by a public agency is necessary as a surrogate for market forces, or to protect interests which would not be adequately protected by the free play of market forces, for example the provision of emergency telephone services. We now move from consideration of the product market to the market for corporate control. What powers has government taken to intervene in this market after privatisation?

In the early examples of privatisation, residual shareholdings were retained by government. Although these were accompanied by undertakings that the Government did not intend to use them to intervene in company decision-making, the right to do so was reserved, and the undertakings were anyway so vague as to be unenforceable (see e.g. the Secretary of State for Energy at 16 HC Debs col. 171, 19 January 1982). This inevitably led to speculation that such shareholdings could be mobilised as a means of intervention if the need arose: indeed, the Labour Party in Opposition announced a clear intention of doing so (Labour Party 1986: 6). However, as the target for asset sales has increased in size, such residual shareholdings have been sold off, and in more recent sales, such as that of British Gas, no residual shareholding has been retained. The most important example, the 49.8 per cent of ordinary shares of British Telecom left in government hands, is subject to a pledge that it be retained only until April 1988, although some recent reports have indicated that it is unlikely to be sold in the wake of mounting criticism of British Telecom (*Financial Times*, 29 September 1987; *Financial Times*, 4 November 1987).

In several cases, government directors are to be appointed to the boards of privatised companies. In the past such directors in other companies have had a very limited role as a means of government intervention. Thus in the case of BP there were both government directors and a substantial shareholding, yet it was widely recognised as possessing an extreme independence from governmental influence, though this was no doubt due largely to its market position and financial success. In the wake of the De Lorean affair, in which the presence of government directors had not prevented serious loss of public funds, the Public Accounts Committee recommended strengthening their monitoring functions, but this is unlikely to herald any major change. Anyway, company law at present places restrictions on the ability of such directors to represent interests other than those of the company as a whole (see HC 33 1985–6 and Cmnd 9755). It thus appears unlikely that government directors will play any important part as a means of influence on privatised companies; at most, they will ease the communication process between company and government.

Thus residual shareholdings and the appointment of government directors do not appear to provide important means for government intervention in privatised concerns. The next technique is, however, of the greatest importance. This is the retention of 'special' or 'golden' shares giving government a veto over major decisions—in particular, the power to block unwelcome takeovers. Aside from the product market, the other major form of market discipline is theoretically exercised through the market for corporate control and the threat of takeover. Golden shares replace this discipline by limiting shareholdings and preventing undesirable takeovers. This is accomplished through a variety of devices written into the articles of association of the privatised companies. Why different forms were used is not always obvious, and there has been little public debate over the choice of devices.

When a special share is created, held by a government nominee, the effect is to ensure that certain parts of the company's articles of association are only alterable with the consent of the special shareholder. The object is to prevent the limitations on shareholdings being avoided by altering the articles. In five cases (Amersham International, British Airports, Jaguar, Cable & Wireless, Rolls-Royce) the disposal of a material part of the assets of the group of companies, roughly 25 per cent of the assets, is also deemed to be a variation of rights of the special shareholder. in these cases, any substantial restructuring of the company can only take place with the consent of the government, and how such negotiations will be conducted is left entirely to the discretion of the parties involved.

The central object of the scheme is the prevention of undesired takeovers. As regards Britoil and Enterprise Oil this is accomplished in a relatively simple way. In summary, if any person, alone or acting in 'concert', controls more than 50 per cent of the votes then, from the date that occurs, the special share has, on any resolution in a company general meeting, one more vote than the total number of votes which are not controlled by the Secretary of State. So voting control remains with the government.

Subsequent flotations have utilised different special share arrangements which operate at a lower level of shareholding, domestic or foreign. Whenever any person, alone or in 'concert', obtains control of more than 15 per cent of the voting shares, then the directors shall (may, in Cable & Wireless) serve a notice on such a person requiring them to dispose of their excess shares. If this is not accomplished within a specified time limit, then the directors shall arrange for the disposal of the excess shares. After service of the first notice, the excess shares confer no voting rights at general meetings, these rights vesting, usually, in the chair of the meeting. Due to problems with European Community law, British Airways has a unique provision, as regards foreign shareholdings.

The significance of these provisions is that they replace the market for corporate control with the protective presence of government. The logic of the corporate market argument is that if managers are inefficient, the share price of a company will be lower than it could be with an efficient management. This provides the opportunity for an outsider to make a takeover bid and, if successful, to reap a profit. The mere threat of takeover is enough to encourage efficiency among managers. However well or badly this market may work in the ordinary case, it is simply non-existent when a golden-share scheme is in operation.

This is illustrated by the Enterprise Oil and Britoil experiences. When Enterprise Oil was floated, oil prices were depressed and the issue was undersubscribed. Rio Tinto Zinc (RTZ) applied, in secret, for 49 per cent of the shares, which meant that they would have gained *de facto* control without bringing into play the special share. This proved politically embarrassing for the Government, which had promised that Enterprise Oil would remain an independent company. It refused to allocate RTZ more than 10 per cent of the shares but, following the allocation, RTZ established a 29 per cent stake in Enterprise Oil. In the ensuing Parliamentary debate it was clearly stated that the special share would be used to block any takeover bid (62 HC Debs col. 20, 2 July 1984).

More embarrassment has been caused by BP's takeover bid for Britoil, which demonstrated that golden shares *will* be used as a means of policy intervention. Although the Treasury, and later the Chancellor of the Exchequer (125 HC Debs cols. 13–16, 11 January 1988) announced that the golden share would be used to prevent a takeover bid, this did not discourage BP, which has launched a full bid, after receiving clearance from the Takeover Panel, apparently hoping that either the Government will back down out of embarrassment or that company law will prevent the board acting against BP's interests. The problem has been exacerbated by the build-up of the Kuwait Investment Office's stake in BP which, according to one report (*Independent*, 8 January 1988), received the tacit blessing of the Government. BP has no protection in its articles of association because it declined the Government's offer of a golden share (*Observer*, 20 December 1987). Although the Government has powers under the Industry Act 1975 to limit foreign control of BP, it is not clear why the Government should be hostile to the BP bid for Britoil and yet, at worst, neutral about the Kuwait stake. There seems to be a lack of clarity about policy aims and, in particular, a lack of thought about the purposes for which the protective provisions of the golden share should be used (*Financial Times*, 15 December 1987 and 5 January 1988).

Other means of government intervention

The Government has retained, or created, important means of intervention in relation to privatised industries. These operate to modify the product market through government involvement in the regulatory process and in the market for corporate control through possession of golden shares. There are, however, a number of other ways in which government will be intimately involved in the affairs of privatised concerns. The first of these is through contracting. We have dealt with this issue at greater length elsewhere and will briefly summarise it here (Graham and Prosser 1987: 41–9).

Many of the most important privatised concerns are necessarily dependent on government contracts. Thus government is British Telecom's largest customer, but contractual relations are of particular importance in the defence field where particularly close links will continue to exist in the case of the warship yards of British Shipbuilders, Rolls-Royce, British Aerospace and the Royal Ordnance factories. Indeed, the saga of the Royal Ordnance sale illustrates the close interdependence of privatisation and contracting. Originally the company was to have been floated as a single entity. However, at a late stage the flotation was abandoned; it appears that this change of plan was for two reasons. First, it seems that the Ministry of Defence had refused to allow Royal Ordnance's accounts to be published or to provide an opening balance when it had become a limited company some months earlier: the inevitable lack of information as to its potential would hardly have produced a successful flotation (Veljanovski 1987: 125). Secondly, strong protests had been received from its sole British competitor in tank manufacture about the non-competitive award of a contract for the supply of tanks to Royal Ordnance shortly before the proposed flotation.

When the flotation was called off, it was decided to sell the tanks factory to the competitor at a price based on a confidential formula, thus creating a domestic monopoly in tank manufacture. Bids were invited for the remainder on the basis of a selling memorandum, and eventually the company was sold to British Aerospace, itself only recently privatised, thereby creating the largest Western defence manufacturing company outside the United States. In addition to the placing of the tanks contract, the Ministry of Defence had intended to prepare the company for flotation by placing with it all small arms ammunition contracts for three years, and virtually all explosives and propellant orders for seven years. After the cancellation of the flotation, the exclusive supply arrangements were retained, although the duration of that for explosives and propellant orders was scaled down to three years (*Financial Times*, 5 March 1987). The sale was heavily criticised both because of allegations that the method of sale had resulted in a serious undervaluation of the company, and because of the competition implications discussed above.

In other examples of privatisation, it has been alleged that the award of contracts has been used by government to 'fatten up' firms about to be privatised and to ensure their continued existence after sale, notably in warship procurement (Graham and Prosser 1987: 44–5). The key point is that, as we have seen in the case of Royal Ordnance, both before and after privatisation major links exist between government and the industries concerned through contracting. This would be less important if the contracts in question had any resemblance to the classic private law model of bargaining between equal parties in a competitive environment. In fact, such a contract—for example, the Joint Understanding between British Coal and the CEGB—will be 'more than a technical device for securing the wanted goods and fixing the reward of the supplier; it is also a kind of treaty, by which the conditions of a relationship of interdependence are established' (Turpin 1972: 264).

Despite the current moves towards a greater use of competitive tendering and fixed-price contracts in defence procurement, much contracting in the defence field is by its nature not susceptible to competitive bidding; currently, about 60 per cent by value of the Ministry of Defence's purchases are non-competitive, and this is expected to reduce only to about 40 per cent as a result of the current reforms. Where competitive tendering is not possible, a profit formula based on recommendations from the highly-corporatist Review Board for Government Contracts is employed (for more details see Ch. 3 above [see *Waiving the Rules*, Ch. 3]. This formula and its application have recently been subject to heavy criticism from the Public Accounts Committee as allowing the defence industry to fare much better than the rest of British industry during the recession: 'we feel bound to conclude that the profit formula has, in recent years, been applied in a very one-sided manner in favour of defence contractors'. The Committee has concluded that control of expenditure on defence equipment has been 'one of the most conspicuous records of failure in the whole field of Public Accounts' (HC 390 1984–5; HC 56 1985–6; HC 406 1985–6).

Apart from the question of whether contractual arrangements are such as to maximise value for money, contracting may provide a means of effective government involvement in private industries. In the 1960s the ship-building industry was reorganised largely through the use of government contracts (B. Hogwood 1979: 79, 87, 168, 171, 189–90 and esp. 264–5). More recently, a Public Accounts Committee Report on the supply of gases to the National Health Service revealed a process of negotiation over price setting very similar to that between government and nationalised industries, but without the equivalent minimal requirements of outside consultation (HC 67 1984–5). In the United States, government through contract is more fully recognised as a means of control; after a major Anglo-American conference on areas of public and private interdependence it was reported that there was 'general agreement that the US Government has achieved a greater degree of *de facto* management control over the aerospace industry through the contract device than the British Government has achieved by nationalising certain industries' (B. Smith 1971: 19). The danger is that a network of links between government and privatised concerns might develop through private law techniques largely immune to public law scrutiny.

Other opportunities will doubtless arise for the exertion of informal pressure by government on privatised concerns. A well-known example occurred during the Westland affair, when it is quite clear that the Secretary of State for Trade and Industry expressed considerable concern to the Chairman and Chief Executive of the wholly privatised British Aerospace about the company's involvement in a rival rescue plan for Westland to that favoured by the Secretary of State. Accounts of the key meeting between the Secretary of State and the Chief Executive differ, but it is clear that, at mildest, pressure, both explicit and implicit was exerted by the Secretary of State in an attempt to persuade British Aerospace to withdraw. It was assumed by, amongst others, the

Chairman of the company, that behind this lay a threat to withdraw government financial support necessary for the company to participate in the separate airbus project (for the different views see HC 519 1985–6, paras. 206–12; HC 169 1985–6 paras. 365 and 369 and Qs 781–800, 833–5; HC 193 1985–6 para. 121 and Qs 144, 869–83, 546–53, 558–85; Linklater and Leigh 1986: 145–8).

Openness and accountability

It is thus clear that there are a variety of legal devices available through which government can intervene in the decisions of privatised industries. In other words, to see privatisation as a straightforward example of the rolling back of the state would be misleading; although it limits the ownership of industry by the state, it is compatible with an extensive governmental role in the economy. Indeed, in an economy as complex and interdependent as that of modern Britain, it should not surprise us that no government can stand aloof from strategic industrial decisions.

This raises a new question of the greatest importance. If state intervention in industrial matters is inevitable even after privatisation, what degree of openness and accountability exist in relation to such intervention? This of course far transcends the issue of privatisation. Two central issues are the effectiveness of ministerial responsibility to Parliament and the degree of openness in government generally. In this sense the problems raised by privatisation are part of the more general problems endemic in British government, and are discussed more fully in Chapters 1 and 10 [see *Waiving the Rules*, Chs. 1, 10]. However, there are particular problems of openness surrounding privatised concerns, and these will now be examined.

Openness in regulation

The first issue is the openness of the regulatory arrangements. In particular, is regulation by government and the new agencies likely to be more open to public scrutiny than regulation by government (and to an extremely limited degree, the consumer councils) under nationalisation? It must first of all be said that privatisation has in some respects the potential to increase openness. For example, where licensing has been chosen as the mode of regulation, the licences must be published (see e.g. Department of Trade and Industry 1984; Department of Energy 1986) and these will set out the key regulatory provisions. In addition, the central problem to plague the consumer councils was their inability to acquire information from nationalised industries. Arrangements for obtaining information by the new Directors-General of Telecommunications and of Gas Supply are in principle superior, both by virtue of statutory provisions and through licences issued to regulated industries (Telecommunications Act 1984, s. 48; Department of Trade and Industry 1984, conds. 16, 20, 52; Gas Act 1986, s. 38; Department of Energy 1986, cond. 7). However, the arrangements for the provision of information to the Gas Consumers' Council also established as part of the privatisation of the British Gas Corporation retain many of the deficiencies of the nationalised industry consumer council provisions (Department of Energy 1986, cond. 8; for problems under nationalisation, see Prosser 1986, ch. 8).

Nor has the acquisition of information by the regulatory bodies always proved easy in practice. The reason lying behind this is that, despite the apparently liberal powers to acquire information given in licences, there are limited requirements for the regulated industries to collect information in a form which will be useful to the regulator. As one commentator has put it:

> It was open to the Government to diminish the risk of regulatory capture by defining in legislation or licences the type of information required—identifying the form of accounts by laying down principles and by defining cost categories and cost centres. In the case of both British Telecom and British Gas, these opportunities have been largely overlooked. Detailed reporting requirements have not been specified with the monitoring functions of OFTEL and OFGAS in mind. As a result the ways in which the industries allocate their costs are not as transparent as they ought to be.
>
> (Helm 1987: 51)

This can be partly attributed to the decision not to enforce the splitting of the industries into separate cost and profit centres on sale which would have enabled the performance of different regions to be compared (Helm 1987: 48).

The Director-General of Telecommunications has expressed himself satisfied with his access to information from British Telecom, though he would prefer more information to be provided as a matter of course rather than on request and has criticised the lack of regular accounting information to enable him to deal effectively with pricing complaints (HC 15 1985–6, Minutes of Evidence Q254; HC 461 1985–6 para. 1.15; cf HC7 1987–8 paras. 2.17–2.19). His major problems have occurred in relation to performance indicators. The nationalised industries had been encouraged to publish such indicators as a central means of accountability and, although their quality varied, they provided one of the most important tools through which consumer councils could assess performance. After privatisation, British Telecom refused to provide them. The Director-General then took steps to monitor services and commissioned public opinion surveys. At this point, British Telecom agreed to provide indicators, although there has been lengthy wrangling about the timing and form of publication; almost a year after the agreement to publish the indicators, the Director-General was reported as complaining that British Telecom had not yet 'agreed a plan of action with me, nor has it made any public statement about its intentions' (*Guardian*, 7 July 1987). Shortly afterwards, British Telecom started publishing performance indicators (*Financial Times*, 27 October 1987; OFTEL 1987).

The major problem of the supply of information to the regulator has arisen in gas. A lengthy wrangle occurred between the Director-General of Gas Supply and the British Gas Corporation over the information supplied to justify its first tariff change after privatisation. The Director-General considered that insufficient information had been given to satisfy him that forecasts forming part of the process had been properly prepared, and had to threaten court action to obtain more details, which were handed over the day before the company's annual general meeting (*Financial Times*, 29 July 1987; *Guardian*, 28 August 1987).

As regards access to information, then, the provisions for disclosure are wider than under nationalisation, but in practice licences have been drafted, and privatisation has taken place, in such a way as to limit the amount of information actually available. What about the procedures adopted by the regulators? The Director-General of Telecommunications has already shown an impressive degree of openness in approaching his task. He has:

made a commitment, in public statements, to be as open as possible in the discussion of issues arising out of my functions and duties. I intend to make public statements about major issues under review and to invite representations from any interested parties; I intend to establish contact with individuals, companies and representative bodies with interests in telecommunications so that I may become fully aware of their views on important issues; and I intend to give the fullest possible explanation of the basis for my conclusions, subject only to the need to respect commercial confidentiality.

(HC 457 1984–5, para. 1.27)

He has taken a variety of steps towards implementing this commitment, for example by publishing a number of consultative documents and receiving representations on them, and taking the initiative in establishing a Telecommunications Forum representing a variety of interests and organisations for consultative purposes. He has also committed himself to making public his advice to the Secretary of State in so far as matters of commercial confidentiality do not arise. This degree of openness is most encouraging, especially in comparison to the dearth of information available under nationalisation, though it is unfortunate that it is dependent on the liberal instincts of this particular Director-General rather than being required from all regulators.

When one examines the regulatory functions exercised directly by government, the picture is less impressive. In some areas procedural duties are attached to powers given to ministers; for example, in drawing up traffic distribution rules for airports, the Secretary of State is obliged to consult the Civil Aviation Authority, which in turn is to consult airport and aircraft operators and organisations representing them, and similar provisions apply to rules limiting aircraft movements and allocating airport capacity (Airports Act 1986, ss. 31–3). This no doubt reflects the tradition of relative effective participative provisions in civil aviation (for which see Baldwin 1985). As regards the licensing of telecommunications systems and gas suppliers, the procedural duties are minimal. Thus in the case of telecommunications licensing, all that is required is that the Secretary of State must consult the Director-General of Telecommunications, and in the case of public telecommunications systems (the major operators such as British Telecom and Mercury), the Secretary of State must give notice that the licence is to be granted, state reasons and consider representations and objections. Originally, the Telecommunications Bill did not contain any provision for the laying of licences before Parliament, but eventually an amendment was accepted providing for the laying of licences for public telecommunications systems (Telecommunications Act 1984, ss. 8(5), 9(2)). The provisions are similar for gas, except that no authorisations need be laid before Parliament. It has already been noted above that the vitally important price-restraint formula for British Telecom was the product of private negotiations between Government and the company. Such lack of procedural constraint stands in marked contrast to the arrangements for regulation of the United States utilities, where a range of devices including open hearings and highly sophisticated provisions for disclosure of information have been adopted (see e.g. Henney 1986).

Shareholder accountability

Thus the arrangements adopted for regulation after privatisation have on the whole produced limited improvements as regards the openness and accountability of the regulated industries. However, a major argument offered by government to justify privatisation is that it introduces a new and more direct form of accountability—accountability of the privatised companies to their shareholders:

> The existence of large numbers of shareholders who have both paid for their shares expecting a reasonable return *and* are customers interested in good service at a fair price is an irresistible combination and a powerful lobby in favour of both efficiency and price restraint.
>
> (Moore 1986a: 95, original emphasis)

However, when one comes to examine the mechanisms of shareholder accountability which actually exist, the claims seem less impressive. First, the number of individual shareholders in privatised industries has often declined rapidly and markedly (Buckland 1987: 254). Second, on the basis of the reports of the first few annual general meetings of the privatised companies, the scrutiny of management performance can only be described as perfunctory (for British Telecom see *The Economist*, 13 September 1985, 89–90; for TSB, *Financial Times*, 2 May 1987; for British Airways, *Financial Times*, 30 June 1987). In the case of British Gas, an attempt by industrial customers to have Sir Ian MacGregor appointed to the board of directors as their representative was heavily defeated at the annual general meeting (*Financial Times*, 28 August 1987). Indeed, the TSB has complained that it costs too much to send out its annual report to shareholders and is seeking permission to send out a shorter, less informative, document to them. As for private legal action by a disgruntled shareholder, this is unlikely given the current state of British company law (see Graham and Prosser 1987: 39–40).

As for takeovers, we have already discussed the effect of 'golden' shares on them. Even ignoring golden shares, there must be some doubts about the openness of the market for corporate control. Under certain circumstances, takeover bids are potentially subject to a reference to the MMC. It is worth re-emphasising what a discretionary and secretive process this is, particularly on the question of whether to refer or not. Although the present guidelines, for the exercise of the Secretary of State's discretion, state that references will primarily be made on competition grounds, this is not an inflexible rule, The Elders IXL bid for Allied Lyons was referred because of its unique financing (109 HC Debs col. 373, 28 January 1987). This can give rise to controversy. When BTR made its bid for Pilkingtons there were rumours that three of the ministers in the Department of Trade and Industry were not consulted and did not approve of the non-referral. Given our current constitutional conventions, only one side of the case was presented to Parliament (*Financial Times*, 16 January 1987).

Even when government is not directly involved, the takeover process can be very secretive. One of the best recent examples occurred in the proxy fight over Westland. At the last minute some 20 per cent of the shares were acquired on behalf of six unnamed beneficiaries, three operating through Swiss banks. This assistance proved crucial to the success of the Sikorsky bid. At the end of their investigation the Stock Exchange's committee concluded:

[We] found it difficult to understand why overseas buyers should consider it worth their while in order to gain voting rights to pay substantially above the prevailing market price. It is not beyond the bounds of possibility that there are six such ingenuous foreigners in the world, but [our] credibility was sufficiently strained to be sceptical as to the absence of a concert party of some sort.

(Stock Exchange 1986: 7; HC 176 1986–7)

The Council of the Stock Exchange accepted the committee's view that the law should be changed so that companies should be empowered to disfranchise shares registered in nominee names but where the ultimate beneficiary is not disclosed—a proposal which has not been implemented.

The notorious Guiness affair revealed equal problems with the operation of the takeover system. In its bid for Distillers Guinness offered its own shares in exchange for shares in Distillers. Therefore the higher the Guinness shares price, the more valuable and attractive the bid. It was arranged that allies of Guinness should purchase its shares, having been given an indemnity against any subsequent loss, thus raising the price and creating a false market (and being in breach of the Companies Acts) (Kochan and Pym 1987).

The point we wish to make is not that the takeover process can never be open and above board (indeed, this is the aim of the myriad of rules surrounding takeover bids), but that there are grave doubts whether the present regulatory mechanisms will ensure fair procedures (see Ch. 4 for an overview) [see *Waiving the Rules*, Ch. 4]. These doubts appear to be shared by the Government as a review of the Takeovers and Mergers Panel's work has taken place.

Accountability and contracting

Finally, accountability for government contracting with the industries is attenuated in the extreme. The complex systems of review of contracting by Federal Government in the United States are notable by their absence here, as is the sophisticated regime for the review of government contracts by the administrative courts which exists in France. In the United States the Federal Government's discretion is structured by procedural devices, and its decisions are subject to review by a complex network of independent institutions, including the federal courts. There are procedural protections for the debarment and suspension of prospective bidders, and a formal disputes procedure exists for dealing with grievances over awarded contracts (Nash and Cibinic 1977, Vol. 2, ch. 30; Calamari 1982; Steadman 1976). In the making of procurement regulations, although not subject to the 'Notice and Comment' requirements of the Administrative Procedure Act, the Office of Federal Procurement Policy requires that the views of interested non-governmental parties be given due consideration in the formation of federal procurement policy (Nash and Cibinic 1977, Vol. 1:42). In France there is a large case-law on review of government contracts by the public law courts (Brown and Garner 1983: 125–30). More recently, the increasing intervention of the state in the economy through contractual and quasi-contractual devices has given rise to a lively debate on whether the traditional methods of control and conceptualisations of the problem are adequate (Delmas-Marsalet 1969; Truchet 1980; Nitsch 1981).

The major constraint in Britain on central government is the work of the Public Accounts Committee in conjunction with the National Audit Office, and the Committee has undertaken extensive work in this area. However, it only examines individual cases on an *ex post facto* basis. For example, the report referred to above, on the application of the profit formula in defence contracting, came too late to prevent the payment of 'windfall profits' to contractors of between £220 million and £360 million. In a different context, the highly critical report on payments to De Lorean Motor Cars Ltd came too late for the recovery of over £70 million of wasted public money (HC 390 1984–5; HC 127 1983–4). Very recently, the Committee has complained about the failure to learn from previous experience and from its own reports, going back to the Ferranti affair of the early 1960s, in the development of major equipment for the Ministry of Defence from outside contractors (HC 104 1986–7). Moreover, the Comptroller and Auditor-General will not have access to the books and records of contractors but only to information in departmental files; this lack of access has been criticised by the Committee in relation to the books of the contractors who will manage the Royal Dockyards under the privatisation arrangements (HC 286 1985–6). (Compare these limitations with the remedies given to private contractors under the Local Government Bill.)

Conclusion

On close examination, then, the privatisation programme does not represent so major a change as appears at first sight. Of course, important and no doubt irreversible changes have been made to the ownership of British industry, but this has not achieved the fundamental cutting back of governmental powers suggested by the advocates of the programme. This means that major problems still remain concerning the openness and accountability of governmental intervention in the economy. The major stumbling block is still the secrecy characteristic of British government and the inadequacies of ministerial responsibility as a means of effective governmental accountability; inadequacies illustrated vividly during the Westland affair, which after all, concerned relations between the Government and companies which were privately owned. Thus privatisation does not offer a new constitutional departure, replacing inadequate mechanisms of political accountability with direct accountability to consumers through the play of market forces and to shareholders. Rather, the privatisation programme is one more example of the deficiencies of current British constitutional arrangements.

References

BALDWIN, R. (1985) *Regulating the Airlines*. Clarendon Press, Oxford.

BROWN, L. N. and GARNER, J. (1983) *French Administrative Law*, 3rd edn. Butterworths, London.

BUCKLAND, R. (1987) 'The Costs and Returns of the Privatization of Nationalized Industries', *Public Administration*, 65, 241–57.

CALAMARI, J. (1982) 'The Aftermath of *Gonzalez* and *Horne* on the Administrative Debarment and Suspension of Government Contractors', *New England Law Review*, 17, 1137–74.

DELMAS-MARSALET, J. (1969) 'Le Contrôle juridictionnel des interventions économiques de L'État', *Études et documents du Conseil D'État*, 22, 133–60.

DEPARTMENT OF ENERGY (1986) *Authorisation Granted and Directions Given by the Secretary of State for Energy to the British Gas Corporation under the Gas Act 1986*. HMSO, London.

DEPARTMENT OF THE ENVIRONMENT (1987) *The National Rivers Authority: The Government's Proposals for a Public Regulatory Body in a Privatised Water Industry*. HMSO, London.

GIST, P. and MEADOWCROFT, S. (1986) 'Regulating for Competition: The Newly Liberalised Market for Private Branch Exchanges', *Fiscal Studies*, 7(3), 41–66.

GRAHAM, C. and PROSSER, T. (1987) 'Privatising Nationalised Industries: Constitutional Issues and New Legal Techniques', *Modern Law Review*, 50, 16–51.

HELM, D. (1987) 'RPI Minus X and the Newly Privatised Industries: A Deceptively Simple Regulatory Rule', *Public Money*, 7, 47–51.
HENNEY, A. (1986) *Regulating Public and Privatised Monopolies: A Radical Approach*. Public Finance Foundation, Newbury.
HILLS, J. (1986) *Deregulating Telecoms: Competition and Control in the United States, Japan and Britain*. Frances Pinter, London.
HOGWOOD, B. (1979) *Government and Shipbuilding*. Saxon House, Farnborough.
KOCHAN, N. and PYM, H. (1987) *The Guinness Affair*. Christopher Helm, London.
LABOUR PARTY (1986) *STATEMENTS TO CONFERENCE*. Labour Party, London.
—— (1987) *Britain Will Win*. Labour Party, London.
LINKLATER, M. and LEIGH, D. (1986) *Not with Honour*. London, Sphere.
LITTLECHILD, S. (1983) *Regulation of British Telecommunications' Profitability*. HMSO, London.
—— (1986) *Economic Regulation of Privatised Water Authorities*. HMSO, London.
MOORE, J. (1986a) 'The Success of Privatisation', in J. Kay, C. Mayer and D. Thompson (eds.) *Privatisation and Regulation: The UK Experience*. Clarendon Press, Oxford.
—— (1986b) 'Why Privatise?', in J. Kay, C. Mayer and D. Thompson (eds.) *Privatisation and Regulation: The UK Experience*. Clarendon Press, Oxford.
NASH, R. and CIBINIC, J. (1977) *Federal Procurement Law*, 3rd edn. George Washington University, Washington, DC.
NITSCH, N. (1981) 'Les Principes généraux du droit à l'épreuve du droit public économique', *Revue du Droit Public*, 97, 1549–79.
O'BRIEN, D. P. (1982) 'Competition Policy in Britain: The Silent Revolution', *The Antitrust Bulletin*, 27, 217–39.
OFTEL (1987) *British Telecom's Quality of Service 1987*. OFTEL, London.
PROSSER, T. (1986) *Nationalised Industries and Public Control*. Basil Blackwell, Oxford.
SMITH, B. (1971) 'Accountability and Independence in the Modern State', in B. Smith and D. Hague (eds.) *The Dilemma of Accountability in Modern Government*. Macmillan, London.
STEADMAN, J. M. (1976) 'Banned in Boston—and Birmingham and Boise and . . . : Due Process in the Debarment and Suspension of Government Contractors', *Hastings Law Journal*, 27, 793–823.
STEEL, D. (1984) 'Government and the New Hybrids', in D. Steel and D. Heald (eds.) *Privatizing Public Enterprises*. Royal Institute of Public Administration, London.
STOCK EXCHANGE (1986) *Dealings in the Shares of Westland PLC*. Stock Exchange, London.
TREASURY (1984) *Nationalised Industries Consultation Proposals*. HM Treasury, London.
TRUCHET, D. (1980) 'Réflexions sur le droit économique public en droit français', *Revue du Droit Public*, 96, 1009–42.
TURPIN, C. (1972) *Government Contracts*. Penguin, Harmondsworth.

VELJANOVSKI, C. (1987) *Selling the State*. Weidenfeld and Nicolson, London.
VICKERS, J. and YARROW, G. (1985) *Privatisation and the Natural Monopolies*. Public Policy Centre, London.

Internal Resolution of N.H.S. Complaints

J. HANNA*

The Government has recently announced its intention to create a new N.H.S. complaints procedure to be implemented by April 1996.[1] The Government's proposals are in response to the report of the Wilson Review of National Health Service (N.H.S.) complaints, 'Being Heard',[2] which reported last year with recommendations for a radical overhaul of existing complaints systems. Given widespread dissatisfaction with the present procedures from health professionals, management and patient organisations, the proposals for reform are timely, if not long-overdue.[3]

The Government recommends a new simplified and fairer N.H.S. complaints procedure, embodying the principles recommended by the Review Committee.[4] Closely following guidance of the Government's Citizen's Charter Complaints Task Force, the Wilson Committee promotes the following principles for incorporation into any N.H.S. complaints procedure: responsiveness; quality enhancement; cost effectiveness; accessibility; impartiality; simplicity; speed; confidentiality and accountability.[5] The essential features of the new procedures follow the recommendations of the Wilson Committee. Underpinning the design of the new reforms is a choice by the Wilson Committee to use lessons from the private sector to develop and implement these principles in keeping with other reforms in the N.H.S., in particular the creation of an 'internal market' for health.[6] The public law implications of this more general trend in the market-orientated administration of public services are significant.[7] In the complaints context the adoption of a market model is used by the Wilson Committee to justify one of the core recommendations accepted by the Government in the creation of the new system, namely, the promotion of internal resolution of complaints by providers of services to patients. The aim of this analysis is to examine the assumptions that underpin this particular recommendation by the Wilson Committee to illustrate the need for caution in the adoption of lessons from the private sector.

* Manchester College, Oxford.

I. Learning from the private sector

Central to the Wilson Committee recommendations for reform of the existing system is the view that common lessons can be drawn across the public and private sector in the handling of complaints. The evidence for this was drawn heavily from a management consultancy report by Peter Gibson Associates that sought to draw lessons on complaints handling from other organisations in the private and public sectors.[8] It is clear from its presentation of relevant lessons that the Wilson Committee views most of the good lessons coming from practice in the private sector. Traditional public sector complaints machinery is considered by the Committee as too formal and hierarchical with an unhealthy tendency to publicise and escalate complaints through various stages of appeal up to an ombudsman or tribunal. Wilson comments that 'This no doubt reflects the public sector's commitment to fairness and adherence to published rules'[9] but goes no further in addressing the potential relevance of these values to accountability in the public sector. In contrast to experience from the public sector, the Wilson Committee found the lessons from complaints handling in the private sector instructive and appealing. The common lesson promoted by the report is that in dealing with complaints all organisations should aim to satisfy those who complain as part of a general strategy to retain customers and increase profits as part of Total Quality Management (TQM) within an organisation. TQM was developed in the private sector to promote an organisation-wide effort to improve quality through changes in structure, practices, systems and attitudes and in particular investment in high technology information systems.[10] The Committee is keen to transplant across the TQM paradigm as the key to a radical overhaul of existing complaints machinery.

The particular lesson that is used to justify 'first-line' resolution of complaints is that:

Customer satisfaction with how a complaint is handled goes down dramatically the more contacts the complainant has with the organisation. It should therefore be an objective to resolve as many complaints as possible at the first point of contact.[11]

II. The lesson of 'first-line' resolution for reforming complaints

The Wilson Committee considers that one of the major problems with the existing complaints machinery in the N.H.S. is the hierarchy and formality of procedures external to the providing institution which delivers health care services. Complaints against primary carers and against hospitals and their employees are currently considered under separate procedures which lack any common structure or principle.

Complaints against primary carers may be dealt with informally by a lay conciliator or formally by a Medical Services Committee (MSC) of a Family Health Services Authority.[12] An appeal lies from the MSC to an independent Appeal Unit. In hospital complaints,[13] complaints are divided according to whether they raise clinical or non-clinical issues. Although non-clinical complaints may be resolved by management, clinical complaints are subject to a three stage process involving informal discussion, referral to the Regional Medical Officer and finally, an Independent Professional Review by two specialists from outside the locality. The Health Service Commissioner has a role as external watchdog,[14] but is excluded entirely from clinical complaints and from any complaint which is under the jurisdiction of the FHSA. The Wilson Committee noted that there are approximately 2,000 Medical Service Committees and 300 Independent Professional Reviews annually. The Health Service Commissioner received some 1,200 complaints a year, of which 150–200 are actually investigated.[15]

One of the Wilson Committee's primary recommendations for reform to N.H.S. complaints procedures is to shift the forum for the resolution of the majority of complaints to the actual provider of health services in preference to Medical Services Committees or external reviews. 'First-line responses' by providers of services are promoted by Wilson as the most effective way of fulfilling the principles which underly proper complaints handling. In particular, practice-based resolution is seen as more accessible, responsive, quicker, cheaper and confidential than external systems and more likely to lead to improvements in quality of service.[16]

The Committee recommends that the majority of complaints should be resolved internally. Ideally, complaints should be resolved by a 'listening and first-line response' by practice staff within 48 hours of a complaint being made. Only if the first-line response is unsuccessful will the complaint proceed to investigation and/or conciliation by a senior officer within the practice. Finally, the intervention of the chief executive of a N.H.S. hospital trust, or a 'Complaints Executive' of an FHSA will be reserved for particularly serious cases or where dissatisfaction persists. Practice staff will have responsibility for identifying these serious cases.[17]

If internal procedures fail to resolve the dispute then exceptionally the complaint may be heard by an independent panel. The Committee envisages individual health authorities as perhaps having as few as ten such panel cases in any one year.[18] The final review for complaints envisages external investigation by the Health Service Commissioner who would be at the apex of the reformed complaints system.

In 'Acting on Complaints' the Government has accepted Wilson's recommendation for a two stage complaints procedure within the N.H.S., overseen by the Health Service Commissioner.[19] In stage one the following options would be available for resolving a complaint: an immediate first-line response; investigation and/or conciliation; action by an officer of the health authority or family health services authority or by the chief executive of an N.H.S. trust.[20] One significant qualification to the Wilson recommendations is the Government's proposal that the kind of response at stage one should depend on the nature of the complaint and the wishes of the complainant.[21] A complainant would not be required to go through every option for resolving a complaint sequentially. Nevertheless, the central recommendation that internal resolution of complaints by providers of services should be promoted is retained as a core feature of the design of new N.H.S. complaints procedures.

III. The appropriateness of the lesson

The Committee view the move to provider-based resolution as supported by academic studies. In particular, they rely on the view of Norman Lewis and Patrick Birkinshaw that 'providing, of course, that the citizen obtains the redress which is required, then complaints or grievances are best settled at an immediate and local a level as possible'.[22] Whilst local resolution appears generally attractive its effectiveness does depend, as the quotation suggests, on provider-based resolution actually delivering appropriate redress in most cases This requires further analysis.

The Wilson Committee purports to justify its choice of provider-based resolution in terms of a lesson transferable from the private sector. The issue of the general transferability of private sector institutions and values to the public sector is an important one and has been central to critical commentary on the N.H.S. reforms.[23] More generally, John Stewart and Kieron Walsh have argued against an over-simplified model of the private sector being adopted as appropriate for management in the public sector where public bodies continue to prioritise and ration services and where values of the public domain need to be integrated with any benefit that can be gained from management in the private sector.[24]

The adoption of the lesson of 'first-line' resolution serves as an interesting illustration of the difficulties of transferring simple lessons from a market model to a complex organisation like the N.H.S. Although the Wilson Committee acknowledges that the lessons learnt from other sectors 'may need some adaptation in application'[25] to the organisational structures within the N.H.S. there is a key failure to identify the differences between a private sector organisation and a N.H.S. provider of services. It is significant that the Wilson Committee considers the experience of health complaints systems in other jurisdictions as being of marginal relevance because of organisational differences in health service organisation in other countries.[26] In contrast, the issue of institutional differences between the N.H.S. and private sector companies such as Sainsbury plc is mentioned simply by way of qualification to the general assumption made by the Committee that common lessons apply.

The Wilson Committee assumes that most complainants in the N.H.S. will be satisfied by the sort of response that would be expected when returning a defective product to Sainsburys. Thus, the new procedures should empower N.H.S. staff to give a rapid, often oral, response when a complaint is made.[27] In some cases the Committee states it will be more appropriate to concentrate resources on 'remedying the fault' than on a detailed examination of the complaint.[28] The move to provider-based resolution rests on the controversial assumption that the relationship between complainants and respondent providers in the N.H.S. can appropriately be viewed as equivalent to that between a customer and provider of services in the private sector.

A. *Patients as equivalent to customers in the private sector*

It will be argued below that for economic and more wide-ranging reasons patients are vulnerable as complainants in the N.H.S. internal market and ought not to be treated as equivalent to customers in the private sector.

1. *Economic power.* The first and most obvious difference with the market model is that in the N.H.S. internal market users of health services simply do not have the purchasing power of their private sector counterparts. Commentators on the reforms have argued that the power of choice lies with the purchasers who are health service managers or professional fund-holders rather than directly with patients who are not privy to the contracts negotiated within the internal market.[29] Economic incentives exist for providers to please and respond to targets set by purchasing agencies rather than patients.

2. *Other differences in power.* The analogy between patients and customers assumes that complainants in the N.H.S. will be both able and

willing to complain directly to the provider. The Wilson Committee provides no evidence that this is likely to be so. Indeed, the Committee noted evidence from organisations early on in its report that would appear to suggest that forcing complainants to deal directly with the provider institution may inhibit access.[30] Patients may experience psychological barriers to complaining: gratitude, powerlessness, medical language, the erosion of a sense of entitlement; lack of information on procedures, standards and outcomes. Users of health services may have a long-term relationship as patient and may be dependent on continuity of care and some might feel threatened by the possibility of being struck off a general practitioner's list. Further, as the Wilson Committee noted itself, the proportion of complaints which follow bereavement is high.[31]

Although the Wilson Committee does make recommendations to provide support for complainants there is reason to doubt whether this goes far enough to overcome inequalities in the relationship between complainant and provider. The Committee supports the role of Community Health Councils (CHCs) in aiding complainants, but public participation by the CHCs or other external patient representation is not built into the organisational structures of either the provider institutions or purchasing authorities.[32] The Government is not intending to improve the status or resourcing of CHCs and is content to leave new initiatives for supporting complainants to local developments by providers of services.[33]

B. *N.H.S. providers as equivalent to private sector providers*

The main assumption underpinning the Wilson Committee's recommendations is that N.H.S. provider institutions will operate under the same incentives as private sector companies in responding to complaints. It will be argued that this ignores important constitutional perspectives and is not supported by an empirically informed understanding of the provider's role. Differences in economic incentives, cultural values and organisational constraints will be considered below.

1. *Economic incentives.* It has already been seen that economic incentives in the N.H.S. internal market work directly between purchasers and providers, not between providers and patients. The Wilson Committee does go some way in recognising the role of the purchasers in overseeing provider complaints procedures.[34] All providers will be expected regularly to establish what users think of their complaints handling and make an annual published report to appropriate agencies, including the main purchaser. Purchasers will be responsible for auditing complaints and

incorporating such audits within the contracting process. The Government proposes to review the systems for filing, recording and monitoring complaints in the light of the Wilson recommendations.[35] Nevertheless, there are two reasons that justify concern with the effectiveness of the purchaser's role in complaints in the N.H.S. context.

First, there is a serious question mark concerning the ability of the purchaser to fully oversee the quality of complaints handling. The Wilson Committee prefers to treat the issue of quality in complaints as part of the monitoring of quality in the contractual process between purchaser and provider. The Government has also proposed that performance management arrangements in the N.H.S. will take account of performance in satisfactory resolution of complaints. Yet the TQM management paradigm which treats quality as something which can be measured with the use of high-technology information systems and targets has been contested in the context of reform of the N.H.S. Critics of the N.H.S. reforms have pointed to the fact that quality testing in the internal market has tended to focus on efficiency concerns, such as waiting times or numbers treated, that are easily measured rather than overall effectiveness of health care services which is less susceptible to scientific verification.[36] It is not surprising, perhaps, that the only indicator of poor performance suggested by the Wilson Committee is the measurement of the numbers of complaints that are not resolved internally by the provider at stage one of the procedures.[37] There is a danger that those criteria that are verifiable through quality testing will be given undue priority over less tangible, but potentially more important, quality concerns. For purchaser oversight to be effective, full information about all aspects of quality of complaints handling will need to be available.

Secondly, the constitutional accountability of the purchasing authority in overseeing complaints is vital to the legitimacy of the monitoring role and to the integration of information from complaints. In the N.H.S. where service provision is not demand-led but is limited through public funding, a purchasing agent cannot simply meet the concerns of complainants, but must make decisions based on all information available on quality enhancement. This mediation of different interests in health care is a matter of political judgement which requires constitutional legitimation. Yet, the lack of a developed constitutional framework for the internal market has been an issue of some concern.[38] Indeed, it has been pointed out that a significant feature of organisational restructuring has been a 'purge' of local authority and professional representatives at all levels of purchasing and provider agencies.[39] This has not led to the development of new democratic processes, but to a proliferation of government appointees on smaller business-like boards. Given the focus in the report on complainants as consumers, it is surprising that the Committee only touches on the issue of lay participation in those institutions responsible for listening and monitoring complaints. The Committee suggests that the annual report on provider reviews of their complaints handling 'may' be sharpened by the inclusion of an independent person such as a health council Chief Officer, but even this minimal acknowledgement of public participation does not feature as a recommendation of the report.[40]

A final point on the adoption of the market model relates to the issue of investment in complaints handling by the provider institutions. In the private sector investment in complaints handling is viewed as integral to customer retention and increased profits. It is not at all clear that the Wilson Committee is correct in assuming that the costs of investment in complaints in the N.H.S. will be met by improvements in quality of service. Significantly, the Committee does not cost the actual investment needed or the likely financial benefits from such investment. Costs will include training of all practice staff, investment in new information systems and human resource costs of using practice staff to resolve complaints. The effect on the quality of health services of the addition of new responsibilities on 'first-line staff' such as receptionists and health professionals is assumed to be positive. Benefits to the provider of increased investment in complaints will only accrue to the extent that purchasing authorities are satisfied with the quality of the service delivered by the provider, including the handling of complaints. In the context of limited public funding in the N.H.S. the outcome of the cost-benefit analysis cannot be assumed.

2. *Cultural values.* There is a serious question of the effectiveness of N.H.S. providers responding to complaints. The Wilson Committee notes concerns from agencies giving evidence about the impartiality of provider institutions, in particular professional defensiveness and 'closing of ranks' under existing procedures.[41] In the light of this it is significant, and perhaps surprising, that the Wilson Committee recommends the promotion of provider-based resolution for the majority of complaints.

Although the Wilson Committee acknowledges impartiality as an important principle of complaints handling, it considers that public confidence would be promoted by ensuring that investigation was careful and accurate and by communicating to patients that they could take their complaint further if they remain dissatisfied.[42] The Committee does not, however, consider it necessary to have an independent element at every stage. Although it recommends that each provider must have a complaints officer to whom a complaint can be referred, this complaints officer would be internal to the practice complained about. In primary care this would be the practice manager or the senior partner. The Committee recommends that complaints procedures should include at some stage the possibility of complaints being considered by impartial lay people, but provides no guarantee that this will occur at the early stage of a complaint where most complaints are expected to be resolved. It is also suggested that practitioners or members of staff within the practice may be specially trained for this conciliation role.[43] The Wilson Committee's recommendations rest on the controversial assumption that the promotion of care and accuracy will provide a neutral set of values for the resolution of complaints. A fundamental question left unresolved by the Wilson Committee is the extent to which providers and complainants share a consensus on the relevant criteria to judge quality of care.

Empirical research on patient evaluation of good practice suggests that patient judgments on quality do not always accord with professional or government criteria of quality of care.[44] Further, in a recent study of complaints against GPs, Judith Allsop suggests that complaints can be understood as a story of what happened told from different perspectives.[45] Both the complainant and the GP interpreted the complaint from the viewpoint of their own experience, interests, knowledge and ideas about the handling of illness. For both parties, identities were at stake and conflict was likely.

There are some indications unless values are clarified, professional determinations will be used to resolve ambiguity in the new reforms. Although the Committee rejects the logic of a strict distinction between clinical and non-clinical complaints and makes clear that the investigatory panels at stage two of the procedure will have a lay majority, professional advisers and independent reports from health care professionals will be used at panel hearings of clinical issues.

The ambivalence in respect of values is significant given the Committee's commitment to integrating the concerns of complainants in quality improvements and sociological perspectives which suggest that the potential for conflict between complainant and provider may not be as easily resolved as the Wilson Committee suggests. Accuracy and care does not in itself mediate differences in values that may permeate a complaint. It is significant that the Committee considers that complaints against purchasers in the internal market represent a different context where issues of potential conflict and controversy may be raised. The Committee considers that because decision-making by purchasers is controversial, proper consideration ought to be accorded to complainants' views on policy and should be passed on directly to the Health Service Commissioner if an initial investigation fails. The Committee's treatment of purchasers is based on an assumption that politicised issues arise on the purchaser, rather than the provider, side of the market equation. Given the sociological perspectives mentioned above, that is seriously contestable.

3. *Organisational constraints.* Another assumption underpinning the Committee's recommendations is that the provider institution is capable of providing a neutral setting for the resolution of the complaint. No analysis is presented, however, on the relevance of organisational constraints on practice staff charged with responsibility for resolving complaints. One important issue would be how fairness in handling the grievance is affected by the pull of organisational constraints such as professional culture, internal politics and budget constraints. For example, recent research on managers as third party dispute handlers in hospital complaints argues that managers operate as important legal actors in this context and that an understanding of their role is crucial.[46] Managers often fulfilled a 'chameleon-like' approach to complaints which would be influenced by a sliding scale of factors including the impact on the organisation and the credibility of managers.[47] Further, a comparison of interviews with managers with a file study of complaints suggested that whilst managers often adopted the language of third party resolution, their role was in practice more superficial. Managers were often content to deal with the complaint on the basis of the facts and context as presented and where clinical issues were raised were much more likely to act in an agency role for the complained-against clinician.[48] Significant to the issue of provider reporting of the quality of complaints resolution was the finding that complainants might remain dissatisfied even though on the management account the dispute was resolved.[49]

IV. Conclusion

The Wilson Committee's adoption of 'first-line' resolution as a valuable lesson from the private sector illustrates the need for caution in assuming too readily the transferability of lessons from the private sector. The complexities of a quasi-market like the internal market of the N.H.S. raises issues of institutional structure and organisational culture that require a more sophisticated analysis that draws on constitutional perspectives and an empirically grounded understanding of complaints. A simple market paradigm fails to account for the complex economic, cultural and organisational arrangements in the N.H.S. In particular, there is an urgent need for discussion of the conflicting perspectives and value systems which are likely to be the result of opening up health services to the users' voice and how these conflicts will be resolved at all levels, but especially at the provider level.

In 'Acting on Complaints' the Government proposes that implementation of the reforms will be managed by the N.H.S. Executive. The Government's stated aim is to leave as much flexibility as possible for individual organisations, while ensuring that the essential features of the new system are implemented.[50] Whilst the promotion of a new complaints procedure responsive to complainants is likely to be welcomed by many, it is to be hoped that those responsible for developing and monitoring provider-based complaints procedures will be sufficiently sensitive to the complexity and importance of the issues at stake.

Notes

1. *Acting on Complaints: The Government's new proposals in response to 'Being Heard'*, the report of a review committee on NHS complaints procedures, Department of Health March 1995 (hereinafter *'Acting on Complaints'*).
2. *Being Heard, the Report of a Review Committee on N.H.S. Complaints Procedures*, Department of Health 1994, DO16/BH(1)M (hereinafter noted as 'Wilson').
3. *Suggested Future Arrangements*, National Association of Health Authorities and Trusts Complaints Working Party (London 1992) and *Complaints do Matter* (NAHAT, London 1993); *General Medical Services Committee's report from the Complaints Review Working Party* (GMSC, London 1992) and *A Health Standards Inspectorate: a Proposal by the Association for Community Health Councils of England and Wales and Action for Victims of Medical Accidents* (ACHCEW, AVMA 1992) and *The Way Forward* (Consumer's Association 1993).
4. *Acting on Complaints, op. cit.,* at I.
5. Wilson, at para. 161.
6. Wilson, at para. 118. For a detailed description of the reforms see D. Longley, *Public Law and Health Service Accountability* (Open University Press 1991).
7. D. Oliver, 'Law, Politics and Public Accountability' [1994] P.L. 238 and M. Freedland, 'Government by Contract and Public Law' [1994] P.L. 86.
8. Wilson, at para. 118.
9. Wilson, at para. 121.

10. B. Barnes, *Managing Change* (Pitman Publishing 1992).
11. Wilson, at para. 130.
12. N.H.S. (Service Committees and Tribunal) Regulations 1992 (S.I. 1992 No. 664) and guidance notes for FHSAs. See also the N.H.S. Medical (General Medical Services) Regulations 1992 (S.I. 1992 No. 635).
13. See Hospital Complaints Procedure Act 1985 and circular HC(88)37.
14. Health Service Commissioners Act 1993.
15. Wilson, at para. 294.
16. Wilson, at paras. 165 and 228.
17. Wilson, at para. 246.
18. Wilson, at para. 311.
19. *Acting on Complaints*, op. cit., at IV.
20. *Ibid.*, at 7.
21. *Ibid.*, at 8.
22. N. Lewis and P. Birkinshaw, *When Citizens Complain* (Open University Press 1993) at 9.
23. K. Walsh, 'Citizens, Charters and Contacts', in R. Keat, N. Whiteley and N. Abercrombie (eds) *The Authority of the Consumer* (Routledge 1994) and I. Harden, *The Contracting State* (Open University Press 1992).
24. J. Stewart and K. Walsh, 'Change in the Management of Public Services' (1992) 70 *Public Administration* 449.
25. Wilson, at para. 123.
26. Wilson, at para. 159.
27. Wilson, at para. 190.
28. Wilson, at para. 199.
29. See *supra*, note 23.
30. Wilson, at para. 78.
31. Wilson, at para. 49.
32. Wilson, at paras. 196, 289 and annex H.
33. *Ibid.*, at 16–8.
34. Wilson, at paras. 221–4 and paras. 329–35.
35. *Acting on Complaints*, op. cit., at XVI.
36. A. Coulter, 'Evaluating the Outcomes of Health Care' in J. Gabe, M. Calnan and M. Bury (eds), *The Sociology of the Health Service* (Routledge 1991).
37. Wilson, at para. 333.
38. See *supra*, note 23.
39. R. Klein, *The Politics of Health* (Longman 1989) at 239.
40. Wilson, at para. 222.
41. Wilson, at paras. 96 and 112–3.
42. Wilson, at para. 226–30.
43. Wilson, at para. 258.
44. C. Haigh Smith and D. Armstrong, 'Comparison of Criteria Derived by Government and Patients for Evaluating General Practitioner Services' (1989) 299 *British Medical Journal* 494.

45. J. Allsop, 'Two Sides to Every Story: Complainants and Doctor's Perspectives in Disputes about Medical Care in a General Practice Setting' (1994) 16 *Law and Policy* (No. 2) 149.
46. L. Mulcahy and S. Lloyd-Bostock, 'Managers as Third Party Dispute Handlers in Complaints about Hospitals' (1994) 16 *Law and Policy* (No. 2) 185.
47. *Ibid.*, at 196.
48. *Ibid.*, at 206.
49. *Ibid.*, at 206.
50. *Acting on Complaints, op. cit.*, at 24.

The Contractual Approach to Public Services: Three Examples

I. HARDEN

This chapter examines three areas in which major changes are occurring to the way public services are provided: the National Health Service; compulsory competitive tendering for local authority services and 'Next Steps' agencies. It is not intended to provide a complete explanation of the changes, but rather to draw out the common contractual elements which link developments in the three areas. The first is an institutional separation of the responsibility for deciding what service there shall be from the responsibility for delivering that service. The second is that this division takes the form of an agreed definition of rights and duties, which is intended to be binding, if not necessarily legally enforceable.

In itself, this re-structuring results in an expansion neither of individual legal rights, nor of the role of consumer preferences in decision-making about public services. Its intended effect is to improve the management of public services, though this is intertwined with a number of politically controversial goals. However, it will be argued in Chapter 4 [see *The Contracting State*, Ch. 4] that the contractual approach has a potential constitutional value that is separable from the particular political goals with which its implementation has been connected and which is concealed by the apparently simple notion of 'improved management'.

The NHS internal market

Statutory duties to provide, or ensure provision of, health services are imposed on the Secretary of State by the National Health Service Act 1977. Regional and District Health Authorities (RHAs and DHAs) exercise on behalf of the Secretary of State such of his functions as he directs. In addition to DHAs and RHAs, which are statutory corporations, there exist the non-statutory NHS Policy Board and NHS Management Executive (NHSME). The NHSME, headed by the Chief Executive, has terms of reference which include taking central responsibility for the operation and management of the NHS within Ministers' overall policy framework. The Policy Board advises the Secretary of State on policy formulation and strategic oversight of the health service.

The NHS is almost entirely funded from public expenditure. Most NHS services (prescriptions are the main exception) are provided free of charge to the ultimate consumer, including those of general practitioners (GPs) who act as gatekeepers for hospital services. Hitherto, the provision of such services was the direct responsibility of DHAs. The aim of the reforms introduced under the National Health Service and Community Care Act 1990 is to separate the role of the DHA into two separate functions—that of purchaser and provider—and to create an 'internal market" in public health care services. The new system began formal operation in April 1991, but its full impact will take time to emerge.

The concept of an internal market has three interlocking aspects: the creation of incentives to greater efficiency; the delegation of decision-making responsibilities to lower levels; and the principle of money following patients, so as to link resource allocation to service output (Crafts 1989; Hulme 1990). DHAs' primary responsibilities are no longer to provide health services but to assess the health needs of their populations and to arrange for those needs to be met by purchasing services from providers. GPs with larger practices may also become purchasers of services on behalf of their patients. Such 'fund holding' GPs receive directly from the RHA resources which would otherwise have been allocated to the relevant DHA. Although the fact is not emphasised in the official literature, there is no formal barrier to DHAs and GP fund holders purchasing services from private providers, nor to the NHS providing services to private purchasers.

At first, DHAs will continue to have direct management responsibility for most NHS service providers. However, hospital and other units can apply to become NHS Trusts, operationally independent of district and regional management (NHSME 1990a). The key idea of the internal market is that the relationship between a provider—whether a directly managed unit (DMU) or a Trust—and a purchaser (DHA or GP fund holder) should be in the form of a contract which specifies the agreed quality, quantity and cost of services to be provided.

DHAs are expected to wish to secure the maximum volume and quality of health services for their residents at the lowest available cost. Providers are assumed to want to secure the DHA's commitment to funding the maximum use of their facilities, at as good a price as possible, with as little interference in the way in which the service is delivered as possible (NHSME 1990b: para 3.34). The contractual framework of the internal market is supposed to harness these motivations by structuring incentives so as to achieve cost-effectiveness.

Three different types of contract are possible: 'block', 'cost and volume' and 'cost per case' (Department of Health 1989). Cost per case contracts involve *ad hoc* referrals, with the price being fixed on a case-by-case basis and with no prior commitment by either party as to the volume of cases which might be so dealt with. Although an important role was originally envisaged for cost per case contracts, later guidance says that transaction costs make such 'bespoke' contracts for individual patients uneconomic and that their use is likely to be relatively rare (NHSME 1990b: para 3.40).

The block contract involves payment of an annual fee in return for access to a defined range of services and facilities, but with no commitment to a specified volume of output. In essence, this is little different from previous management practice in the NHS. Cost and volume contracts specify outputs, in terms of numbers of patients treated. The form of such contracts favoured by the NHSME (1990b: para 3.39) involves a fixed price up to a volume threshold of treatment and a price per case being paid above the threshold up to a volume ceiling. The intention is that cost and volume contracts should increasingly replace block contracts, thus enabling DHAs to link resource allocation to outputs and providing clear incentives to DMUs and Trusts to improve their efficiency.

The NHS reforms have three aspects. The first is a revised form of NHS management, in which the contractual framework is intended to effect a cultural change towards a focus on outputs, cost-effectiveness and the clear identification of service priorities and requirements. The purchaser role will make DHAs formulate their requirements more clearly and so generate and use the information necessary for that role. Provider units will also need to adopt a similar attitude towards information about their costs. The second element is the integration of private purchasers and providers into a single health care market. Taken by itself, and whilst the obligations of the 1977 Act remain, purchasing and provision of services across the public/private divide is unlikely to alter public health care fundamentally. The sheer financial muscle of public spending on health will ensure that NHS decisions continue to determine the overall pattern of services. However, integration facilitates the marketing of health care services by public providers, particularly if NHS funding leaves them with surplus capacity.

The third element is competition within the NHS itself. Although emphasised in the earlier working documents, this has been down-played more recently, partly for political reasons but also because of the real difficulties it risks creating. Allocation of resources through a competitive market and through management processes structured as contracts are not different versions of the same thing, but fundamentally different processes. The first leads to outcomes which are not planned by anyone. The second is about making the planning process more effective.

A proliferation of GP fund holders and NHS Trusts, making genuinely independent decisions about the purchase and provision of services, would be incompatible with DHA management of health care resources in its area through purchasing. Hence, competition in the NHS is to be 'managed competition' (Appleby *et al* 1990). Prices, for example, are to reflect costs rather than responding to competitive opportunities (NHSME 1990b: para 4.2). The market players are also subject to discretionary regulation, direction and persuasion from the Department of Health. The Secretary of State has ample statutory powers for this purpose in relation to Trusts. GP fund holders, however, may be a different matter.

Patients will benefit from the changes to the NHS if, whilst other things remain the same, health needs are more accurately identified by DHAs and/or providers become more efficient. However, there is no trace here of greater individual legal rights, nor of the preferences of individual consumers being used to decide what services are provided. At no point do the contracts of the NHS internal market operate with the consumer as the customer. Patients can become agents only by paying for health care privately.

Competitive tendering and local authority services

This section of the chapter will first examine the concept of competitive tendering and its relationship to contracting out. It then considers compulsory competitive tendering for local authority services.

Competitive tendering occurs when a body which undertakes an activity 'in-house' (i.e., through employing and managing its own staff) invites outside contractors to tender for the work in competition with the in-house unit. The outside contractor could be either a public or a private body, though discussions of competitive tendering by public bodies usually emphasise the role of private contractors. Large-scale competitive tendering for services provided in-house has been promoted by government in the fields of defence (e.g., stores, security functions and facilities management); NHS support services (cleaning, catering and laundry); central government support services and local authority services.

Competitive tendering overlaps with 'contracting out', but is distinct in two ways. Firstly, if the in-house unit wins the competition then the service is not contracted out. 'Market testing'—providing information about market costs and an opportunity to get better value for money either by contracting out, or by improving efficiency in-house—has long been the declared basis of government policy towards competitive tendering (Health Circular 83 (18); Treasury 1986). Secondly, it necessarily involves competition, whereas, in principle, a public body could contract out a service without inviting competitive bids. However, it is general government policy that procurement of goods and services should be competitive (Treasury 1988a). There are also specific legal requirements which will be considered in due course.

The particular context may make competitive tendering a sudden death process for the in-house unit. If it loses the competition, it is broken up and the public body relies henceforth on outside contractors for the service. Competitive tendering may thus be simply a stage on the way to permanent contracting out of a service. Alternatively, the in-house unit may be subject to a more graduated version of market discipline, with the opportunity to improve its performance, though it is in the nature of competitive tendering that failure to compete effectively will sooner or later lead to its demise.

Compulsory competitive tendering (CCT)

CCT was first applied to construction and maintenance work carried out by direct labour organizations (Local Government Planning and Land Act 1980) and was phased in for a range of other services under the Local Government Act 1988. The basic principle of the legislation is that competitive tendering is required before a local authority is permitted either to undertake defined categories of work for itself ('functional work'), or to perform work under contract for another authority (under the Local Authorities (Goods and Services) Act 1970). Discussion here will focus on CCT and functional work under the 1988 Act, because some of the services concerned are either direct or straddle the boundary between direct and indirect services. The Act lists a number of defined activities, with power to add to the list and to provide for exemptions from it, by order (see Table 1 below). In-house units providing such services are known as direct service organizations (DSOs).

The legislation is complex, partly because of the element of compulsion and thus the need to avoid loopholes. What follows is no more than an outline of the main CCT requirements. If an authority intends its DSO to carry out work within a defined activity, it must prepare a detailed specification, advertise the work in a local newspaper and invite expressions of interest in carrying it out. Of those from the private sector who respond, at least three (or all, if there are three or less) must be invited to submit a bid. However, if a large number of contractors respond, four to six should be invited to bid (DoE 1991a: para 28). There must also be a written bid on behalf of the DSO. This may only be accepted in preference to a lower bid if it can be shown that there were sound reasons to justify doing so.

Specification of contract requirements is a complex and difficult task. However, as with NHS contracts, it is the process of specification which requires authorities clearly to identify what it is that they are using public money to provide. In practice, local authorities have been able to do so sufficiently clearly to enable them to use contracts analogous to 'cost and volume contracts' in the health service (Walsh 1991: para 5.15). In carrying out the work, a DSO must comply with the detailed specification for which it bid.

As well as regulating the tendering process itself, the legislation imposes duties on local authorities to keep separate accounts in relation to any work done by a DSO which is subject to competitive tendering. It also gives the Secretary of State power to prescribe financial objectives for an authority in relation to work it undertakes within a defined activity. The effect of these provisions is to enable central government to determine the market performance which DSOs must achieve in order to survive (see DoE 1988: Annex A).

Just as significant, however, is the impact which CCT has on the culture of local authorities. The accounting provisions and financial targets are fundamental to the creation of separate purchaser and provider, or 'client' and 'contractor' roles. The Audit Commission has emphasised the importance of re-designing institutional structures so as to provide a clear definition and separation of the two roles and this approach was also endorsed by the main public sector accountancy body (Audit Commission 1989; CIPFA 1989: para 6.5 and see Table 1). Once such an institutional separation of roles comes about, perceptions of interests and patterns of behaviour tend to change as a result.

If there is sufficient interest from the private sector to provide competition for DSOs, local authorities will inevitably come to be identified primarily with the client role, at any rate in relation to services subject to competitive tendering. After all, the client role will continue even if the DSO does not. Furthermore, the best a DSO can hope for is to beat the

TABLE 1. *Services subject to CCT under the Local Government Act 1988**

(a)	Collection of refuse	household and commercial waste.
(b)	Cleaning of buildings	interior and window cleaning but not other exterior cleaning.
(c)	Other cleaning	removal of litter and emptying of litter bins; street cleaning and gully emptying; cleaning of traffic signs and street name plates.
(d)	Catering (education welfare)	providing ingredients for, preparing and delivering and serving meals and providing refreshments. Exceptions include delivery (but not preparation) of meals on wheels
(e)	other catering	
(ee)	Management of sports and leisure facilities.	
(f)	Maintenance of ground	cutting and tending grass; planting and tending trees, hedges, flowers etc.; weed control.
(g)	Repair and maintenance of vehicles	excludes accident repairs and maintenance of police and fire service vehicles.

* A local authority is exempted from the CCT requirements in respect of a particular category if its expenditure on that activity through a DSO is less than £100,000 a year.

Sources: Local Government Act 1988, s. 2(2) and Schedule 1; The Local Government Act 1988 (Competition in Sports and Leisure Facilities) Order SI 1989 No. 2488; The Local Government Act 1988 (Defined Activities) (Exemptions) (England) Order SI 1988 No. 1372. Cm 1599 promises to extend CCT to the field of housing management as well as to further indirect services.

private sector at its own game and according to its rules. The client role, on the other hand, involves distinctively public, governmental functions:

- assessing consumer demand and satisfaction;
- developing new ideas on service provision and quality;
- defining desired levels and quality of provision;
- planning for and securing adequate financial resources;
- managing the competitive tendering process;
- monitoring achievement against policy (Audit Commission 1989).

In practice, DSOs appear at present to be winning a substantial proportion of the work for which they bid (see e.g., Walsh 1991), but the Audit Commission's principal document on CCT is sceptical at a number of points about their long-term future (Audit Commission 1989). The view of central government is that:

local authorities' role in the provision of services should be to assess the needs of their area, plan the provision of services and ensure the delivery of those services.

There are also fields in which local authorities will continue to have important regulatory functions and providing roles. But councils should be looking to contract out work to whoever can deliver services most efficiently and effectively. (DoE 1991b: para 4.)

This formulation is nicely ambiguous, since it is not clear whether 'contract out' is being used in a sense which implies that the contractor is necessarily private. However, it remains the case that although CCT necessarily gives primacy to the client role in defining the purpose of a local authority in relation to those services which are subject to competition, it is not inconsistent with its retaining a continuing contractor role.

Figure 1

The Audit Commission's view of options for the client-contractor split:

	Unacceptable	Undesirable	Possible	Ideal
Members	Same Committee	Same Committee	Same Committee plus DSO plus client subcommittees	Client Committee DSO Board
Officers	Same officers	Different officers in same departments	Client department DSO	Stand alone or umbrella DSO. Client departments plus perhaps contract supervision unit.

Source: Audit Commission (1989) p.6

As with the NHS internal market, CCT may benefit consumers of local authority services, but it does not give them legally enforceable rights to services, still less make them sovereign. The customer/client/purchaser is the local authority itself. Its authoritative decisions determine the supply of services. The 1988 Act creates a general duty not to take into account 'non-commercial considerations' in exercising certain contractual functions. This prevents local authorities from seeking to achieve most of the policy objectives which could be pursued through 'contract compliance'. Subject to this limitation, however, and within the ambit of their statutory powers and duties, exactly what services they buy with their money is for local authorities themselves to determine. CCT does not even require them to explore the comparative costs of different service levels (Hartley 1990).

'Next Steps' agencies

Contracts in the NHS and competitive tendering for local authority services are directed towards separating the client (or purchaser) and contractor (or provider) functions. The broader aim is to effect a cultural change in the management of public services. The 'Next Steps' programme has been called 'the most ambitious attempt at Civil Service reform in the twentieth century' (Treasury and Civil Service Committee 1990). It also involves a separation of the functions of deciding what a service should be and delivering that service cost-effectively. Again, the overall aim is to effect a profound cultural change, in order to improve the management of public services.

The desirability of cultural change in the civil service is a long-standing theme. The Fulton Report (Cmnd 3638) in the late 1960s emphasised the importance of specialist skills, as opposed to the civil service 'cult of the generalist'. It made recommendations to improve efficiency and effectiveness by the creation of accountable management structures with a clear allocation of personal responsibility. The Fulton Committee was favourably impressed by the Swedish model of small central departments making policy which was then executed by autonomous agencies. It recognized that this would raise Parliamentary and constitutional issues, but recommended early and further consideration of 'hiving-off' areas of civil service activity to autonomous public boards or corporations (Cmnd 3638: paras 188–191). No such examination was conducted in public, but 'hiving-off' took place through the creation of quangos such as the Manpower Services Commission and by putting the operation of bodies such as the Royal Mint and the Post Office on a more commercial basis.

Though the term is not used, 'hiving-off' is a central aspect of the 'Next Steps' programme, so called after the report of that name from the Prime Minister's Efficiency Unit (Efficiency Unit 1988). Again citing the Swedish model, the report recommended that executive functions of central government departments be transferred to agencies. These should have operational independence and responsibility, within a framework of policy, resources and performance targets set by departments. The central civil service should eventually consist only of a small core, engaged in the function of servicing ministers responsible for policy matters. However, the report did not address other features of the Swedish agency model. In particular it neglected to mention that Swedish ministers are by law prevented from interfering in matters which are the responsibility of agencies. Furthermore, Sweden not only has a developed system of administrative courts but also four ombudsmen, who carry out inspections on their own initiative as well as receiving complaints from citizens. The Swedish ombudsmen are also prosecutors and part of the authority their office enjoys stems from the fact that the Swedish penal code makes wilful or negligent maladministration a criminal offence (Ragnemalm 1991).

Following acceptance of the Efficiency Unit's recommendations, the first Next Steps agency was established in mid-1988. The pace of change thereafter has been rapid. By May 1991 more than fifty agencies had been established, covering over 180,000 civil servants (see Table 2 below). The largest is the Benefits Agency which has carried out many of the functions of the Department of Social Security since April 1991.

The Next Steps programme is not merely the implementation of Fulton after two decades. In the interim, a major change had occurred in perceptions of the weaknesses of the traditional civil service. The core problem was identified as the exclusive focus on policy-making as its *raison d'être* (Efficiency Unit 1988). Other tasks were regarded as of lower status and, in particular, the effective management of resources was neglected, particularly as a result of the lack of external pressures demanding improvement in performance. Rather than Fulton's prescription of more experts in a variety of specialisms, the aim of Next Steps is to improve the quality and efficiency of government through better management.

This aspect of Next Steps builds on the Financial Management Initiative (FMI), which in turn was the successor to Sir Derek Rayner's efficiency scrutinies of the early 1980s (see Harden and Lewis 1986: 134 and references). A major goal of both the efficiency scrutinies and the FMI was the control and reduction of public expenditure. However, the FMI also focused on accountability for the effective management of resources, requiring 'a clear view of . . . objectives; and means to assess, and wherever possible measure, outputs or performances in relation to

TABLE 2 *Next Steps agencies established by May 1991*

	No. of Employees
Building Research Establishment	690
Cadw (Welsh Historic Monuments)	220
Central Office of Information[1]	730
Central Veterinary Laboratory	580
Chemical and Biological Defence Establishment	580
Civil Service College	210
Companies House	1,150
Defence Research Agency	11,700
Directorate General of Defence Accounts[2]	2,100
Driver and Vehicle Licensing Agency	5,450
Driving Standards Agency	2,050
Employment Service	5,600
Forensic Science Service	580
Historic Royal Palaces	300
Historic Scotland	580
HMSO[1]	3,300
Hydrographic Office[2]	880
Insolvency Service	1,450
Intervention Board	910
Laboratory of the Government Chemist	320
Land Registry	10,400
Meteorological Office	2,250
Military Survey[2]	850
National Engineering Laboratory	430
National Physical Laboratory	820
National Weights and Measures Laboratory	50
Natural Resources Institute	390
National Health Service Estates	120
Occupational Health Service	100
Ordnance Survey	2,500
Patent Office	1,150
Queen Elizabeth II Conference Centre	70
Radiocommunications Agency	500
RAF Maintenance[2]	5,700
Rate Collection Agency (Northern Ireland)	280
Recruitment and Assessment Services Agency	320
Registers of Scotland	1,100
Royal Mint[1]	1,050
Scottish Fisheries Protection Agency	230
Service Children's Schools (North West Europe)[2]	2,300
Social Security Benefits Agency	68,000
Social Security Contributions Agency	6,600
Social Security Information Technology Services Agency	3,350
Social Security Resettlement Agency	510

TABLE 2 cont.

	No. of Employees
Training and Employment Agency (Northern Ireland)	1,700
UK Passport Agency	1,200
Vehicle Certification Agency	70
Vehicle Inspectorate[1]	1,650
Veterinary Medicines Directorate	70
Warren Spring Laboratory	320
	183,460
Customs & Excise[3] (30 Executive Units)	26,800
Total employees	**210,260**

1. Trading Funds
2. Defence Support Agency. Figure does not include service personnel.
3. Moving towards full operation on Next Steps lines following publication of Framework Documents.

Source: Goldsworthy 1991: 40

those objectives' (Cmnd 9058). To this end, it sought to match responsibility for operational decisions with authority for committing resources and to delegate decision-making as far down the line as possible.

Next Steps goes further by separating the managerial or executive role of the agencies from the policy-making role focused on ministers. In addition, each agency is headed by a chief executive who has responsibility for and autonomy in the management of the resources provided to it. The intention is that Next Steps should lead to a federal structure of more autonomous units (Cm 841). Ministers and departments should to the greatest extent practicable stand back and leave agency managers free to manage (Efficiency Unit 1991: 30).

Such independence is necessary to raise the status of management in the civil service, so that being an agency chief executive, for example, is seen as a worthwhile and high-status role for a high-flying civil servant. However, separation of agencies from departments is not just about raising the profile of management skills in the public service, it is also intended to put those skills to work. For agencies to be able to function autonomously, it is essential to be explicit about the objectives of a particular agency, the resources which it will have, the specific targets it is to meet and how its performance will be monitored. Government has accepted that the emphasis in setting targets should be on outputs rather than inputs (T&CSC 1990, Cm 1263).

Framework documents
The key mechanism through which the above principles are put into effect is the 'framework document' (FD) for each agency. The original Next Steps proposal envisaged setting out the policy, the budget, specific targets and the results to be achieved; specifying how politically sensitive issues were to be dealt with and the extent of the delegated authority of management (Efficiency Unit 1988: para 20). No formal definition of a FD has ever been produced but the concept and its perceived importance have developed in the course of implementation of the programme. The Project Manager in charge of Next Steps (Sir Peter Kemp) has emphasised that each FD is a unique document, tailored to the requirements of the particular agency. However, although there is no standard pattern, there are certain common elements. A FD usually:

- sets out the status, aims and objectives of the agency and its services;
- makes clear the relationship between the department and the agency and in particular the rights and responsibilities of the agency chief executive, with specific reference to financial planning and control and to pay and personnel matters;
- sets out, or provides for, the performance targets which the agency must achieve (and the possibility of alteration to performance targets in-year);
- provides for accounts and reports to be published;
- provides for regular review of the agency's performance by the department and for a major review of the FD itself after three years;
- provides for approval of a long-term corporate plan and/or an annual business or operational plan for the agency and for the role of these documents in relation to the above matters.

FDs are fundamental to the relationship between agencies and departments, but the precise nature of that relationship has not been clearly defined. The White Paper on financing and accountability of the agencies refers to the 'department as "owner" and where appropriate customer of the Agency' (Cm 914: para 2.7). The official history of Next Steps says its aim is 'to establish a more contractual relationship between the Chief Executive and the Minister' and speaks of the 'underlying principle of a "bargain" between them' (Goldsworthy 1991: 7). The Treasury and Civil Service Committee recommended that each FD should 'be regarded as a contract' (T&CSC 1988). In evidence to the Treasury and Civil Service Committee, Sir Peter Kemp described the minister and chief executive as being in 'a quasi-contractual position, though I do not particularly like that expression' (T&CSC 1990: Q 170). The description was nonetheless accepted by the Committee and quoted by the Efficiency Unit (1991: para 2.6).

Although emphasising that FDs are not legal contracts for the provision of services, but statements about how services will be delivered, government has accepted that ministers and departments should avoid intervening in matters delegated to an agency. Where intervention is judged, exceptionally, to be necessary outside the terms of the FD, or of the normal planning and resource allocation arrangements, then it should be done explicitly (Cm 841, Cm 1263).

The precise legal status of FDs will be considered in Chapter 5 [see *The Contracting State*, Ch. 5]. However, it is clearly the intention that they should be a binding (albeit revisable) agreement of the respective rights and responsibilities of agency and department. Furthermore, they are meant to separate (i) responsibility for deciding what services an agency should perform and what targets it should meet, from; (ii) responsibility for the agencies' performance in delivering those services and meeting those targets. In other words, there is a split between departments and agencies like that between a local authority and its DSOs, or between a DHA and an NHS provider. FDs thus perform a function which is as 'contractual' as that of the other types of arrangement examined in this chapter. Next Steps is part of the more general move towards 'management by contract' in the public sector (Kemp 1990).

References

APPLEBY, John, ROBINSON, Ray, RANADE, Wendy, LITTLE, Val and SALTER. Judith (1990) 'The use of markets in the health service: the NHS reforms and managed competition', *Public Money and Management*, 27–33.

AUDIT COMMISSION FOR LOCAL AUTHORITIES IN ENGLAND AND WALES (1989) *Preparing for Compulsory Competition*, Occasional Paper no. 7. HMSO, London.

CHARTERED INSTITUTE OF PUBLIC FINANCE AND ACCOUNTANCY (1989) *The Audit Approach to Competition for Public Services*. CIPFA, London.

CRAFTS, Rosemary (1989) *Commissioning Health Services: contract funding in the NHS*. Public Finance Foundation, London.

DEPARTMENT OF THE ENVIRONMENT (1988) *Competition in the Provisions* (sic) *of Local Authority Services*, Circular 19/88. HMSO, London.

—— (1991a) *Local Government Act 1988 Part 1. Competition in the Provision of Local Authority Services*, Circular 1/91. HMSO, London.

—— (1991b) *Local Government Review: the internal management of local authorities in England*. DoE, London.

DEPARTMENT OF HEALTH (1989) *Funding and Contracts for Hospital Services*, Working Paper 2. HMSO, London.
EFFICIENCY UNIT (1988) *Improving Management in Government: the next steps.* HMSO, London.
——— (1991) *Making the Most of Next Steps: the management of Ministers' departments and their Executive Agencies.* HMSO, London.
GOLDSWORTHY, Diana (1991) *Setting Up Next Steps: a short account of the origins, launch and implementation of the Next Steps Project in the British civil service.* HMSO, London.
HARDEN, Ian and LEWIS, Norman (1986) *The Noble Lie: the British constitution and the rule of law.* Hutchinson, London.
HARTLEY, K. (1990) 'Contracting-out in Britain: achievements and problems' in J. J. Richardson (ed.) *Privatization and De-regulation in Canada and Britain.* Dartmouth, Aldershot.
HULME, Geoffrey (1990) 'Contract funding and management in the National Health Service', *Public Money and Management*, 17–23.
KEMP, Peter (1990) 'Can the civil service adapt to managing by contract? *Public Money and Management*, 25–31.
NHS MANAGEMENT EXECUTIVE (1990a) *NHS Trusts: a working guide.* HMSO, London.
——— (1990b) *Contracts for Health Services: operating contracts.* HMSO, London.
RAGNEMALM, Hans (1991) 'The Ombudsman', unpublished paper to seminar of the European Commission for Democracy through Law.
TREASURY (1986) *Using Private Enterprise in Government: report of a multi-departmental review of competitive tendering and contracting for services in government departments.* HMSO, London.
——— (1988a) *Public Purchasing Policy: Consolidated Guidelines.* HM Treasury, London.
TREASURY AND CIVIL SERVICE COMMITTEE (1988) Eighth Report 1987/88 *Civil Service Management Reform: the Next Steps*, vol II HC 494–II. HMSO, London.
——— (1990) Eighth Report 1989/90 *Progress in the Next Steps Initiative*, HC 481. HMSO, London.
WALSH, Kieron (1991) *Competitive Tendering for Local Authority Services: initial experiences.* HMSO, London.

Command Papers

Cmnd 3638, *The Civil Service*, 1968.
Cmnd 9058, *Financial Management in Government Departments*, 1983.
Cm 841, *Developments in the Next Steps Programme*, 1989.
Cm 914, *The Financing and Accountability of Next Steps Agencies*, 1989.
Cm 1263, *Progress in the Next Steps Initiative*, 1990.

The Next Steps Initiative: An Examination of The Agency Framework Documents

P. GREER

Introduction

In 1988 the Prime Minister's Efficiency Unit produced a report, 'Improving Management in Government: The Next Steps', which has since become the manifesto of change in Whitehall. The main aims of the initiative are to improve management in government and to deliver services more efficiently and effectively within available resources. The rationale behind the initiative is the recognition that 'the Government machine is too big and its activities too diverse to be managed as one unit'. The solution is the current move to create executive agencies from the operational arms of Whitehall.

The issues raised by the initiative are fundamental, questioning existing perceptions of the functions and nature of the British executive and of the executive's relations with Parliament. The first evidence of the likely resolutions of such issues is available in the framework documents which, along with the annual business plans and the five yearly corporate plans, define the framework in which agencies are to operate. This article examines the first 34 framework documents of the agencies established up to the end of March 1991. A list of these is shown in appendix 1. Essentially this article covers two areas: first, it considers the similarities between the documents and the unresolved dilemmas apparent therein; second, it examines the differences and considers the reasons for these variations and their implications.

The author wishes to thank Rudolf Klein for his helpful comments on this paper, and the Leverhulme Trust for funding her work on the Next Steps initiative.

The similarities between the framework documents

The framework document is crucial in setting out the contractual responsibilities of the various parties involved in an agency arrangement so it is not surprising to find that the Next Steps team at the Office of the Minister for the Civil Service (OMCS) have taken a central lead in providing guidance on its contents. The bare bones of the guidance are clear from the framework documents. First, the five ingredients the OMCS regard as key in defining the structures in which agencies are to operate are identified: aims and objectives; relations with Parliament, ministers, the parent department (unless the agency is a separate department), other departments and other agencies; financial responsibilities; how performance is to be measured; and personnel issues including the agency's delegated personnel responsibilities and the agency's role and flexibilities for pay, training and industrial relations arrangements. Second, the guidance offers direction on how to step around dilemmas that agencies cannot be left on their own to tackle.

A first dilemma: the division between 'policy' and 'operational' matters

Despite the framework documents' (remarkably similar) attempts at clarification, there are potential overlaps in the responsibilities of ministers, departments' headquarters and agencies. This obfuscation stems from the lack of a clear dividing line between policy and operational matters. For example, with regard to the respective responsibilities of departments' headquarters and agencies' accounting officers, the Veterinary Medicines Directorate's framework document is fairly typical; 'The departmental accounting officer remains answerable for the general structure within which the Directorate is required to operate . . . The agency accounting officer is answerable on the Directorate's detailed operations and its performance against targets' (Department of Agriculture 1990, para. 4.9). If the headquarters accounting officer is responsible for all 'policy' matters then how does he know where his responsibilities begin and end? This issue is likely to raise more difficulties where the operations of an agency are politically sensitive, for example, in the case of the Social Security Benefits agency. In such instances it is more likely that 'policy issues' will extend much further down into operational matters than would otherwise be the case.

Similarly, on the issue of parliamentary questions, the framework documents are all very clear; ministers will reply on matters concerning policy or where Members of Parliament specifically seek a ministerial reply, and chief executives will reply on matters concerning the day-to-day operations of the agency. In this example 'policy' issues are likely to push down into 'administrative' issues so as to avoid contentious questions being raised in the House and published in *Hansard*.

A second dilemma: the aims of accountability and flexibility

There are essentially two types of accountability; first, there is the question of internal accountability, that is, the way in which agencies are held accountable for their performance by senior management and parent departments (Day and Klein 1987) and second, there is the question of external or parliamentary accountability. Difficulties arise in twinning the notions of parliamentary accountability and flexibility. The Next Steps premise is that the nature of parliamentary accountability will remain unchanged. Indeed, the government response to the Treasury and Civil Service Select Committee report (HC 481, 1990) concluded that the establishment of agencies does not diminish ministerial accountability (Cm. 1263, 1990, p. 13). If 'flexibility' is to be at all meaningful, however, it will involve experiment and thereby risk. Examples include the plans for a number of agencies to expand in their existing or maybe in new markets and the specified plans of a number of agencies to move from gross to net cost accounting systems and then perhaps to trading fund status (the details of these plans are explored below). If the National Audit Office and the select committees continue their traditional roles of reporting to Parliament on the economy, efficiency and effectiveness of the use of the resources in specific areas, then, particularly at the early stages of agency development, they are likely to be accused of stifling innovation and, consequently, the spirit of the Next Steps initiative. The potential incompatibility of increased flexibility with parliamentary accountability, therefore, provides agencies and departments with a rationale for attempting to limit the scope of the parliamentary watch-dogs.

The differences between the framework documents

A typology

Agency status, despite the overall uniformity of Next Steps' language and aims, is likely to hold very different meanings and provide very different opportunities for agencies. This is reflected in their framework documents. The natures of the executive functions being transferred to agency status are diverse and consequently the differences between the framework documents result, in part, from this diversity. A typology categorizing by nature of executive function is therefore a useful tool for exploring the extent to which differences between framework documents relate to function and the extent to which they relate to other factors. Such a typology is shown in table 1.

The typology develops that of Dunleavy and Francis (1990) who identify eight main agency types but do not develop this categorization to consider the main features of those agency types. The typology developed in table 1 is a three-tier typology identifying the main agency types, dividing those charging for services from those not charging and those with monopolies over their markets from those with no such monopoly. The 'self-funding agencies' and in particular, the 'non monopoly self-funding agencies' (especially those carrying out work for the private sector) clearly have the greatest potential for development as autonomous business units.

Reasons for agency status creating improvements in effectiveness and efficiency

All the agencies adopt the central Next Steps objective of achieving improvements in effectiveness and efficiency. A main difference between the framework documents, however, is in the expressed reasons for why such improvements should occur. Many of the 34 framework documents cite the new freedoms and flexibilities as enablers of improvements in effectiveness and efficiency. Examples of this include the Hydrographic Office where greater management freedoms and flexibilities are to make it easier for it 'to seize the opportunities which its new status offers for delivering progressive improvements in performance' (Ministry of Defence 1990, foreword); the Land Registry where 'greater managerial freedom is expected to achieve progressively improving performance targets' (Lord Chancellor's Department 1990, foreword); and the Employment Service agency where 'agency status will allow the management and staff freedom to manage and deliver their services more efficiently and imaginatively, and so improve the whole range of their service to clients' (Department of Employment 1990, foreword). Other framework documents single out further factors such as improvements in internal management (Patent Office), increased motivation for managers and staff (Patent Office), a sense of corporate identity (Vehicle Certification Agency, Driving Standards Agency) and the businesslike framework within which to operate (Civil Service College). The variations in expressed emphasis as to why agency status should result in improvements in effectiveness and efficiency have no correlation with agency function as defined in table 1.

TABLE 1 A typology of Agencies

Agency type		Agencies	Accounting system	Recruitment	Industrial relations
NOT SELF-FUNDING	WELFARE SERVICES				
Monopoly		Resettlement Agency	AS3	R5	IR2
		NI Training and Employment Service		R6	IR3
	PUBLIC SERVICE				
		Employment Service	AS3	IR6	IR3
		Meteorological Office	AS3	R1	IR2
		Ordnance Survey	AS3 (may be AS2 depending on Treasury discussions)	R6	IR2
SELF-FUNDING	REGULATORY				
Monopoly		Land Registry	AS1	R6	IR2
		Registrars of Scotland	AS1	R5	IR2
		National Weights and Measures Laboratory	AS3	R5	IR2
		Veterinary Medicines Directorate	AS2	R2	IR3
		Vehicle Certification	AS1	R6	IR1
		Radiocommunications Board	AS1	R6	IR2
		Intervention Board	AS3	R5	IR2
		Companies House	AS1	R5	IR2
		Patent House	AS3	R5	IR2
		Insolvency Service	AS3	R5	IR2
		Vehicle Inspectorate	AS2	R5	IR2
		Driving Standards Agency	AS1	R5	IR2
		Driver and Vehicle Lic.	AS1	R2	IR3

The Next Steps Initiative

	PRODUCTION	The Royal Mint	AS1	R3	IR2
Not Monopoly		National Physical Laboratory	AS1	R5	IR2
		HMSO	AS1	R4	IR3
		Hydrographic Office	AS1	R6	IR2
	CONSULTANCY	Occupational Health Ser.	AS2	R4	IR2
	To Govt. Depts. and other agencies	Civil Service College	AS1	R5	IR2
		Natural Resources Instit.	AS1	R5	IR2
		Central Office of Info.	AS1	R5	IR2
		Information Technology Services Agency	AS2	R1	IR1
	To Govt. Depts. and agencies and private sector	National Engineering Lab.	AS1	R2	IR2
		Central Veterinary Lab.	AS2	R1	IR3
		Building Research Estab.	AS1	R5	IR2
		Warren Spring Laboratory	AS1	R5	IR2
		Lab. of the Govt. Chemist	AS1	R5	IR2
	LEISURE	QEII Conference Centre	AS1	R5	IR1
		Royal Historic Palaces	AS1	R1	IR1

KEY
AS1: Trading fund or plans to become trading fund
AS2: Operating a net cost accounting system
AS3: Operating gross cost accounting system

KEY
R1: Agency delegated to recruit up to grade 6
R2: Agency delegated to recruit up to grade 7
R3: Agency delegated to recruit up to grade 6
R4: Agency delegated to recruit up to Executive Officer
R5: Agency delegated to recruit Administrative Assistants and Administrative Officers
R6: Framework document non-specific on this issue

KEY
IR1: Appropriate means for communicating with staff to be developed
IR2: Existing systems or adaptation of these
IR3: Vague

Differences in the intended development of agencies

The framework documents show that agencies will increasingly diversify as they develop. The 'self-funding agencies' all emphasize the importance of becoming more commercial and profitable and of increasing the proportion of their costs covered through fees. Even at their conception, some agencies are more commercial than others. Two of the 'Consultancy' agencies are useful examples of 'developed agencies', the Central Office of Information and the Civil Service Occupational Health Service. The Central Office of Information became a full repayment department in 1984 and since 1987 departments have been free to choose whether or not to purchase certain services from the Central Office of Information. Similarly, the Civil Service Occupational Health Service already recovered its full costs from charges to customer departments, who have the choice of purchasing the services provided by the Civil Service Occupational Health Service from elsewhere.

In addition, many of the 'self-funding agencies' have the aim of expanding both in existing and in other markets. One example of this is HMSO where it and the Treasury will periodically review the scope for HMSO expanding its client group and range of services; another, Warren Spring Laboratory, where 'the size and shape of the Laboratory will depend on the orders it can win from customers' (Department of Trade and Industry 1989, foreword). The only framework document which explicitly takes these commercial values a step further to talk of future privatization is that of the National Engineering Laboratory. As an aside, both the National Engineering Laboratory and Vehicle Inspectorate were earlier candidates for privatization. It appears they are again being presented but this time under a different guise. Clearly there are other potential candidates for privatization in particular among the 'self-funding agencies'. On the whole, the framework documents are explicit in areas where issues are under review and change is likely. The fact that only the National Engineering Laboratory framework document talks explicitly of privatization would therefore suggest either that privatization is not an issue for the other agencies or that a cautious approach is being adopted where decisions on privatization depend on a number of factors including the success of agencies in achieving their aims and on political priorities.

Differences in financial arrangements

Agencies' accounting arrangements are important in determining their flexibilities, freedoms and future. Gross accounting is basically where all receipts and expenditures are presented in the accounts and net accounting is where receipts are netted off against expenditures and only the final figures are shown. Trading funds are basically net accounting systems operating independently of the Supply system, that is, the system by which Parliament provides and receives money. Table 1 shows that the framework documents divide a most cleanly between the 'self-funding' agencies which, on the whole, are, or intend to, adopt net cost accounting systems or trading fund status and the 'not self-funding' agencies which will retain gross cost accounting systems. Three exceptions are 'regulatory' agencies with no stated intentions to change from their current gross accounting systems—the Insolvency Service, the Patent Office and the National Weights and Measures Laboratory.

The move to net, not gross, running cost control and, for some, to trading fund status, will provide agencies with much greater freedoms as it is likely to allow them greater control over their own resources. However, this assumes that other controls, for example the consultation procedures (with parent departments and Treasury) for agencies wishing to increase charges for services are not so bureaucratic as to invalidate the new 'freedoms'. Related to this is the technical but important point that the price for the 'self-funding agencies', new freedoms is, again, reduced parliamentary accountability. The move to net accounting means that Parliament is relinquishing control of the right to see all receipts except where special dispensation has been granted to net off certain receipts against expenditure.

Personnel delegations: recruitment and industrial relations arrangements

The transformation to agency status has considerable potential for significantly transforming what it means to be a civil servant because of the wider organizational changes and because of changes in personnel practices, varying between agencies and departments. Personnel areas where differences between the framework documents occur include the arrangements for recruitment and industrial relations negotiations.

The powers for an agency to carry out its own recruitment provide the agency with flexibility to recruit with greater speed as and when the need arises, the 'type' of staff required both for carrying out the task in hand and also to fit the self-image. Two points are clear from an analysis of the recruitment responsibilities of agencies. First, table 1 shows that the initial delegated powers of recruitment as stated in the framework documents are cautious. The highest level to which agencies can make permanent appointments is up to grade six or seven and only seven out of the 34 agencies can do this. Of the 34 agencies, 17 can only make permanent appointments at administrative assistant and administrative officer level. Second, those agencies with specialist staff have greater freedoms—the 'self-funding' agencies (in particular, the 'Consultancy', 'Production' and 'Regulatory' agencies). This reflects less their functions than the nature of their staff.

All of the framework documents examined outline in some detail the proposed industrial relations structures and all delegate responsibility for agency industrial relations to the chief executive (see table 1). Some of the framework documents say that existing arrangements or an adaption of existing arrangements will continue for the present but will be kept under review (management consultation through trades unions within the Whitley system). For example, the Occupational Health Service Framework Document states, 'The existing arrangements with trade unions will continue to apply initially, but will be subject to review and development over time' (Office of the Minister for the Civil Service 1990, para. 7.8).

Others say that new arrangements for communicating with staff will be developed, and outline their plans, for example, the Historic Royal Palaces executive agency's framework document states, 'Committees, including the trades unions (both industrial and non-industrial) as appropriate, will be established for the main palaces, and for the agency as a whole' (Department of Environment 1989, para. 5.8).

Others are quite simply vague, for example, the Employment Service's framework document states, 'The chief executive is responsible for conducting effective employee relations within the agency including consultation, as appropriate, with recognised trades unions within the agency' (Department of Employment 1990, para. 7.4).

The differences between the framework documents on this issue appear to relate less to the function of the agency than to a confusion about what arrangements are and will be appropriate. There is therefore a paradox that while industrial relations structures are likely to be important throughout the progress of this major reform of the civil service, particularly where the reform affects staff's terms and conditions of employment, the structures themselves are also experiencing change and uncertainty. Looking further ahead the question arises of if and how, the two main civil service unions (the Civil and Public Servants Association and the Society of Civil and Public Servants) will adapt to ensure they continue to represent the interests of *all* their members when the nature of those members' jobs, working environments and consequently interests becomes increasingly diverse as agencies develop.

Conclusions

The framework documents therefore raise many, as yet, unanswered questions although they also provide some sense of the direction and potential impact of the Next Steps initiative. First, analysis by 'type' of agency shows that what it means to be an 'agency' is largely dependent on function. Most agencies who rely on exchequer funding aim to become more effective, efficient and distinct executive arms of government (the exception being the Resettlement Agency which is unlikely to continue its role as an agency when it has achieved its aim of closing the resettlement units). The future of many of the 'self-funding agencies' however, seems more uncertain and at least some will probably move into the private sector. Second, the analysis of the documents suggests that the uniformity of the civil service and of the prevailing image of the civil servant is likely to change. In addition, the agencies' ethos are likely to become increasingly diverse. Third, the framework documents provide a number of indicators that parliamentary accountability may not be fully upheld. The lack of a clear dividing line between policy and operational matters can result in 'policy' being incrementally defined to suit ministers, department headquarters and, to a lesser extent, agencies. The flexibility which is core to the aims of Next Steps, involves risk, and risk is a concept which Parliament's watch-dogs traditionally aim to minimize. Flexibility can therefore only progress if the watch-dogs are called off or marginalized. Finally, the move to net cost and trading fund accounting systems will provide those agencies involved with greater flexibility but at a cost to parliamentary accountability. Such a move involves Parliament relinquishing its right to see all receipts except where special dispensation has been granted.

Appendix one: List of the 34 framework documents examined

Department of Agriculture
Central Veterinary Laboratory Executive Agency Framework Document, April 1990.
Intervention Board Executive Agency Framework Document, April 1990.
Veterinary Medicines Directorate Executive Agency Framework Document, April 1990.
Department of Employment
Employment Service Executive Agency Framework Document, April 1990.

Department of the Environment
Queen Elizabeth II Conference Centre Executive Agency Framework Document, July 1989.
Historic Royal Palaces Executive Agency Framework Document, October 1989.
Building Research Establishment Executive Agency Framework Document, April 1990.
Ordnance Survey Executive Agency Framework Document, May 1990.

Department of Social Security
Resettlement Agency Framework Document, May 1989.
Information Technology Services Agency Framework Document, April 1990.

Department of Trade and Industry
Companies House Executive Agency Framework Document, October 1988.
National Weights and Measures Laboratory Executive Agency Framework Document, April 1989.
Warren Spring Laboratory Executive Agency Framework Document, April 1989.
Insolvency Service Executive Agency Framework Document, March 1990.
Patent Office Executive Agency Framework Document, March 1990.
Radiocommunications Agency Framework Document, April 1990.
National Physical Laboratory Executive Agency Framework Document, July 1990.
Laboratory of the Government Chemist Executive Agency Framework Document, October 1990.
National Engineering Laboratory Executive Agency Framework Document, October 1990.

Department of Transport
Vehicle Inspectorate Executive Agency Framework Document, August 1988.
Driver and Vehicle Licensing Agency Framework Document April 1990.
Driving Standards Agency Framework Document, April 1990.
Vehicle Certification Agency Framework Document, April 1990.

Lord Chancellor's Department
HMSO Executive Agency Framework Document, December 1988.
Central Office of Information Executive Agency Framework Document, April 1990.
Royal Mint Executive Agency Framework Document, April 1990.

Land Registry Executive Agency Framework Document, July 1990.
Ministry of Defence
Hydrographic Office Executive Agency Framework Document, April 1990.
Meteorological Office Executive Agency Framework Document, April 1990.
Northern Ireland Central Office
Training and Employment Agency Framework Document, April 1990.
Office of the Minister for the Civil Service
Civil Service College Executive Agency Framework Document, June 1989.
Occupational Health Service Executive Agency Framework Document, April 1990.
Overseas Development Administration
Natural Resources Institute Executive Agency Framework Document, April 1990.
Scottish Office
Registers of Scotland Executive Agency Framework Document, April 1990.

References

Cm. 1263 (1990) Government reply to the Eighth Report from the Treasury and Civil Service Committee, Session 1989–90, HC 481, *Progress in the Next Steps Initiative*. HMSO, London.
DEPARTMENT OF AGRICULTURE (1990) *Central Veterinary Laboratory Executive Agency framework document*.
DEPARTMENT OF EMPLOYMENT (1990) *The Employment Service Executive Agency framework document*.
DEPARTMENT OF ENVIRONMENT (1989) *Historic Royal Palaces Executive Agency framework document*.
DEPARTMENT OF TRADE AND INDUSTRY (1989) *Warren Spring Laboratory Executive Agency framework document*.
—— (1990) *The National Engineering Laboratory Executive Agency framework document*.
DAY, P. and R. KLEIN (1987) *Accountabilities*. Tavistock, London.
DUNLEAVY, P. and A. FRANCIS (1990) 'The development of the Next Steps programme 1988–90', appendix 5 in HC 481 Treasury and Civil Service Committee Eighth Report, *Progress in the Next Steps Initiative*. HMSO, London.
EFFICIENCY UNIT (1988) *Improving management in government: the Next Steps*. HMSO, London.
LORD CHANCELLOR'S DEPARTMENT (1990) *HM Land Registry Executive Agency framework document*.

MINISTRY OF DEFENCE (1990) *Hydrographic Office Agency framework document.*
THE OFFICE OF THE MINISTER FOR THE CIVIL SERVICE (1990) *The Occupational Health Service Executive Agency framework document.*
TREASURY AND CIVIL SERVICE COMMITTEE (1990) HC 481 Session 1989–90, *Progress in the Next Steps initiative.* HMSO, London.

THE POLICY-MAKING PROCESS

Discretionary Justice and the Development of Policy

R. BALDWIN

Cargo charter policy

The CAA played a major part in reducing anti-competitive restrictions in scheduled versus charter competition in transporting freight. It is perhaps in this area that we see most clearly how the CAA moved away from policy-making via trial-type adjudication to a mixed system of hearings and informal policy statements.

After 1945, 'split' cargo charters (i.e. those allowing more than one consignor to use an aircraft) were not allowed under IATA rules. Britain, however, allowed splitting on the cabotage routes of the Empire. In the late 1960s, air freight grew rapidly with charter development and, after 1966, the ATLB's Class EJ licence allowed split charters for up to four journeys between named places. No more than four consignors were to share the aircraft and no consignment might be less than 1000 kgs weight: these conditions were designed to protect scheduled operators' markets. By 1969 such limitations were found to be 'unduly restrictive' by the Edwards Committee which advocated that scheduled and charter operators should compete for large consignments, that the consignors limitation should be dropped and that the weight limit should be retained or increased from 1000 kg.[1]

In spite of Edwards, the ATLB retained the consignor limitation on its standard licence but it liberalized by granting 'specific' licences of a more generous nature where demand on individual routes was shown.[2] The ATLB would licence a cargo charter operation even if this affected the scheduled service—provided that demand was growing rapidly.

When the CAA took over, it continued to licence on proof of demand. Scheduled operators became increasingly concerned that their position was being eroded by the provision of regular services by charterers and by the evasion of regulations on consolidation. In 1972 BOAC claimed that over half of its cargo weight came in consignments over 1000 kg and that its traffic was imperilled.[3] The CAA undertook to make a general study of the problem. Pressure from the charterers mounted in 1973 and, at a hearing on 26 and 27 February 1974, five major operators applied for the relaxation of cargo charter conditions.[4] They stressed their need to be allowed more consignors and more journeys in order to compete with foreign opposition and, in response, the CAA announced that it was to undertake a comprehensive review of cargo chartering.[5]

The Economics and Statistics (ECS) and Economic Policy and Licensing (EPL) divisions of the Authority conducted an extensive investigation into the whole cargo charter process in 1974, consulting users and providers of services, airline agents, wholesalers, shippers and their customers. Memoranda and forecasts were submitted to the CAA and discussions were held on these. CAA officials met airline staff on a number of occasions to discuss the study, the aim of which was to find ways of decreasing restrictions on charters without unduly prejudicing scheduled operators.

When airline representatives went to a private meeting with the CAA in early 1975 they were given a three-page document entitled 'International Air Freight Charter Services' which stated that, although the Authority had not completed its review of cargo policy, it had reached some conclusions which 'would guide its licensing decisions for the coming year'. It saw no further justification for restricting the numbers of consignors under the Class 6A licence and stated that, from 1 March to 31 December 1975, the standard 6A licence would only be subject to a minimum consignment weight (1000 kg) and a maximum number of journeys (10) between two places. The prohibition on consolidating consignments under 1000 kg was to be retained but after the end of 1975 the CAA said that it would:

expect any airline wishing this condition to be retained . . . to put forward convincing arguments. . . .

The CAA gave notice that, in 1976, it would consider applications for Class 6 licences of up to five years duration, it would discuss with scheduled freight carriers the possibility of their adopting more competitive rates, it would consider whether capacity restrictions should be placed on charterers and it would produce a further consultative document based on its reviews.

The 1975 document gave a clear indication of the CAA's change of policy on the consignors restriction. It was clear to operators that those opposing this change at public hearings would be placed in a difficult position. On 10 and 12 March 1975 all major freight charterers applied for 'standard' class 6A licences and BA and B.Cal. objected to all applications involving relaxation of the consignors limitation.

At these hearings not all parties agreed to disclose confidential information[6] and so this was restricted by the CAA. The Chairman of the panel explained that the CAA policy statement was 'intended to provide guidance and assistance' but that the panel 'would be prepared to hear arguments on the proposals'.[7] Mr David Beety of B.Cal. asked the CAA to disclose the detailed figures behind the Authority's policy statement and the CAA circulated some information following this request. The panel admitted that the figures given were sparse and, for reasons of confidentiality, less informative than they might have been. B.Cal. were worried that the CAA had adopted a policy in advance of the case and had not disclosed figures sufficient for examination of this policy at the hearing. B.Cal. and BA considered that the CAA had chosen to experiment in this area and that the public hearing would make little difference to a policy already adopted. All applications considered at the hearings were granted without limitation on numbers of consignors. The scheduled operators felt that between public hearing and private consultation the CAA had somewhere avoided the need to justify its policies. They did, however, see the hearing as having some value as an opportunity of publicly expressing their views.

In September 1975, as promised, the CAA published a 'consultative document', 'International Air Freight Services'.[8] This thirty-eight page booklet went into far greater detail than the earlier policy statement and was meant to assist 'in the further discussion of . . . options and of . . . policy issues facing the air freight industry'.[9] The CAA rejected complete liberalization in favour of allowing further competition. It proposed to end the 1000 kg consignment limit, saying that belly capacity on scheduled flights, when costed in relation to charters, did not merit protection but that schedule services were to be protected during the initial development of a route since belly capacity provided the most efficient method of transporting small amounts of freight. The airlines welcomed the discussion document as an exposition of the CAA view and as an analysis of the regulatory issues. The editor of the magazine *Airtrade* commented that it was the most concise, lucid and logical review of airfreight that he had ever read.[10]

To follow up this document, the CAA held 'a programme of consultations and discussions' with users and providers of air freight and continued its researches. In February 1976 it used its 'Official Record' to make an announcement on 'Freight Regulatory Policy'[11] and set out a new set of principles which were to 'guide' it in freight cases in the near future.

The new policy entailed reducing the minimum consignment size for both 'standard' and 'specific' 6A licences to 500 kg but consolidation of loads below this size was still to be prohibited. An experiment was proposed for the Hong Kong route: there would be no minimum consignment size or prohibition on consolidation and results would be monitored in order to test the effect on scheduled service freight tariffs. (Being a cabotage route, and so free from IATA controls, scheduled operators could compete freely.) No new limitations on flight numbers were to be imposed and 'specific' licences would continue to be allocated on the basis of shown demand. The Authority promised that the policy would not preclude arguments for or against liberalization in particular cases.

Applications by charter operators for three Class 6A (Standard) and eighteen 6D (Specific) licences were heard on 13 and 14 May 1976. All applications had been amended in accordance with the CAA's proposals. Class 6A licences were granted for five years but the CAA considered that no case had been made for extending Class 6D licences (for specific routes) beyond one year since the policy review and the Hong Kong experiment were to continue. As for the number of flights asked for on specific routes, the Authority commented on BA's failure to produce evidence of diversion and saw no reason to limit these except where the scheduled service appeared vulnerable.

Further intensive studies and consultations resulted in another statement on freight regulatory policy, published in the CAA's 'Official Record' on 1 March 1977.[12] This proposed to allow consolidation of loads to make up the minimum of 500 kg weight and to modify licence conditions so as to replace capacity control defined in terms of specific destinations with aggregate control defined by the number of pairs of points served.

Following the above statement, a freight hearing originally planned for May 1977 was postponed until November of that year. A further statement reiterated the above policy on 28 June 1977.[13] A number of operators requested that a fuller explanation of CAA policy be supplied, and, in response, the CAA produced and circulated to parties a paper setting out the reasoning behind the latest proposals. This paper, together with a parallel statement on scheduled air freight policy was published in August 1977 as 'Air Freight Policy—a Consultation Document'.[14] This sixteen-page 'brown book' dealt with regulatory options and philosophy in great detail and the CAA's proposals were argued for rather than merely set down. On scheduled air freight policy the Authority proposed that where demand did not justify the operation of a scheduled freighter the use of bellyhold capacity on passenger planes should be encouraged. To this end, freight rates, it said, should be allowed to find their own level, and the use of bellyhold capacity maximized. With respect to charter freight operations, it proposed to replace capacity control in terms of specific destinations by aggregate control defined by the pairs of points served. This was to leave the achievement of balance between supply and demand on individual services to commercial rather than regulatory factors. Control of consignment size was to be retained at 500 kg but consolidation would be allowed.

When operators applied for freight licences in November 1977 they did so not so much with regard to prior case law but on the basis of 'the brown book'. The CAA decided those applications collectively in four pages of reasoning and stressed the need for a precise regulatory framework where control was to be applied. Though accepting the need to distinguish scheduled from charter traffic on long-haul routes, the CAA accepted that this would not be so necessary if scheduled operators could adopt a competitive rate structure and it urged them to do this. In accordance with the discussion document, it retained the 500 kg minimum consignment size but discontinued the prohibition of consolidation within that limit. It said that it would control charter capacity on longer routes for the time being but would do so in terms of weight carried rather than numbers of flights. Since the Authority had decided to issue 'standard' 6A freight licences for five year periods, it envisaged a change in licensing procedure:

large annual freight hearings will no longer be needed; instead it should be possible to consider applications for 6D licences for particular markets together as need arises.[15]

After November 1977, the hearing, as was intended, further diminished in importance in charter freight licensing[16] as a liberal regime was applied.[17] A further indication that cargo charters did not include the most contentious issues in licensing came with the publication in November 1979 of the CAA's 'Statement on Air Transport Licensing Policies': no outline of cargo policy was given.

Although this area has rarely occupied the spotlight in air transport regulation it did prove significant in the 1970s. It was concerning cargo charter policy that the CAA developed most rapidly a combination of different regulatory procedures. In the 1960s, policy had emerged case by case, it lacked continuity and a consistent strategy. The first detailed process of consultation followed by administrative rule-making took place in 1975 and resulted in the document 'International Air Freight Services'. That publication signalled the start of a new way of developing regulatory policy. Instead of argument at public hearings, there came debate via research and consultation. This was followed by publication of 'proposals' or 'discussion documents' which were to serve as the bases for licensing decisions that dealt with large numbers of applications collectively. With liberalization of licensing and the use of longer term 'blanket' licences, the public hearing further diminished in importance. The adjudicative process had, with the development of regulatory expertise, been replaced with a system in which administrators, in the main, made decisions, policies and rules by informal and more flexible means.

Conclusions: the role of public hearings in policy development

Unlike the Restrictive Practices Court[18] or the ATLB, the CAA was equipped to supplement policy-making via the trial-type process with other procedures. It has done so and what has been striking has been the variety of procedures used in the different regulatory areas. Although the CAA started off life relying very much on public hearings as the ATLB had done, there were soon changes. In the field of competition policy it published its own criteria to supplement the Act and the guidance; charter policy involved 'policy announcements' and 'public consultations'. Combined research work such as the 'Cascade' study and the 'Fare-Type' analysis were conducted with the airlines and a series of 'brown books' were produced across the regulatory arena. On some topics, such as ABC charters, the public hearing was shunned, on others, such as part-charters, it was deliberately employed. Hearings played a prominent role in airports policy development but in the cargo-charter area they were secondary to policy-announcements. More recently the CAA has taken major steps towards the structuring of its own discretion and the description of its strategy following the general statement of licensing policies of November 1979: what had formerly been set out in discussion or consultative documents had concretized as policy. How then has the CAA coped with the problems highlighted at the start of Chapter 10 [see *Regulating the Airlines*, Ch. 10]?

(a) Polycentricity

CAA experience indicates that polycentricity may not threaten the integrity of the decision-making process provided that agency expertise is sufficient to lay bare the context within which discretion operates[19] and provided that undue emphasis is not placed on the trial-type process. Jeffrey Jowell has argued that urban planning decisions are so polycentric as to 'threaten the integrity of decision by adjudication'[20] but in aviation licensing meaningful arguments can be made at public hearings on a range of issues[21] in spite of the CAA having to exercise a considerable degree of judgement and having to balance a host of different considerations in taking most of its decisions. The commission type of agency can juggle with a far greater number of inter-related issues than the tribunal— this is a function of expertise and is achieved by applying that expertise in a number of ways that Davis would call the 'structuring' of discretion.[22] As Gifford puts it:

the guiding, structuring or confinement of discretionary decision-making is ultimately a function of the process of information collection and evaluation.[23]

What danger there is comes from the experts and professionals using the hearings process as 'something to hide behind'. One staff member commented on this point:

The quasi-judicial procedure means that in finely balanced cases where the issue turns on professional judgement you can use the discretion to exercise expertise that is built into the system. Wearing a more legalistic hat you might say we shouldn't strictly do that but what we're talking about here is marginal cases or cases where the evidence doesn't give a clear-cut pointer in any direction. You have to make the quasi-judicial procedure fit the system.[24]

Since the evidence indicates that decisions are 'glossed' in comparatively few cases this may be accepted as a price that has to be paid in a system that seeks to combine disparate functions to achieve the best of a number of worlds. CAA decisions indicate that whether the best of those allegedly incompatible worlds can be achieved and whether polycentric issues can be dealt with acceptably depends largely on two things: the balance that is effected between the public hearing and other procedures and the success of the agency in making decisions that are perceived to be coherent and legitimate. To these we turn.

(b) Case law or rulemaking?
It is clear that many of Shapiro's advantages of rule-making have been sought by means of various formal and informal CAA procedures. Standards have been articulated in the 'short-haul criteria', participation has been provided in seminars and consultations, the planning of

longer-term strategies has gone ahead by means of the 'brown books' and this has allowed not merely wide consultation to take place but has disseminated information to the industry and has led to open discussion of policy. Except in limited areas, the CAA has not waited for applications to be submitted so that it can formulate policies but has used its resources to look forward to what it sees as potential regulatory problems—as in its analysis of the European fares market.

It might be questioned whether the Authority has always taken positive decisions to develop certain procedures rather than others. In fact, the determining factor here has been the development of the CAA's regulatory expertise. Thus the short-haul criteria and new cargo charter policies arose out of the period 1974–75 when the CAA's economics branch were perceived as having developed their ability to apply economic theory to regulatory issues.[25] At the same time, following the fuel crisis of 1973–74, the CAA board and senior staff were concerned especially with low domestic profitability and were determined to strengthen the airlines' economic position. One CAA executive told the author how in 1975, and leading up to the publication of the short-haul criteria, there was 'a coming together of a developed theory and a pragmatic recognition that we ought to do something about the industry'. When asked whether the application of more rigorous standards to case-decisions had been a motive for producing the criteria, he stated that the criteria 'did serve to supplement the guidance', but that 'the need for consistency was definitely a second-order factor'. Legalistic values were clearly placed lower on the scale than economic and pragmatic considerations.

One factor, however, that did directly influence the choice of whether to consult, publish policies or adjudicate by public hearing was the anticipated degree of airline or other resistance to proposals. Thus, the three year profitability rule set out in the short-haul criteria was considered uncontroversial and so was not the subject of prolonged consultation or a special hearing. In contrast, when the 'part-charter' facility was introduced in 1974, CAA staff arranged to hold a public hearing specifically because they envisaged a dispute between charter operators and the larger scheduled airlines. In the case of the 1979 review of regulatory policy entitled 'Domestic Air Services', both economic and organizational motivations played their part. By that time the airlines' financial position had improved and the CAA had reached 'a new stage in the evolution of CAA thinking' in which the Authority took a step backwards to examine in depth what the policy guidance stood to achieve. By this time also, departmental specializations in the CAA had blurred to produce a more co-ordinated approach. The economists had gained a greater appreciation of, and a greater involvement in, policy issues and the policy makers were more able to suggest avenues of economic investigation to their colleagues. One person closely involved contrasted the operation of a 'multi-disciplinary team' to procedures in the CAA's earlier days when some administrators would isolate a problem and immediately 'find out what the economists say about it' rather than work on the policy issues themselves and then turn to the specialists. By 1979, under the more co-ordinated system that had been encouraged by the hiring of both policy and specialist staff from industry rather than the Civil Service, there were three main influences behind the policy review: the CAA's desire to look more closely at its objectives; the fact that less severe economic conditions encouraged a less restrictive approach to regulation and the advent of a new CAA chairman with a wish to see 'regulation with a light touch'. Revisions in the use of the public hearings system were by-products of such changes rather than objectives. On the aims of rule-making and policy-statements another senior CAA staff member said:

We aim to create a policy framework based on the best knowledge available and on the Act, one that allows us to put proposals to the airlines which can be challenged but a framework that also allows the airlines to put new proposals back to us: you always have the fall-back of the quasi-judicial hearing.[26]

Of published research and policy documents, Sir Nigel Foulkes commented:

The CAA aims to simplify and minimize regulation. The brown books look at dark areas, at underlying policies. I think the CAA can't just lurch from one decision to another. The statistical, research and consultation work is part of a process of seeing if we can understand the industry, regulation and the airlines. They constitute the raw material for reassessing policies.[27]

If Gifford's test is applied and it is asked whether the CAA has matched its structuring devices to the different types of regulatory issue, the answer is that, whether the perfect balance has been achieved or not, the agency has been sensitive to variations in issue-type and has been highly adaptable in dealing with these.

(c) 'Managerial' or 'regulatory' decision-making? Muddling-through or coherence?

In looking at how an agency can make coherent decisions it is helpful to move from K. C. Davis's focusing on the need to structure discretion to an examination of the amenability of decision-types to structuring.

Examination of the CAA shows, moreover, the folly of dealing with the discretion of any body as a single, constant factor. To assume that all the decisions of a particular agency such as the Supplementary Benefits Commission (SBC), the CAA or the IBA are of the same kind, is rash. It makes little sense to question whether the decisions of an agency are 'justiciable' or 'polycentric' or anything else without first examining what varieties of decision the body makes. The kinds of control that may usefully be applied to discretions may in practice only be assessed in relation to quite particular circumstances. That is not to say that we should ignore general analyses of the advantages and disadvantages of 'legalizing' (imposing substantive rules on) or 'judicializing' (applying adjudicative procedures to) agency activities: this will aid consideration of individual discretionary powers but it is no substitute for analysis on the ground. As Jeffrey Jowell has stated:

> an understanding of the limits as well as the merits of legal techniques will allow us to provide for the control of administrative discretion in a manner that is sensitive to the nature of the administrative process and the constraints operating upon decision-makers, and is hence more likely to prove successful.[28]

Davis himself has stressed that 'almost every proposal (must) be examined in each of the thousands of specific contexts of particular discretionary powers'.[29]

Within one agency there can be, as Gifford has pointed out, a massive array of issues that have to be decided in the course of regulation. In the case of the CAA some hearings are concerned with complex non-recurring test cases and new policies (e.g. should 'part-charters' be allowed on scheduled passenger services?). Others deal with one case within a larger policy framework (e.g. Skytrain applications) and yet more are routine issues being decided not in terms of new policy lines but of relatively well established and recurring standards (e.g. simple questions of diversion). Some policy issues are politically contentious, others give less cause for concern. There are cases involving just two parties and others that entail debates between a multitude of operators. Even the concept of the public hearing is indefinite: many important policy decisions are made not at formal licensing hearings but on the basis of informal consultations and semi-formal consultative hearings that may resemble public inquiries or even academic seminars. The CAA employs a wide variety of procedures both to decide a multitude of issue-types and to 'structure its discretion by openness'.[30]

Appreciation of the range of discretionary activity within each agency should discourage unqualified generalizations of the kind 'the CAA should adopt more precise standards' or 'the SBC should operate on a "rights" basis', since it is plain that such prescriptions cannot apply to all the decisions of such agencies.[31] Similarly, another temptation may be resisted. Bodies such as the IBA, CAA and SBC are all independent and specialist and so there is a tendency to overplay parallels between authorities of similar institutional status. When details are taken into account, however, we may be dealing with very different forms of decision and contrasting decision-making contexts. The CAA's regulatory clients are an expert few enjoying considerable resources: they are interested in a small number of decisions involving substantial economic values. Such a position contrasts with the SBC whose clients are on the whole inexpert, huge in number and, by definition, lacking resources.[32] The SBC is concerned with a plethora of decisions which, taken individually, tend to involve small amounts of money. If, therefore, we are examining the advantages of rule-making as a method of structuring in relation to the powers of the CAA and SBC, there is a limit to useful generalization. Legalization in the one field may be based on perceived needs to focus planning and to increase political legitimacy. In the other, it is more likely that rules are used to regulate clients' demands and to increase the speed with which cases are processed.[33]

If an attempt is made to break down agency functions into different classes of decision-making it becomes clear that in evaluating agency rule-making and procedures it is unhelpful to apply tests of 'predictability', 'efficiency' etc. across the board.[34] Apart from espousing what might be called 'the fallacy of homogeneous decision-making', it implies that there exists some sort of consensus on the optimal levels of predictability or efficiency in the regulatory scheme. It would be mistaken, however, to assume that even the airlines all want predictable and formally rational decisions: their interests vary. The large operators, who are concerned to preserve their existing routes, to rationalize investment decisions and to plan safely for the future, may advocate such a system even at the cost of delay, but (as was indicated to the author) certain private operators of an aggressive disposition would prefer CAA decisions that place more emphasis on speed and responsiveness to economic change than to any dictates of consistency or precedence: they hope to benefit from any degree of chaos that results from changing existing route structures. The fact that a wide spectrum of such demands is made of the CAA does not, however, indicate that 'anything goes' in CAA decision-making. In the end, credibility still has to be preserved and the Authority has to justify decisions not merely to the operators but to the public and to the Secretary of State as appeal authority.

The implication of this account of the CAA is that it may be a misapplication of emphasis generally to favour certain models of decision-making and particular forms of control over discretions. Within an agency there may be something more important than a willingness, say, to create standards across the board or to decide issues openly or to give the appearance of rationality: administrative justice may, more than anything, depend on the skill of the agency in adjusting its procedures to fit the kinds of decision-making it is required to indulge in.[35] The more a body is involved in 'polycentric' issues, and the greater the number of constituents to be satisfied, the more that the choice of procedures and rule-making devices becomes a political art rather than an administrative or judicial science. The test of success in such issues may be public confidence—whether politicians, regulatees and consumers have faith in an agency's control. To this extent the regulatory agency's tendency to direct its activities towards self-justification is understandable. For those studying the agency, the lesson is to limit the premium attached to general analyses of procedures and to consider the match between procedures or issues and the processes used in adapting those procedures to ends. Not only academics such as Gifford have stressed the importance of the adaptive process: it was no other body than the American Bar Association that said:

administrative procedures, initially developed as a safeguard against the threat of regulatory abuse, have come to mimic the judicial process, with inadequate regard for the flexibility available under existing statutes. Regulatory procedure requires a new flexibility which respects traditional concerns for accuracy, fairness and acceptability, but meets the need for more efficient administration.[36]

There is a difference, however, between advocating that formal licensing procedures should be more responsive (in the sense of less time-wasting and rigid) and appreciating the stronger sense of 'flexibility' that involves a preparedness to consider all the varieties of agency decision-making and the need to adapt these to changing circumstances. A body's 'discretionary power' is a package containing disparate elements that interact (as we have seen public consultations and public hearings do in CAA procedures). This rules out the divorcing of discretionary activity from policy-making and demands that the relationship between the two is explored. Thus we could not account for CAA decisions on major route applications without looking at relevant CAA research and policy development through the 'brown books'. K. C. Davis would advocate neither the separation of policymaking from discretion nor the abolition of discretion[37] but his argument is open to narrow interpretation, especially at the hands of lawyers. He says that we should eliminate all unnecessary discretion and control what remains by confining, structuring and checking it. He inserts qualifications (e.g. that we should aim for the 'optimum breadth' of discretionary power and that discretions be looked at in context) but the thrust of Davis's argument is towards an unexplored concept of justice. His central concern is:

> How can we reduce injustice to individual parties from the exercise of discretionary power.[38]

It should be repeated: focusing concern in this way biases any analysis of discretion in favour of legalistic values. The broader the responsibilities of a government agency, the less useful it is to treat discretion as an issue of justice rather than of judgement, administrative art or politics. The real questions are sidestepped when Davis states:

> Whenever any agency or officer has discretionary power, rule-making is appropriate. The general objective should be to go as far as is feasible in making rules that will confine and guide discretion in individual cases.[39]

He adds, however, a footnote:

> Unfortunately how far is feasible is a question that must be determined for each discretionary power in each particular context.

We may agree that the particular context must be looked at but the point underemphasized by Davis is that, when we do examine the exercise of discretion in individual cases, we find that not only is it unhelpful to prescribe techniques of structuring and confining across the board but that we cannot ignore the political dimensions of discretionary activity. The heart of the matter lies in his footnote: the real question is not about *justice* but about *feasibility* and the balances that have to be established between predictability, efficiency, openness, political self-justification and many other factors. To concentrate on justice to the individual in any case fails to take on board the public's interest in regulatory activity[40] and, from the agency's point of view, justice to the individual must be a secondary consideration when put next to political survival.

Here again Gifford proves helpful. Soon after Davis had published *Discretionary Justice*, Gifford argued that those involved with agency decision-making should concentrate their attention not on the development of standards but on 'decisional referents'.[41] Most theorists had 'avoided asking why one standard is employed rather than another',[42] but decisional referents were 'whatever a decision-maker deems significant in deciding'[43] and they included and went beyond rules, criteria factors, considerations, principles, policies, goals and reasons. They thus allowed a broader and more policy-oriented approach to decision-making. The notion of decisional referents could, for instance, explain the taking into account of potential states of affairs—matters of special concern to regulatory bodies with their vague mandates and prospective orientation.

Decisional referents, for Gifford, had to 'possess more content' than generally-stated goals or directives and he gave as examples: studies and analyses; specialized disciplines and programme plans. Davis might stretch 'rule-making' to cover the consideration of hypothetical instances but Gifford looked further. CAA regulation suggests that Gifford was right: 'rules' do merge with non-rule decisional factors, there is a continuum of tentative through to firm considerations operating in agency decisions, factors taken into account by individuals do blur into those handed down by organizations, informal relationships are of relevance in decisions and decisional referents are in a constant state of flux. What might be added to Gifford is this statement: in areas of complex economic decision-making the attempt should be made not merely to develop standards and a range of decisional referents but to express the latter *in a layered fashion so that each layer is reinforced and made intelligible by others*. These layers should extend from the most general to the particular and cover not merely legal and policy but political considerations.

In the case of the CAA, the development of a 'layered' discretion was made possible because the Authority did not restrict itself to the trial-type process. A range of procedures had emerged by 1980 with the effect of both substituting for hearings and of harnessing them to the regulatory process. This allowed legislative and trial-type adjudicative methods to be applied to a wider range of activities than would otherwise have been possible and meant that public hearings could be used both to adjudicate on particular issues and to promulgate major rules. In addition, the mixed system allowed hearings to be used in relation to comparatively rarely-recurring cases and to regulatory areas subject to rapid change.[44] Whereas in the ATLB's days policy only existed at two levels, the most general (the Act) and the specific (the case decision), there emerged under the CAA a wide-ranging system of policy-making and accountability. A licensing representative could look not only to the 1971 Act but also to the policy guidance and this provided a crucial link between statutory and administrative principles, standards and rules. The guidance attempted, with limited success, to describe policy in terms of some precision but of sufficient breadth to allow the CAA operational flexibility. It influenced CAA rule-making without compelling any one approach. The system that emerged by 1980 was one in which the absence of a rule or policy on any one point or level might be balanced against the existence of a broader policy or principle[45] and all policies of detail might be interpreted against a background of published CAA statements of regulatory philosophy. As a final reference, representatives, in preparing cases, were also able to consider relevant case law against their background knowledge of the views or attitudes of CAA members and staff as expressed at the various forms of consultative or public meetings.

On the question whether the CAA simply 'muddles through' in an *ad hoc* or incrementalist fashion, it is useful to refer to Charles Lindblom's contrast between rational-comprehensive (root) and successive limited comparison (branch) styles of developing policy in relation to complex issues.[46] The former is said to rely on the clarification of values and objectives, policy-formulation is approached by selecting the means to achieve defined ends, analysis is comprehensive, all values are considered, all relevant factors are taken into account and theory is relied on. The branch method, on the other hand, links the selection of value goals with empirical analysis: it relies very little on means end analysis; policies are based on short-term choices; a high degree of selectivity ignores important alternatives and values and policy-making by a succession of comparisons reduces reliance on theory.

Lindblom's point is that the more complex a problem, the less useful the 'root' method becomes and the more appropriate 'branch' policy-making will be: one could take into account all the relevant factors involved in directing traffic on a bridge but to suppose that this could be done in making national foreign policy would be misleading. Branch policy-making is therefore said to be flawed but 'it is not a failure of method for which administrators ought to apologise': it often makes sense to simplify choices, to rely heavily on past policies and only to change these incrementally.

Taken to its extreme, the argument for 'muddling through' would seem to rule out the need for truly expert agencies that differ from tribunals in taking a more comprehensive view of 'the factors attending a decision or policy'. Quick and cheap decision-making, it could be said, is therefore preferable to expertise. The contrary view, however, is supported by this review of the CAA. Lindblom's arguments are at their strongest in relation to the 'one-off' (e.g. foreign policy) decisions of individual administrators: they are weaker where, as in the case of civil aviation licensing, mere 'muddling through' would fail to yield the longer-term policies that are needed to render particular decisions coherent. Incrementalism is also incapable of responding to rapid economic and technological changes with radical new policies.[47] The Lindblom thesis, furthermore, minimizes the role of expertise (in his view it is impossible and so misleading to attempt to prioritize values): in aviation licensing, however, the contrast between the CAA and the ATLB shows how an approach that is more ambitious than mere pragmatism can avoid many pitfalls.

The CAA, for its part, falls between the Lindblom models: it acts incrementally in some areas (e.g. airports policy) but in others its staff have attempted to take a 'new look' at a whole problem (e.g. at ABC's, 'blanket' charter licenses, cargo charter liberalization) and it has both relied on theory and a prioritizing of values. Lindblom points out that the objectives of many agencies are limited by statute. This is so in the CAA's case and, to some extent, this restricts the breadth of approach that the CAA can take. A developing expertise has, nevertheless, played an important role and the ATLB's ad-hoccery has been avoided. What counts in aviation licensing is, as we have seen, the potential to make policies that are intelligible and responsive, lasting and politically secure.

(d) Politics and the regulatory process

The political legitimacy of the CAA is closely dependent on its policy-making and its case decisions being seen as coherent, expert and efficient. In so far as the CAA has succeeded where the ATLB failed it has avoided gross confusion of policies and has established a framework to render most of its decisions intelligible within the real world of political contention. Trial-type procedures have been combined with other devices without turning the public hearing into a charade and lawyers have been used in hearings without causing undue rigidity. Arguably because of the policy guidance and the resulting relationship between CAA and the Department, the CAA has managed to develop its own policies without suffering the kind of interference via ministerial *ad hoc* and political inconsistency that was seen in the ATLB era. The benefits of 'layered' policy-making have been clear. Those seeking in the CAA a source of both longer-term and day-to-day policies have, if necessarily imperfectly, been offered both. The extent to which the CAA's achievements in combining powers might have been (and will in future be) possible in the absence of the written policy guidance system is, however, a matter to be returned to in Chapter 13 [see *Regulating the Airlines*, Ch. 13].

Notes

1. Cmnd. 4018, paras. 742–3.
2. See e.g. Decisions E9486/1, 25 March 1970; E11022/4; and E12518, 30 November 1971.
3. Decision E12530/5, 24 October 1972.
4. Decision 6A/8006.
5. See also CAA Annual Report (1973–4), pp. 22, 32.
6. Under s. 36 of the Act.
7. Decision Class 6A licence, 12 March 1975, para. 2.
8. CAP 379.
9. *Ibid.*, p. 16.
10. *Airtrade*, October 1975.
11. See ATLN No. 203, 24 February 1976, p. 15.
12. ATLN No. 255, pt. 4.2.
13. ATLN No. 270, pt. 4.2.
14. CAP, 405.
15. Decision on Applications for Class 6A (General) etc., licences, 30 December 1977.
16. Decision 5A/70096, etc., 4 July 1978.
17. Decision 5A/70098, etc., 11 August 1978; para. 20.
18. R. B. Stevens and B. S. Yamey, *The Restrictive Practices Court*, London (1965), e.g. p. 140; see also the Confederation of British Industry's criticisms of the RPC, in J. A. Farmer, *Tribunals and Government*, Weidenfeld and Nicholson (1974), pp. 31–2.
19. See D. J. Gifford, 'Decisions, Decisional Referents and Administrative Justice' (1972) 37 *Law and Cont. Prob.* 3.
20. Jowell, *op. cit.* (1975), p. 153.
21. *Ibid.*, p. 154.
22. See P. Weiler, 'Two Models of Judicial Decisionmaking' (1968) *Can. Bar Rev.* 406 at 423–5.
23. Gifford, *loc. cit.*
24. Interview with author, February 1982.
25. For discussion of legally controlled discretions in relation to particular topics, see e.g. R. Titmuss, 'Welfare "Rights", Law and Discretion' (1971) 42 *Pol.Q.* 113; Stevens and Yamey, *op. cit.*; G. Ganz, 'The Control of Industry by Administrative Process' [1967] *P.L.* 93.
26. Interview with author, February 1982.
27. Interview with author, 1 March 1982.
28. Jeffrey L. Jowell, 'The Legal Control of Administrative Discretion' [1973] *P.L.* 178 at 183.
29. Davis, *op. cit.* (1969), p. 232.
30. *Ibid.*, pp. 99–111, 226.

31. See E. Gellhorn, and Glen O. Robinson, 'Perspectives on Administrative Law' (1975) 75 *Col. L.R.* 771.
32. See M. Adler and A. Bradley (eds), *Justice, Discretion and Poverty* (1976).
33. On processing by legal rules, see J. Bradshaw, 'From Discretion to Rules: The Experience of the Family Fund' in *Discretion and Welfare*, M. Adler and S. Asquith (1981), Heinemann. For an examination of reasons not to control discretion see R. Baldwin and K. Hawkins, 'Discretionary Justice: Davis Reconsidered' *P.L.* (forthcoming 1984).
34. See e.g. the treatment of tribunals as bodies with homogeneous needs in the *Justice/All Souls Review of Administrative Law in the UK* (1981); for comment, see R. Austin and D. Oliver (1981) *P.L.* at 441–52.
35. See G. O. Robinson, 'The Making of Administrative Policy: Another Look at Rulemaking and Adjudication and Administrative Procedure Reform' (1970) 118 *U. Pa. L. Rev.* 485, pp. 535–9; Baldwin and Hawkins *loc. cit.*
36. ABA 'Federal Regulation: Roads to Reform' (1979), p. 92.
37. Davis, *op. cit.* (1969), pp. 5–6.
38. *Ibid.*, p. 216.
39. *Ibid.*, p. 221.
40. See e.g. J. T. Winkler, 'The Political Economy of Administrative Discretion', in M. Adler and S. Asquith (eds), *op. cit.*
41. D. J. Gifford, 'Decisions, Decisional Referents and Administrative Justice' (1972) 37 *Law and Cont. Prob.* 3.
42. *Ibid.*, p. 12. But see: Colin S. Diver, 'The Optimal Precision of Administrative Rules' (1983) 93 *Yale L.J.* 65.
43. *Ibid.*, p. 12.
44. On the problems of rapidly changing circumstances, see Jowell, *op. cit.* (1975), p. 135.
45. On principles as supportive of rules, see R. Dworkin, 'The Model of Rules' (1967) 35 *U. Ch. L.Rev.* 14; R. Hughes, 'Rules, Policy and Decisionmaking' (1968) 77 *Yale L.J.* 44; and Galligan, *loc. cit.*
46. Charles E. Lindblom, 'The Science of Muddling Through' (1959) 19 *Pub. Admin. Rev.* 79.
47. See Y. Dror, 'Muddling Through—Science or Inertia?' (1964) 24 *Pub. Admin. Rev.* 153. Dror argues that the Lindblom approach presupposes the generally satisfactory nature of present policies, the absence of a need for radical change and continuity in both problems and the available means for dealing with them.

Regulatory Agencies: An Introduction
Conclusions: Regulation and Public Law

R. BALDWIN AND C. MCCRUDDEN

1 Why agencies?

Under which conditions may regulation by agency be seen to be preferable to government action through a Whitehall Department, to nationalisation, self-regulation, or to control by the ordinary courts, tribunals or local government? On a pragmatic level there are managerial reasons for resort to agencies. Thus, Frans Slatter[1] has noted how the rise of administrative agencies has paralleled a substantial increase in the overall size of government. He argues that one of the most obvious reasons for the creation of these bodies has been to take some of the workload off the more traditional branches of government, either by hiving-off old functions or giving new functions to administrative agencies.

A number of factors may influence the placing of new governmental burdens on agencies, rather than allocating them to the existing departments of the public service or giving more work to the courts. In some cases, the new activities would have been seen as awkward or burdensome bedfellows if added to already existing duties of departments or courts. In other cases, functions are thought likely to be better administered if they are the sole or central interest of the organization, and not just a peripheral matter dealt with by someone whose attentions are primarily directed elsewhere. A division of labour in this way may also help to develop specialization and expertise.

Greater government involvement with the economy combined with new and increasingly complex forms of technology, has led to the establishment of administrative agencies which are seen as experts in these substantive matters. The required expertise might have been developed inside existing departments or courts by appointing or retaining the appropriate personnel. However, the need for expertise is sometimes found in combination with a rule-making, decision-making or adjudicative function that is thought to be inappropriate for a government department or a court. Sometimes a department is seen as not able to provide the independence from government needed in some of these applications of expertise. Sometimes it is thought to be difficult to develop all the different kinds of expertise needed in courts without increasing their number and size. In any event, if new courts are needed, or if the existing courts have to function in divisions based on expertise, a new administrative agency might as well be used. In other cases the expenditure of the time and prestige of highly paid senior judges may be thought to be unjustified.

The advantages agencies are said to have over traditional courts are numerous. The sheer volume of decisions may call for a separate structure. Economy, speed in decision making, ability to adapt quickly to changing conditions, and freedom from technicality in procedures are other commonly cited advantages. Agencies are also thought to be able to relax the formal rules of evidence when appropriate, to avoid an over-reliance on adversarial techniques, and to avoid strict adherence to their own precedents. Administrative agencies are thought not to be as restricted to formulating policy on a case-by-case basis as are the courts.

As government expands into unfamiliar territory it is often hard to set fixed criteria that will adequately anticipate marginal cases. Sometimes it is simply not possible to foresee what circumstances will arise. In these cases a greater or lesser degree of discretionary power may need to be left to the administrator. A rule-making power is often found where regulation of a highly complex or technical nature is required. Delegation of rule-making powers may also be needed where constant fine-tuning of the rules and quick changes to meet new circumstances are required. Often the Cabinet cannot justify devoting the time needed to these matters, or else it is felt that it simply cannot act quickly enough. It would, of course, be wholly impractical to take the numerous amendments needed through Parliament. This is not to say that the impossibility of fixing criteria in advance and insufficient Cabinet and Parliamentary time lead inexorably to the establishment of an independent agency. Placing a wide discretion on the Minister or department has, in Britain, been a more common reaction.

Many of these factors have also been considered to represent advantages of agencies over the traditional central government department. The opportunity for consultations by way of a public hearing is often denied to departments because of the conventions under which they operate. An agency structure, however, may enable interested members of the public to participate in the making of the decision, rather than merely to react to the decision once it is made. Being able to use a modified adversarial system to decide between competing parties may be another advantage agencies have over departments.

Agencies' separateness from government may also make them a preferred mechanism for giving various interest groups a place in government policy-making through the appointments process. Agencies are also argued to be more suited than courts and probably than Whitehall departments to collegial decision-making. This facilitates the bringing together of many and diverse ideas, and can be used to encourage openness in decision-making. Groups that are coopted into the system in this way may be more likely to support government activity in the field. An agency may be able to provide for greater continuity and stability in policy-making and implementation, being one step removed from the vagaries of Cabinet reshuffles and changes in government.

The exercise of a policy-making function by an administrative agency is said to provide flexibility not merely in policy formulation but also in the application of policy to particular circumstances. Agencies are said to be able to react to new circumstances more quickly than the departments of state, and are able to be more flexible in the application of standards than the judiciary. Somewhat paradoxically, as we have seen, they are also said to provide greater continuity than courts and greater stability than cabinets because the administrative agency is one step removed from the election returns.

In the discussion up to this point, one characteristic of administrative agencies has recurred: their independence from, or arm's-length relationship with, central government. In many cases this is the only reason why an agency is used rather than a more conventional department. Departments could generate advice, build expertise, develop new procedures, or manage programmes and projects. It is only an administrative agency, however, that can provide this added remoteness. In other cases the very function itself calls for independence. Adjudications, arbitrations, and the exercise of some statutory discretions are examples of this need. It is here that agencies are said to play an important role as insulators between government and the public.

There are several facets to the agencies' role as insulators. They may be used for dealing with 'hot potatoes', i.e. politically controversial decisions. Agencies have sometimes been used to give the appearance that the government was 'doing something' about some sensitive matter. The agency can then serve to deflect criticism and give the government time to examine the alternative solutions available. An agency may also be used where the administration of a particular policy is viewed as being politically dangerous. The use of an agency means that the Minister is not directly responsible for decisions taken, and so the government is protected.

Where valuable benefits such as licences or rate increases are at stake, an agency may be used to deflect lobbying and attempts at invoking political favours. It is easier for a government to say 'no' when it can say that the responsibility does not lie with it. The hope here is not that pressure groups will have no influence on government, but rather that their arguments will be evaluated more technically. An impartial decision-maker is useful whenever it is hoped to free government administration from partisan politics and party political influence. Adjudication and arbitration are examples where this is necessary, as are the granting of funds for research or cultural purposes. Where economic regulation interferences with vested private rights, the business community will usually be less antagonistic to government intervention if it is not a matter of party politics.

Finally, agencies have been justified as being able to 'raise and clarify arguments upon which political decisions depend'.[2] Regulatory agencies may help 'focus public attention on alleged harms and bring out issues in a way that enrich[es] political debate'.[3] In short, irrespective of their ability to command expertise they may be useful simply to enlarge the scope of political discussions.

These, then, are the main functional justifications for and the major roles performed by administrative agencies. While few of these functions could not be performed within more conventional government structures, agencies usually bring something to the particular role which departments and courts cannot. No agency is likely to perform only one of these roles. There are therefore technical reasons and justifications for using regulatory agencies, but there is also a variety of political motives for resort to such bodies. Thus Hood[4] has noted that in addition to a government's desire to distance itself from certain decisions or to co-opt certain groups into the decision-making process, it may often resort to agency control so as to create 'space' for a new policy within the machinery of government or to conceal the real size of state bureaucracy or to give the impression of taking firm action on an issue of concern.

2 Operational choices

In establishing an agency, the government faces a number of choices. Where these are not made before legislation is passed, then they devolve, at least initially, on the agency itself. Three choices are particularly important. First, should the agency proceed by way of education, negotiation and bargaining, or by enforcement of established standards through the legal process? This may involve deciding whether the agency is to act as a pressure group and thus adopt a primarily representational role, or whether it is to remain 'neutral'. The government or agency may also have to consider whether a self-regulatory or more interventionist approach is more desirable.

A second choice is whether the agency is to adopt a reactive or a proactive enforcement strategy.[5] A proactive strategy would mean that the controlling body does not allow its priorities to be decided by outside forces (e.g. by those cases that happen to come into the agency of their own accord) but rather develops its own priorities (by way of, e.g., a strategic enforcement plan). Involved in this choice is the question whether the agency is to favour an 'incrementalist' or a 'comprehensive' approach to bringing about change. The characteristics of the former approach are a piecemeal development of policy, with a limited consideration of options and consequences. Regulation under this regime is a series of small-scale experiments and continual adjustments of differing interests and values. A comprehensive approach, often called 'comprehensive rationality',[6] emphasizes a clear specification of goals, choices and the consequences of alternative regulatory decisions. Systematic analysis (as represented by, for example, cost-benefit studies) is central. While some writers have doubted the feasibility in practice of this approach, and have drawn attention to the 'bounded rationality' of administrators,[7] it has had an important influence on attempts to improve the quality of policy-making.[8]

A third choice is necessary where some element of 'strategic' enforcement is adopted. This may involve concentrating on what might loosely be called 'rule-making', i.e. securing the promulgation of rules by the legislature or by the body itself through, e.g., codes of practice. Alternatively it may involve adjudicatory methods of enforcement: encouraging, planning and funding individual cases, or initiating investigations. Courts and tribunals tend to rely on adjudicatory and adversarial procedures rather than engage in consultative rule-making processes. Central government departments make many rules, but seldom adopt procedures that resemble trials. It is, as a consequence, rare to have to ask whether, in the development of policy or in deciding between parties, the appropriate procedure for courts or tribunals is rule-making or trial-based (or indeed some other form of adjudication). With regard to agencies, however, the advantages of rule-making procedures (wide consultation, prospectivity, research, etc.) have to be balanced against those of trial-type adjudication (the protection of individual rights, the 'day in court', etc.). Where rule-making is preferred, other choices are relevant: Is the purpose of the rule to be an aid to prosecution, or a means of promotion or consciousness raising? Which parties are sought to be addressed by the rule or code, and who is to participate in its making?

The different regulatory mechanisms implicit in these three choices may not formally conflict at the level of theory, and indeed may be complementary. However, scarce resources (in staff time, energy, and expertise) mean that these mechanisms may well conflict in practice. An agency cannot in the real world be all things to all people.

3 Pitfalls

A number of (sometimes inconsistent) criticisms have been made of the exercise of power by agencies. They have been accused, inter alia, of being slow, inefficient, unfair, unpredictable, corrupt, ill-managed, badly staffed, undemocratic, and unresponsive to changes in political opinion. It has been argued that it is inappropriate for those who make rules to be permitted also to sit in judgment on their application to individual cases;[9] that agency processes are run by people who do not have a sufficient appreciation of the problems of those who are subject to its rulings; that regulators make their decisions before the issues are sufficiently formalized for requirements of consultation to apply; and that agencies (like all bureaucracies) desire growth, seek new business to justify this growth, and thus produce ever increasing regulation.

Perhaps the most common form of attack is that agencies are too easily subject to influence by various outside interests; 'clientalism' and 'agency capture' are familiar terms. Several capture theories have been advanced. One common version is that capture is the process whereby agencies created in the public interest are subverted to the ends of those supposed to be regulated. The best known example of this theory argues that agencies go through a life-cycle of gestation, youth, maturity and old-age, ultimately becoming 'captured', so that the agency becomes 'the recognized protector' of the industry supposedly being regulated, rather than the champion of the public interest. This theory has been criticized as having surprisingly little empirical backing, and is inconsistent with specific industry studies in the United States and the growth of social regulation in the 1970s. It has also been criticized as apolitical in its concentration on the influence of the regulated industry to the exclusion of other political pressures.

A second variety of capture theory states that agencies are created in the first place in a manner that benefits existing industrial interests. The older 'public interest' theory conceived government intervention as a beneficial corrective for inefficiencies and inequities in the market-place. The 'public interest' was equated with remedying a list of 'market failures' which both explained and justified government intervention. Experience of regulatory programmes has suggested to some that they more often serve private rather than public interests.

A more promising variation on the theme of regulation serving private interests is the interest-group theory. This involves the familiar argument that the political process is a market-place in which rational, self-interested individuals attempt to maximize their utility. Regulation is a valuable commodity sold by politicians to the highest bidders. Thus, for example, consumers might 'buy' protection in return for votes. Given these assumptions, it is unlikely that the outcome of the regulatory process will have any necessary connection with efficiency, since the role of the state is essentially that of mediating distributive claims. Within this market there will be a variety of 'players': bureaucrats, politicians, voters, pressure groups and the media. The resources (legal, informational) available to each group and the strength of incentives to obtain favourable regulation will have a decisive impact on the creation and implementation of regulatory schemes. The problems involved in organizing diffuse and fragmented groups such as consumers, and the limited incentives which individual consumers have to develop informed preferences on policy issues affecting them, suggest that they will be weak players in the regulation game. This economic theory has the merit of simplicity but it is as yet largely untested.

It is often unclear whether such accounts are intended to assess the *origins* or the *means of implementing* a regulatory programme. A theory of regulatory origin may not explain regulatory output. Moreover, the contrast of the 'public interest' and 'interest group' theories suggests different conceptions of rationality in the regulatory process. Within the former, technocratic rationality will be both necessary and possible. The interest-group theory seems, however, to deny the possibility of long-term planning; expert knowledge will be relevant primarily as support for particular political interests. Other theories argue that capture results from the bureaucratic and organizational make-up of agencies. Yet others argue that decision-making in these agencies 'is largely the result of coalitions of diverse participants—career bureaucrats, political appointees, agency professionals, public activists, and corporate managers—all with different motives and all interacting within complex institutional settings which provide changing patterns of incentives'.[10]

We will be concerned, therefore, to examine the extent to which the enabling statutes sufficiently control administrative discretion. Do regulators act as 'professionals' whose discretion is held in check by the tenets of their discipline? Is there effective control of the exercise of discretion through an internal administrative law of the agency? Questions with which public lawyers have been concerned for some years are thus posed in peculiarly acute form. The following are some of the issues on which the book will focus. What are the appropriate criteria for determining the use of rules and adjudicatory mechanisms, the exercise of discretion, the degree and role of participation and accountability, the external scrutiny of specialists and experts? Is there a relationship between the competence and the legitimacy of a decision-maker? How does procedure interrelate with substance? What makes (or should make) an issue justiciable? What is the appropriate relationship between the various agencies and the courts? A central issue is thus the basis upon which the courts should interfere with the decisions, actions and policies of regulatory agencies. Judicial review is only one form of control over agencies, but such influence must itself be based on justifiable criteria if it is to be legitimate. Is there, indeed, a legitimate basis for judicial review of agency action?

* * * * *

Modern British administration is concerned with at least five different governmental functions. One is the prevention of undesirable behaviour and the securing of desirable behaviour in particular areas of public and private life. An example of this is the regulation of health and safety at work, or the prevention of race discrimination in employment. A second is providing state facilities for the reaching of compromises between parties. An example of this is provided by the institution of ACAS. A third is the provision of services and the redistribution of goods. Examples of this are the provision of social security benefits, education, and the health service. A fourth, standard-setting function is involved in attempting to settle those disputes with respect to which the law is unsettled. A fifth, more mundane function is grievance remedying, i.e. settling those disputes with respect to which the law is more settled. Many of the disputes arising in the social security area would also fall into this category. Many of these different government functions have been the focus of discussions in this book.

Academic administrative law has adopted at least three approaches to the study of these functions: first, a conceptual or doctrinal approach; second, a more theoretical approach. A third approach, which concentrates more on the empirical context of law has, in the administrative law area, given rise to at least two major variants. One is a more functional variant; another is an institutional variant. In this book, we have concentrated on the extent to which governmental functions are carried out by one particular institution of government: the regulatory agency. However, our case studies and the earlier chapters do not aim to view regulatory agencies primarily from the institutional approach only. Rather we have attempted to meld the three approaches together into a more coherent view of agencies in their diversity, grounded in law and fact but conscious of theory.

We have also avoided concentrating on the study of these bodies primarily from a grievance-remedying approach. There is a danger of distorting the way we view administrative law, of skewing it towards grievance-settling to the detriment of research on the other functions of government and law. Concentration on those institutions mainly devoted to the settling of grievances would lead, for example, to some rather significant exclusions. Once other functions are considered, other institutions must surely be deemed as important as those more traditionally considered. The function of preventing undesirable behaviour brings the HSE and the Gaming Board to mind. The facilitative function brings ACAS into prominence. The provision of services raises the importance of the CA. We have aimed our selection of case studies to illustrate the wide variety of governmental functions associated with regulatory agencies.

This book has been concerned to examine the strengths and weaknesses of agency performance and to consider the place of these bodies within the British constitutional structure. We are concerned, in particular, with whether agencies can be more fully integrated into the constitutional scheme. The earlier chapters examined why agencies are chosen rather than other institutions. We outlined the operational choices frequently faced by such bodies and pointed out a number of failings to which agencies are often said to be susceptible. The growth of agencies in British government was traced and we considered whether there were criteria which could be used to evaluate agency performance. We argued that the legitimacy of specific agency actions, or the agency itself, could be assessed under at least five criteria: adherence to its legislative mandate; the degree of accountability and control exercised over it; the degree of due process accorded by it; its expertise; and its efficiency. The limitations of these criteria were considered and illustrated.

In Part II, our contributors addressed many of these issues in detailed accounts of eight regulatory schemes. The variety of agencies, tasks, operational styles, schemes of oversight and problems which they illustrate is considerable, and it may be argued that each agency operates in such a highly specific context that it is fanciful to draw lessons from, say, the Gaming Board and attempt to apply them to the CRE. We think that this scepticism goes too far. There are a number of problems peculiar to each agency, of course, but common themes do emerge from the studies and broad lessons can, we think, be drawn.

One theme which emerges is the vagueness of the mandate under which most agencies operate. In the case of the GB, ACAS, OFT, MMC and the CA, our contributors laid particular stress on this point. It might be thought extraordinary that the majority of agencies discussed in this book are not told what they are supposed to be doing. On the other hand, vagueness of mandate tends to reflect Parliament's allocation of a complex, specialist and flexible function to that body. Indeterminacy of mandate may thus be seen not only as not unusual for regulatory agencies, but likely to be the norm. Such indeterminacy makes assessment of agency performance (at least by reference to statutory criteria) highly problematic. It should be recognized, nevertheless, that indeterminacy of practical objectives does not justify lack of clarity about arrangements for control and accountability, or participation. Increased efforts should therefore be made to spell out of whom, in what respects and by which procedures, agencies are to be held accountable. Such provisions could also increase the openness with which controls over agencies are operated. We discussed whether administrative rule-making might provide a means of rendering statutory objectives more precise. In the case of agencies such as the MMC, CAA, HSE and ACAS, we saw that important tasks have been carried out through codes of practice and administrative rules of differing types. Rule-making powers and the legal effects of rules should, however, be more fully dealt with in the statute establishing the agency.

The absence of secure agency legitimacy in the British constitutional structure emerges as a second theme. The consequent attempts at a variety of methods of accountability and control over the agencies were discussed by a number of contributors. In the discussion of the GB, a highly discretionary scheme of regulation subject to low levels of accountability and control was noted. The GB's task has not been made significantly more difficult by the interventions of government, Parliament or the courts. With most other agencies the story is considerably different. In contrast with the GB, ACAS has faced political changes that have put its survival at issue and judicial review (particularly in relation to recognition procedures) has severely cramped one of its most important tasks. The HSE has managed to limit areas of contentious review but in large part by adopting a consensual approach which has arguably restricted agency innovation. The CAA's scheme of written ministerial policy guidance may have provided a means of effective compromise between accountability and agency initiative but this system of control itself failed to survive a combination of judicial review and ministerial scrutiny. In the case of the OFT and the MMC, ministers retained a degree of control over agency actions. The study of the CA illustrated the difficulties of establishing clear controls in an area of conflicting objectives and incoherent governmental strategies. Where deregulation has taken place, or, as in the financial services field, a degree of 'self-regulation' has been relied on, external controls are almost necessarily a weak form of legitimation and the benchmarks of expertise, efficiency and effectiveness have to be relied on instead.

Also in the context of accountability, we have sought to illuminate the relationship between the courts and regulatory agencies, but without concentrating unduly on it. As we have seen, the courts are important, indeed vital, for their standard-setting function. However, they are worthy of study by administrative lawyers not only as a source of standards but also as institutions themselves. Too little work has been done on the procedural mechanisms of the courts, partly, we suspect, because civil procedure has seldom been taken seriously enough as an academic subject. This defect has a continuing effect on the quality of studies of other institutions. In Part I, we emphasized a number of institutional characteristics of the courts and the potential effect these have on their competence in operating judicial review. We considered why institutional design and tradition tend to make the relationship between courts and agencies a difficult one. We pointed to the need for the judiciary to recognize the value of forms of control and accountability other than judicial review (internal, ministerial and parliamentary), when reviewing the activities of agencies. Courts, it was suggested, might not always be the most competent bodies or in the best position to review agency actions or decisions. A call was made for a more developed and explicit approach to judicial review. The studies of specific agency regimes, we think, reinforce these points. Lawyers, we argued, should resist the temptation to focus exclusively on legal values, to assume the omni-competence of the judiciary and to ignore the role of non-judicial institutions in interpreting the law.

Lawyers should be more willing to trade-off legal and other forms of effective oversight and to recognize the value of non-judicial procedures of scrutiny.

A third theme which emerges from the studies is the difficulty which agencies have encountered in arriving at procedures that are seen to achieve an acceptable balance between the participation of parties affected by a decision, and the effective pursuit of statutory objectives. Such a tension may be seen to be particularly likely in the case of bodies such as agencies which possess specialist policy-making powers and exercise judgment, yet are often expected to act in a manner modelled on the courts. In the case of the GB, the kinds of decision made in gaming regulation are not wholly consistent with the supply of full information to applicants. In the case of ACAS, the judiciary's notions of due process were not consistent with ACAS's view of what constituted effective agency action. The useful device of administrative rule-making also gives rise to a host of participatory dilemmas. For the HSE, the tension between due process and effectiveness emerged in the course of agency rule-making. Large, organized interests tended to be consulted rather than the small, disorganized and less willing employers and workers. Yet there are indications that the very group not consulted may be the group creating the greatest health and safety hazards.

The problem of how best to reconcile expertise with other values is a fourth theme which runs throughout the studies. Most of the agencies discussed in the case studies were created as experts or specialists. The need to preserve powers of expert judgement free from undue constraints was frequently mentioned. Particularly in relation to ACAS and the CRE, however, the judiciary have not been conspicuously sensitive to agency realities in exercising review. They have not been seen as intervening in a manner which enhances effective agency action. They appear to have failed to establish a clear basis for intervening (or failing to intervene) in the decisions of expert bodies. One alternative proposed may be for the courts to develop a more coherent notion of respective institutional competence.

A related issue raised by the studies is the problem of how to render polycentric issues manageable by an agency. An agency such as the CAA which employs trial-type procedures to decide complex regulatory issues has a particularly severe problem in that respect. We saw that the CAA, by using policy statements and rules in combination with case decisions, has had some success on this front. Justiciability is a theme which runs through the discussion of the MMC. The agency has been able to consider the economic effects of a firm's behaviour, but only after that behaviour has been tested against a standard of legal form. Craig suggested that the use of guidelines might provide a means (albeit not an unproblematic one) of reconciling these factors. Agencies, then, can employ processes commonly associated with the courts when involved in polycentric decision-making. Success demands, however, astute use of ancillary procedures such as rule-making and disclosure of policy guidance.

A fifth theme encountered in the studies centres on the problem of assessing effectiveness in the context of agency action. The problem which recurs most frequently is that of assessing effective *enforcement*. The HSE, OFT and CRE have had to balance promotional and campaigning activities with enforcement through more formal legal processes. How to target effectively the limited agency resources which agencies have at their disposal, and how to establish clear regulatory priorities are issues which recur repeatedly in the studies. It would seem that agencies could do more to co-ordinate case-by-case enforcement with rule-making strategies. Rules could thus be related more closely to those forms of activity that are to be anticipated in the field, and greater efforts could be made to identify those subject to regulation who are to be the particular targets of agency action.

The issue of effectiveness is also closely related to the level of regulation and the overall regulatory strategy that is chosen in the particular field. The degree of 'self-regulation' relied on was an issue raised particularly in relation to ACAS, HSE, CA and the financial services agencies. Where 'self-regulation' prevails there is a special case for developing means by which its effectiveness can be assessed and demonstrated. However, the need for clarification of agency aims and objectives emerges as a precondition for the adequate testing of effectiveness by an agency in such circumstances. Where, as with the CA, such clarity is not provided by the parent statute, the agency or overseeing Department might take steps to provide such clarification by more open disclosure of rules and policies.

Finally, a theme emerges which could well come to dominate academic administrative law in Britain in the 1990s, as it has already done in the United States. This is the issue, not of *how* to regulate, but of *whether* to regulate at all. It is an issue which is likely to percolate more and more explicitly into the field of public law. Though our contributors have not sought to focus on this issue systematically, except in the context of the regulation of cable television, it is an issue which underlies many of the discussions. Whether to regulate should not, however, be divorced from consideration of the methods to be used when Government reassesses the scope and stringency of regulation. The themes which emerge from this book are thus intimately connected with the question of whether to regulate. Just as public lawyers can no longer ignore economics and political science, so economists and political scientists can no longer afford to neglect public law.

Notes

1. F. Slatter, *Parliament and Administrative Agencies* (1980), pp. 1–19. Slatter's work has influenced this section considerably. See also Law Reform Commission of Canada (LRCC) Working Paper No. 25, *Independent Administrative Agencies* (1980), pp. 34–5. See also C. Hood, 'Keeping the Centre Small: Explanations of Agency Type' (1978) 26 *Pol. Stud.* 30.
2. R. B. Reich, reviewing J. P. Wilson's *The Politics of Regulation*, in *The New Republic*, 14 June 1980, at p. 38.
3. *Ibid.*
4. Hood, *loc. cit.*
5. See the discussion of the pros and cons of both in R. A. Katzmann, *Regulatory Bureaucracy* (1980), chapter 3.
6. Incrementalism is associated with the work of C. Lindblom. See 'The Science of Muddling Through' (1959) 19 *Pub. Admin. Rev.* 79. A useful discussion of the relative merits of these models in their application to regulation may be found in C. S. Diver, 'Policymaking Paradigms in Administrative Law' (1981) 95 *Harv. L. Rev.* 393.
7. H. Simon, *Administrative Behaviour*, 2nd edn (1957, Free Press), pp. 39–41, 80–84.
8. For example, as reflected in the White Paper, *The Reorganisation of Central Government 1970–71*, (1970), Cmnd 4506.
9. LRCC, *op. cit.*, p. 9.
10. Reich, *op. cit.*, p. 36.

"Which Arrow?": Rule Type and Regulatory Policy

J. M. BLACK[1]

The decisions involved in rule making have been described by one regulator as finding answers to the question: 'which arrow is to be taken from the quiver?'[2]. What type of rule, in other words, should be adopted? The issue of rule type is not one which appears on its face to be a matter which should be of great concern in the formation of rules. *What* the rule should say is clearly a policy decision; *how* it should say it does not immediately appear to be significant. However, as the quote indicates, the choice of rule *type* may be a particular regulatory policy in itself. It raises three main questions: what different types of rules may be used; what targets are being aimed at; and is the process as strategic as the quote suggests? This article seeks to address these questions through an analysis of rule making by financial services regulators. Its concern is to understand the nature of the rule making process and the reasons rule makers adopt different types of rules.

Rule making has not been a central focus of attention for those concerned with the operation of regulatory agencies or bureaucracies. The rise and rise of delegated legislation has been documented,[3] and concerns about the control of the process and the legal status of the rules raised,[4] but there have been few extensive studies of the way in which agencies exercise their rule making powers once delegated to them. The focus of those studies which have been done has generally been on overall policy formation and implementation,[5] although more recent studies have indicated that the choice of rule type may be of significance.[6]

Fundamental reasons for the choice of rule type are suggested by the work on rules and discretion. While again the focus of this work has not been rules or rule type, in highlighting the advantages and disadvantages of rules and discretion it illustrates why the choice of rule type matters. Further, the association of rules or standards with different forms of bureaucratic rationality, substantive and formal, suggests that in forming rules rule makers are constantly addressing the tensions which exist between them.

However, neither the empirical studies nor the theoretical debate provides a comprehensive framework in which rule making, can be analysed. In order to begin to develop such a framework, the article argues for a dimensional analysis of rules which can be applied to any rule system. Rules should be seen not as monoliths, but as comprising separate facets or dimensions. This understanding of rules can then be used in the analysis of rule making decisions in order to illustrate and understand what different types of rules may be adopted and why. The article identifies three main objectives which financial services rule makers have aimed to achieve in adopting rules of particular types: a balance between flexibility and certainty; a change in the perceptions of the regulation held by the regulated and the wider community; and the exercise of control over the regulated and other regulators. The article shows that the process of rule making contains both strategic and non-strategic elements. While it exhibits some elements of rationality, it is not a wholly rational process. Moreover, even if the decision is to an extent strategic, constraints operate on it, arising both from the inherent nature and limitations of rules and from factors external to the rule: the political, legal and regulatory context. Finally, rules cannot be seen as discrete decisions. Rules are bargained over, and they are built.

The first part of the article thus proposes a dimensional analysis of rules which can be applied to any rule system. The second part discusses the few existing studies on rule making and points to a need to move beyond their concerns and approaches to begin to develop a theoretical framework for the analysis of rule making. This is developed and applied in the third part of the article in which the rule making process of financial services rule makers is analysed, focusing principally on the change from detailed rules to a three tiered rule system of Principles, Core Rules and Third Tier rules. Finally the article examines the strategic and non-strategic aspects of the rule making process and the constraints which operate on it.

Types of rules

Rules are often seen as relatively precise formulations: 'a legal precept attaching a definite legal consequence to a definite detailed state of fact'.[7] As such they are contrasted with standards, which are seen as less precise, requiring some judgment in their application.[8] The definition of a rule that will be used here is one which is more general: 'a general norm mandating or guiding conduct or action in a given type of situation'.[9] Within this general category rules may vary in their type, but not simply in the manner which the usual rules/standards distinction suggests. A dimensional analysis which recognises the different aspects of rules illustrates more clearly the range of rule types which may exist. Rules have four dimensions, and may vary in each of these[10]: substance (what the rule says); character (whether it is permissive or mandatory: may or shall); status (its legal force and the sanction attaching to it) and structure. The last is the most complex, and has four aspects; the scope or inclusiveness of the rule; precision or vagueness (the degree to which behaviour under the rule is prescribed); simplicity or complexity (the degree to which the rule may be easily applied to concrete situations) and clarity or opacity (the degree to which the rule contains words with well-defined and universally accepted meanings).[11] Rule type is a function of only three of the dimensions: character, status/sanction and structure.

These dimensions are descriptive rather than normative. Further, with the exception of the clarity of the rule, they are objective. Clarity is subjective: it is dependent upon the readers' ability to understand what the rule requires. So a requirement of 'reasonableness' may be clear in a particular community, as a consensus exists as to what conduct would be seen as reasonable. If no such consensus exists, then the rule is opaque.[12] The interrelationship of these dimensions is complex. As will be seen below, the rule's substance may affect its structure, as may its status or the sanction attaching to it. A rule may be precise and simple: *e.g.* 'Send information within fourteen days', but the greater the degree of precision the greater the likelihood that the rule will become more complex: *e.g.* 'send information on costs and expenses, surrender values, projected returns and cancellation provisions within fourteen days'. This will increase further if 'costs and expenses' are separately defined and prescribed. Rules may be precise, complex and clear: clarity is a function of the manner in which the individual requirements are defined. So, for example, a rule which states that licences may be granted to persons fulfilling a list of conditions is complex, and it is clear if those conditions are well understood. However if they are not—being defined in terms of reasonableness, for example, or requiring a person to be 'fit and proper'— then the rule is both complex and opaque. Similarly, the status of the rule or sanction attaching to it may vary: precise, complex rules may be of the status of guidance or they may have legal force. Breach of that rule may lead to disciplinary action, or to a civil action or prosecution.

Rule making by regulators

Rule making is an aspect of policy making and rules have been described as the '"skin" of a living policy'.[13] The choice of rule type, this would suggest, is part of a regulatory policy. But why would rule makers pay attention to the type of rule which they formulate? What difference does it make if a rule provides that 'sufficient information must be given' rather than: 'the following information must be given', followed by a long list of precise requirements? Why does it matter whether the rule is only a guidance, or the sanction is disciplinary (i.e. imposed by the regulator with no court involvement)?

Why rule makers make the rules they do has not been the subject of extensive research. There has been considerable development in the last decade or so in studies on the use of discretion,[14] the enforcement of rules,[15] and the reactions of the regulated to particular types of rules, or 'creative compliance'.[16] This work looks essentially at *responses* to rules. It is concerned with the reactions of different actors to rules: officials within organisations charged with the implementation or enforcement of a particular policy, or those actors outside the organisation whose behaviour the rules are meant to regulate. Rules are considered, but incidentally. The existence of the rule is taken as given. The interest and focus of this work is on what happens next: how is that rule applied; to what extent does it constrain the exercise of discretion; to what extent is its implementation or application affected by other considerations, and what form do those considerations take? The work on discretion and enforcement, particularly, questions the assumptions which are made in the rules/discretion debate, criticising the monolithic conceptions of discretion which are employed, the decision making processes which it assumes, and the assumption that the debate makes about the effectiveness of rules in confining discretion. It is not concerned with why the rule provides what it does, adopts a particular structure or has a certain sanction attaching to it.

One of the implications of the relatively limited focus on rules is that despite the advances in the understanding of the nature and use of discretion, there has been little advance in understanding why regulatory agencies make the rules that they do.[17] There are two exceptions. The first is the work by law and economics writers which is concerned with the 'optimal precision' of rules.[18] 'Optimality' is assessed according to an efficiency criterion in which four categories of costs and benefits are weighed: the rate of compliance, the over and under-inclusiveness of the rule, the costs of rule making and the costs of applying the rule.[19] Rules which are precise are seen as more costly to draft, but less costly to enforce. They are also seen as providing greater predictability, and so encouraging compliance, by increasing the probabilities as perceived by the person subject to the rule that non-compliance is punishable, and compliance is not. These assumptions lead to certain specific recommendations of optimal rule type. Diver, for example, argues that where the rule is directed at an audience which is large, diverse and remote, or where incentives for non-compliance are otherwise high—for example where compliance with the rule requires a person to forgo considerable benefits, or where non-compliance is difficult to detect—then more detailed rules should be used.[20]

The second exception is the few studies which have been done on rule formation, which have been concerned to examine empirically the formation of rules and have not sought to apply the law and economics writers' predictions or suggestions. Nor have they all been concerned with the issue of rule type. They indicate the impact the structure of the rule making process has on rule formation as a whole,[21] and the extent to which the perceived legitimacy of the rules or the issuing body affects the content of the rule.[22] They do also indicate, however, that the question of what type of rule to use can be a significant issue of regulatory policy.[21] Normative assessments have been made as to the nature of rules that should be produced. Rules have been criticised on the basis that they are insufficiently precise, and so do not aid enforcement.[24] Baldwin, for example, argues that rule makers should adopt a compliance-oriented approach to rule making, taking into account the enforcement strategies which will have to be used to achieve compliance.[25] In the area of health and safety Baldwin found that on the whole enforcers prefer more specific rules.[26] However, rule makers prefer across the board regulation in the interests of uniformity and consistency. The fact that rule makers have adopted the rule types they have is attributed by Baldwin to their different preferences, and failures in the rule making process arising from organisational and political constraints and the nature of the process itself. However, while precise rules may aid enforcers,[27] work in the area of tax and accountancy indicates that they have other drawbacks.[28] It further shows that regulators do pay considerable attention to compliance, if not to enforcement, when forming rules, but that the rules which they perceive to be necessary are not those which in fact Diver would recommend. McBarnet and Whelan look at the dynamic interplay between regulated and regulator in the formation of rules. The regulated adopt a strategy which they term 'creative compliance': the use and manipulation of rules as a combination of specific rules and an emphasis on legal form and literalism to circumvent or undermine the purpose of regulation.[29] The regulatory response has been to reinstate control by adoption of anti-formalistic approaches which are more flexible, open-textured and policy-orientated. This includes an emphasis on purpose, avoiding use of statute and courts, using professions to formulate guidance and regulation of standards of practice, reflecting the substance of transactions and relationships and implementing spirit rather than letter of the rule.[30] The strategy is undermined by the regulated's continual demand for certainty, which leads to the formation of regulatory guidance and precedents, and so to the re-introduction of formalism. They thus posit cycles of formalism in which detailed rules are replaced by purposive ones, which in turn are replaced by detailed requirements, facilitating creative compliance behaviour.

Formalism and anti-formalism; rules and discretion: finding the 'right balance'

These theoretical and empirical studies indicate that the choice of rule type matters, either because it can affect enforcement, or because it affects the compliance behaviour of the regulated. However, if we are to understand why rule makers choose to make different types of rules we need to understand at a more fundamental level the nature of the consequences arising from that choice. We can then begin to develop a framework for understanding the rule making decision, seeing whether rule makers pay attention to these consequences, and why.

The debates concerning the nature of rules and discretion and the types of rationalities which are exhibited by administrators at any particular time are instructive as they indicate some of the reasons why the adoption of different types of rules is significant. The pattern identified by McBarnet and Whelan of cycles of formalism echoes the writings of sociological theorists concerned with identifying and explaining the changes in the form that law has taken in society.[31] The use of purpose orientated rules or precise detailed stipulations is associated, drawing on Weber, with different forms of rationality.[32] Formal rationality, or decision according to the literal letter of the rule, is contrasted with substantive rationality, where decisions are made according to the substantive goals of the policy. However, as Galligan notes, a tension exists between the two forms,[33] and 'Weber's ideal types serve as focal points around which those [decision making] strategies are centred, with each exerting a pull and influence on that process, so that in practice decision making is likely to be a complex mixture of the two.'[34]

One of the implications of this tension is that one of the central issues in the exercise of discretion is to discover configurations of these rationalities which are the most suitable to achieve certain goals, whilst respecting other values.[35] The resolution of the competing demands of certainty and flexibility, uniformity and individualisation, is seen to be the combination of different types of rules (or in the language of this literature, rules and standards) to achieve the 'right balance' of discretion at any one time.[36]

The issue of the 'right balance' arises from the perceived advantages and disadvantages of rules and discretion, in which the advantages of one tend to be the disadvantages of the other.[37] The issue has been seen as one best assessed in the context of individualised decision making.[38] However, rule makers may try to achieve that balance through the use of rule type. In choosing the type of rule to use, or arrow to fire, the rule maker may thus be trying to resolve the tensions between rules and discretion, or formal and substantive rationality. The dimensional analysis of rules posited above is particularly valuable in illustrating how this may be achieved. In the rules/discretion debate a simplistic model of rules tends to be used, in which rules are contrasted with standards, but beyond an intimation that one is more precise than the other, no further analysis is done. The dimensional analysis of rules set out above indicates that 'formalism', or rigidity within a system of rules, may be a function of a number of the dimensions: the legal status of the rule, its structure and its character. A detailed rule is given 'flexibility' if it is purely recommendatory, or its breach leads to only a discretionary sanction. Moreover, rules which are detailed and precise may prompt a formalistic approach to their interpretation and application. Those which are more purposive and require judgement in their application facilitate a purposive approach, or substantive rationality. Finally, the structure of the rule affects the amount of discretion the person subject to the rule has: rules which are less specific and more vague confer greater discretion on the interpreter of the rule as to how the rule should be applied than rules which are clear and precise.

While suggesting that rule makers are in their choice of rule type responding to the tensions between different types of rationalities takes us further towards an answer, it is not complete. We need to ask how this is manifested. Is it reflected in a desire simply to balance flexibility and uniformity, or are rule makers trying to affect the interpretation (formalistic or purposive) which is given to the rule? We still need to know why the rule maker wanted to strike that particular balance in that particular case. Is it simply a question of overriding preference: do rule makers always prefer a certain type of rule, as Baldwin's study suggests? Is the choice of rule type a particular tactical choice, for example to counter the 'creative compliance' behaviour of the regulated, or is the process less strategic? Is the rule maker unconstrained in that choice? Finally, given that the different rationalities involve competing notions of justice and accountability, can the tensions between them in fact be resolved? We will turn now to examine the rule making decisions of financial services regulators to see what types of rules they have used, and the extent to which the choice of rule type has been a conscious and strategic decision.

Rule making by financial services regulators

Rule making has been a continual exercise in financial services regulation.[39] The Financial Services Act 1986 established a system of regulation for all those conducting investment business.[40] It requires all persons carrying on investment business to be either authorised or exempt.[41] Authorisation may be obtained through membership of one of the recognised regulatory bodies.[42] There are essentially three levels of regulators: the executive, on which rule making and enforcement powers were initially conferred, the Securities and Investments Board or SIB, to which those powers were delegated,[43] and the self-regulatory organisations, or SROs.[44] SIB and the SROs are private companies limited by guarantee, established solely for the purposes of regulating investment firms. SIB exercises statutory powers; the SROs operate on the basis of contract. There are currently five SROs, each of which has been authorised by SIB to regulate a different area of the market, and each has its own disciplinary and membership processes and its own rule book.[45]

SIB has a range of rule making powers, most of which were conferred in amendments made to the 1986 Act. SIB has the power to issue principles, designated (or core) rules and codes of practice which apply to its own and SRO members.[46] Its rules have to conform to broad statutory criteria, which require that they make proper provision, for example, for disclosure, require firms to act with due skill, care and diligence and promote high standards of integrity and fair dealing.[47] The power of the SROs to make rules arises from their constitutions, and not from the legislation. Each SRO had to have a complete rulebook in place before it could be authorised, however, and SRO rules, like SIB rules, are subject to scrutiny by the Director General of Fair Trading to ensure that they do not restrict or distort competition to a significant extent. SRO rules are otherwise required to provide an 'adequate' level of investor protection.[48] The sanctions for breaches of the rules of SIB and the SROs vary, but not, as one would imagine, with their legal source. Breach of the Principles is subject to discretionary disciplinary action by the person's regulator.[49] Breach of the Core Rules and SRO rules is subject to both disciplinary action and to civil actions on the part of a private investor.[50]

The rule making process involves the formation of policy proposals, which may or may not be accompanied by draft rules, which are published for consultation and sent to all those who hold a copy of the rules.[51] Responses are then reviewed internally, and in the case of significant changes to the rules further proposals are made. Once the rules are formed, the OFT reviews them to assess their effects on competition, and then reports to the Treasury. The Treasury may then require SIB and the SROs to amend the rules if they affect competition to a significant extent.[52]

While rule making has been continual, two main phases can be identified: the formation of the initial rule books in 1987–8 and the change to the New Settlement in 1990–1, which was enabled by the 1989 statutory amendments. SIB has used the powers to issue 10 Principles and 40 Core (designated) Rules.[53] The New Settlement, as the new rule system was termed, represented a striking change in the type of rules used by SIB. It was a move from detailed, specific rules to vaguer, more purpose-oriented rules, with little change in their substance.[54] The SROs' rules, or third tier rules, are more detailed, expanding or clarifying the Core Rules, and stipulating their application to the market area which they regulate.

Which arrow? The choice of rule type

The rule system as a whole, therefore, is made up of rules which vary in their precision, which have different sanctions attaching to them, and which apply to different areas of the market. Why has such a complex combination of rule types been used? Analysis of the rule making decisions indicates that the process involves a complex interaction of both strategic and non-strategic behaviour. Rule making has in part involved conscious policy decisions as to the choice of rule type. However, as we will see below, there are limits to the rationality of the process. Three principal aims have been pursued through the choice of rule type: to provide both certainty and flexibility, to induce compliance and to affect the degree of control exercised. Analysis of the reasons for the pursuit of these aims reveals the nature and interaction of deeper objectives which go far beyond the substantive objectives of investor protection and market efficiency that one might expect.

Certainty and flexibility

The rule making decisions of the financial services regulators indicate that they are concerned to use rule type to address the tensions between formal and substantive rationality by addressing the most immediate tension between certainty and flexibility. The demands of the regulated in relation to rule type are consistent in their inconsistency. The continual demand in responses to the consultative papers has been that the rules should be sufficiently certain to enable firms to know what is expected of them, but sufficiently flexible to allow business practices to develop. The initial rules of SIB and the SROs were criticised as being too complicated, too detailed, too long, and too difficult to understand. Their length, detail and complexity meant that their application was at once both uncertain and inflexible.

The initial impetus for Principles arose from the conviction of the second chairman of SIB, amongst others, that the underlying rationale of the individual rules was not clear. SIB thus drafted a replacement rule book which contained nearly one hundred Principles. Although at this stage the idea of a three tier rule system had not been formulated, and the Principles were later adapted to meet a further objective, discussed below, the aim to combine the benefits of different types of rules within one rule system remained. The Principles are relatively imprecise and vague; they are purposive, and so provide greater flexibility in their application. The SROs' rules are far more detailed, so providing greater certainty ad predictability as to the rule's application in a particular instance. The Principles override these rules, so that issues of application and interpretation are to be resolved in the light of their provisions. The combination of rules of different structure, detailed and general, is thus aimed at aiding the regulated by providing both certainty and flexibility. The Principles also aid the rule maker in that they reduce the pressures for precision which arise out of a fear of creating gaps in the rules. One SRO official in interview described the Principles as a 'safety net', relieving the pressure for the SRO to ensure that all situations were covered in the SRO rules.

Not only the structure but the sanction attaching to the rule can be used in an attempt to resolve the tension between certainty and flexibility. The use of detailed provisions, for example, has been given flexibility by changing their status from rules to guidance. The use of guidance is a direct result of the contradictory demands of certainty and flexibility. As one official explained: 'there is a tension in the desires of members—they want vague, general rules like the Principles and detest the detailed rule books, but they also want certainty and detailed information of what is expected of them. Guidance is an attempt to resolve this tension.' Guidance thus indicates what type of conduct will be accepted as compliance, but removes the irritant of prescription.[55]

Changing the sanction can also mean that rules which have a structure which is less precise and detailed can be used on their own, not necessarily in combination with detailed rules, and still provide certainty. Here we see the nature of the interaction of the different rule dimensions. In the financial services context, the restriction in the application and availability of the right of civil action for breach of the rules[56] had the greatest impact on rule structure, and enabled the refinement from the one hundred Principles to ten. Many of the demands for certainty which were prevalent when the initial rules were being formed arose from the provision of this right of action to all investors. In the uncertain climate of the time there was a strong fear on the part of the regulated of extensive litigation between professionals. As the chairman of SIB explained,

[Section 62] had a serious adverse effect on the SIB rulebook. The problem was that it focused the attention of practitioners, and more particularly their lawyers, on amending the rules, not to improve them generally, but simply to minimize the possibility of claims. This led to long and complicated provisions, attempting to draw fine distinctions to provide safe harbours for legitimate industry practice, while maintaining the essence of the original objective of the rule.[57]

SIB did not initially proposed changing the sanction attaching to the Principles,[58] but their structure was not considered by the SROs and regulated to be acceptable unless the Principles were subject to discretionary disciplinary action only.[59] Why should the change of sanction matter? Why were opaque rules perceived to be too uncertain and so unacceptable unless there was no possibility of court action for their breach? The conviction was strong, but the reasoning for it was unarticulated. The reasons for the perceptions become clear however if we consider how uncertainty can arise. Uncertainty can arise from opaque rules, as those subject to the rule cannot be sure that all others, including enforcers, will interpret the rule in the same way. However, as was noted earlier, opacity is a subjective dimension: whether or not the rule is clear depends on the ability of the reader to understand what the rule requires in a particular circumstance. Opaque terms, such as a requirement to 'exercise due care', may have a particular meaning in a particular community. The requirement is thus clear to that community, although not to persons who are outside it. Further, those who are not part of that community may not share this interpretation. If outsiders are to be involved in interpreting and enforcing the rule, therefore, the community cannot rely on its own consensus as to the meaning of the rule being adopted. Certainty is thus reduced, and has to be regained through detailed rules. The system must therefore be closed off from outside involvement in rule interpretation and application for opaque rules to be capable of providing certainty. One of the simplest ways to effect this closure is through the rule's sanction: to state that breach will give rise to disciplinary action by regulators (insiders) only.[60] To apply this analysis to the financial services context, excluding the right of action from those most likely to use it (professional investors), and from the most opaque rules (the Principles), effectively closed the regulatory system, eliminating the courts (outsiders) as potential interpreters of the rules.[61] This closure of the system, effected by the change in the sanction attaching to the rule, thus attempted to ensure that opaque rules would receive a common interpretation, so increasing certainty in their operation.

Changing perceptions; inducing compliance

Attempting to resolve the tensions between certainty and flexibility has involved the use of the rule's sanction and the combination within one rule system of rules of different structure. Rule structure has also been used to induce compliance, and rules which are relatively simple, vague, opaque, in other words more purposive in their effect, have been introduced. The clearest manifestation of this approach was the initial formation and refinement of the Principles. The aim of the Principles was not simply to act in combination with more detailed rules in an attempt to resolve the tensions discussed above, but to play a more informing and influencing role in enabling and inducing compliance with the rules.

Compliance inducement has been prompted by a set of motivations far more complex than the formal/substantive rationality debate would indicate. In part it is a fairly 'hard nosed' concern with regulatory effectiveness. The formation of the Principles indicates that this had two aspects. The first was an attempt to ensure complete coverage of the rules:—'to remove the unmeritorious technical defence of a lacuna in the rules'.[62] The second was an attempt to emphasise that mere observance of the literal meaning of the rule, while flouting its purpose, would not be accepted as compliance. That compliance should be 'with the spirit and not just the letter' was frequently stressed in the period prior to the change in rule structure.[63] The adoption of more opaque rules, of disciplinary status only (see above), was in large part aimed at pre-empting or foiling the 'creative compliance' behaviour of the regulated.

The strategy was also more subtle: the change in structure and status of the rules manifested in the Principles was addressed not simply and directly at 'creative compliance' behaviour but at what that behaviour symptomised. 'Creative compliance' symptomised a loss of regulatory goodwill, a breakdown in relations between the regulated and regulator, relations which the system had been meant to foster. it was feared that the mass of detailed rules was alienating the regulated and as a result undermining the legitimacy of the regulation. The change was not simply a response to sectoral demands, however: in changing the rule type, the regulators aimed to change the perceptions the regulated had of the regulation in an attempt to alter their internal attitude towards it. The greater the acceptance of the regulation, the greater compliance would be.

The nature and strength of this reason for the adoption of the Principles can be more clearly assessed by looking briefly at the background to the regulation. Prior to the Financial Services Act most areas of the market had not been regulated. The rhetoric which surrounded the Act's formation was that the regulation would combine the advantages of the sensitivity and flexibility of self regulation with the sanctions of a statutory scheme. The regulation which appeared in the form of the initial rules was not what had been expected. The Bank of England in particular felt that the regulation was harming business, and did not have the acceptance of the regulated, or more specifically, the City.[64] The fear of alienation, manifested in non-compliance or creative compliance, thus led to a strategy of co-optation. Co-optation was not attempted simply through involving the regulated in the rule making process,[65] although this was done, but by changing the type of rules used.

The attempts to change the internal attitude to the regulation were expressed through the idea of a 'regulatory contract', an idea adopted by the last two chairmen of SIB. Regulation, according to SIB's second chairman, Sir David Walker, represented a contract between the regulator and regulated, in which the regulator agreed not to interfere in the detailed operation of the regulated, but in return expected compliance which went beyond mere obedience to the letter of the law.[66] In other words, if you, the regulated, act properly, we, the regulators, will not need to be so specific.

The notion of a 'regulatory contract' was accompanied by an emphasis on a particular interpretation of self-regulation as 'regulation of the self'.[67] In other words, each person and each firm should act as its own internal monitor, mentally assessing behaviour against the basic principles of the regulation. There was a fear that the regulation was seen by firms as a matter for compliance officers only, no one else knowing, or caring, what the regulation provided. However, detailed rules are difficult to absorb as they are hard to assimilate. Ten Principles, on the other hand, would be easily memorable and so be able to act as internal benchmarks against which conduct could be assessed. They would enable chief executives to understand the basic tenets of the regulation, and to create a climate of compliance within the firm. They would thus improve compliance in a complex area by lifting the issue into the boardroom and reminding the chief executive that there were basic principles that should govern the conduct of his business. Literal compliance with the rules was not to be a substitute for management integrity: the Principles would help chief executives to see the moral wood for the technical trees. Moreover, it was stressed that they required no more than good business practice, which if followed would increase confidence in the firms and markets, so aiding business. Regulation, in other words, was good for you.

Exercising control

The third aim sought through the use of rule type has been the exercise of different degrees of control. The addressees of this strategy have not only been the regulated, however, but the regulators. Rules of different structure afford different amounts of discretion to those being regulated by them. The sense in which discretion is being used here is that advocated by Galligan and Hawkins: there is no clear dividing line between rules and discretion; rather the two are inextricably interwoven. Discretion in this sense is interstitial: it is 'the space . . . between legal rules in which legal actors may exercise choice'.[68] The rule may vary the amount of discretion which that legal actor has.[69]

Studies of discretion have indicated that how that discretion is used, and indeed the extent to which that discretion is constrained by the rule system as a whole, cannot be determined by an analysis of the rule alone. In applying the rule, enforcers and officials are subject to a range of extraneous factors, with the result that they may be both more and less constrained than the rule would indicate. The regulated may manipulate them; their compliance may be determined by a complex cost/benefit calculation. Moreover, detailed rules in one area may simply displace discretion to another.[70]

However, rule makers have used detailed rules in an attempt to control behaviour by reducing the amount of discretion that the person has as to how to act. The question which is being asked is essentially, to what extent can the rule makers trust the regulated to behave as they want them to. The issue here is not that the regulated are too clever and will manipulate detailed rules to serve their own ends (as it is in creative compliance). Rather it is that in the absence of detailed rules the regulator feels that there will be insufficient compliance: the regulated will simply not act according to the standards of behaviour that the regulator requires unless that behaviour is specified. Contrary to the implications of the creative compliance literature, therefore, the regulator does not want to use a more substantive approach. Rather it is felt that this approach cannot be used. Rule makers cannot afford to confer that discretion as they do not trust how it will be used. This confidence in such rules may be misplaced,[71] but this does not alter the fact that that confidence exists, or at least that such rules are seen as marginally more effective than other types.

Rule makers therefore consider the extent to which they feel they can rely on the person applying the rule to act to further the rule's purpose: detailed rules are, to an extent, a sign of distrust. The regulators have explicitly referred to this aspect of rules. The current chairman of SIB, Andrew Large, emphasised recently that the more SIB felt able to rely on rising standards of competence and business ethics of the regulated, the less prescriptive the rules might be able to become.[72] The type of rule used can thus vary with the person to whom the rule is primarily addressed. Where regulators feel they can rely on the object of the rule, then a rule which requires the regulated to exercise judgement will be used to replace detailed rules.[73]

While the objects of the decision have in these cases been the regulated community, the use of detailed rules in an attempt to exert control has also been evident in relations between all levels of regulators. The highly detailed nature of SIB's initial rule book is attributed by SIB largely to the requirements of the DTI. The DTI's approval of the rules was necessary before the statutory powers could be transferred to SIB. Once they have been transferred, the Secretary of State has no power to order their amendment unless they are significantly anti-competitive. The DTI was in part responding to pressures from the regulated in requiring detailed rules (see above). However it was also uncertain as to how SIB would use its rule making powers once they were transferred. The imposition of detailed rules can be seen in part as an attempt to control that use, even though subsequent amendments would be out of their control.

Conversely, the introduction of the vaguer, more opaque Principles and Core Rules was at one level a deliberate conferral by SIB of a greater degree of discretion on the SROs. One of the main criticisms of the initial SIB rules had been that they did not leave SROs room to write their own. A move to vaguer, more purposive rules was thus a way of affording SROs that room for manoeuvre. SIB stressed that it would be 'standing back' from third tier rule making,[74] recognising that the previous rule system had involved an 'unsustainably intrusive role' for SIB.[75]

However, the greater discretion conferred by the change in rule structure was in part negated by the change in the application of the rules and the sanctions attaching to them. Under the 1989 amendments the Principles and designated rules (Core Rules) of SIB applied directly to all members of SROs. The application of these rules to members of the SROs enabled SIB to assert control over the central provisions of the SROs' rule books. Previously, SROs had to follow SIB's rules simply because of the requirement that their rules provide equivalent protection; now SROs are effectively bound by SIB's rules because their members are. SIB's rules are now more clearly legislative in nature, and centralising in their effect.

Whilst the objectives sought via the Principles were largely a change in the perceptions of the regulation with the aim of inducing greater compliance from the regulated, those sought by the Core Rules relate to the relations between the regulators themselves. The stated aim of the Core Rules was to set what was referred to as the 'common core' or 'harmonising backbone' of the regulation,[76] ensuring consistency between SROs' rule books, and so meeting the criticisms of divergence and inconsistency between the SROs' rule books which had prevailed.[77] However, the reasons behind their introduction lie more deeply, in the institutional structure of the regulation. The New Settlement, and particularly the existence of the Core Rules, makes little sense unless one understands the extent and nature of the pressures which the institutional structure exerts.

The regulatory structure is essentially federal and exhibits consequent tensions between the role of the centre and the autonomy of the outlying bodies. It is the result of the compromises made between differing views of what the role of the regulation, and hence its structure, should be. As the Act was being formed, market efficiency claims pushed simultaneously for self-regulation, which it was felt offered sensitivity and flexibility, and for oversight lest self-regulation degenerated into cartels and restrictive practices. Investor protection claims pushed for public interest regulation with tough enforcement (central statutory body) and awareness of market practices to ensure abuses were addressed (practitioner regulation).[78] The result was a network of agencies whose relative roles were ill-defined.

This structural ambiguity became a source of severe friction between the two tiers of regulators, SIB and the SROs.[79] Many of the criticisms of the initial SIB rules were essentially directed at the role the rules oriented SIB was playing in the regulation. It was felt by many in the City, the SROs and the Bank of England that the 1987 rules represented an over-centralisation of the regulation and the Governor of the Bank of England publicly stated that the 'practitioners' contribution had been overshadowed'.[80] SROs also publicly urged SIB to adopt a less intrusive role, threatening that self-regulation would otherwise come to an early end.[81]

Rule type was thus used to define the nature of that relationship: less specific and detailed rules would confer greater discretion on the SROs, but at the same time changes in the application of SIB's rules and their extension to the members of SROs would ensure greater control and uniformity in the rules of the SROs, which were now more clearly subsidiary regulators.

Summary

Rule type has thus been used to achieve three main aims of resolving the tension between certainty and flexibility, inducing compliance and affecting the degree of control exercised. The dimensional analysis of rules enables us to see how different dimensions have been used to achieve these aims. These dimensions interact: changes in the sanction may confer flexibility on detailed rules, as in the formation of guidance, or certainty on opaque rules through closure of the system, as in the formation of the Principles. It is notable however that detailed rules have remained a central aspect of the rules to which firms are subject. There have been no cycles of formalism as there has never really been a move away from formalism. Vague or purposive rules have been used not to replace detail but to complement it.

The analysis indicates not simply the 'first order' reasons, or aims of the rules, but the deeper, underlying reasons why these aims were pursued at all. These include a desire to improve the effectiveness of the rules in terms of comprehension, accurate application and compliance with them, an attempt to secure acceptance of the regulation, and to educate the regulated in the underlying principles of the regulation. 'The object of the rules has not simply been the regulated. The introduction of the New Settlement, particularly the Core Rules, represents an attempt to resolve the tensions between the role of SIB and that of the SROs which arose from the ambiguities of the institutional structure through the adoption of a particular type of rule. The policy behind them was one of simultaneously conferring discretion, through changing the rule structure, and centralisation through changing the application of the rule. As the relationship between the institutions changes, this affects the rules. This dynamic continues to be important as is illustrated in the current policy of 'de-designation', in which particular core rules are no longer applied to the members of particular SROs.

Constraints on rule making

The attention paid to the types of rules which should be formed, manifested particularly in the New Settlement, indicates that rule making and choice of rule type has been to a significant extent a self-conscious and tactical exercise. Rule type has of itself been an important part of regulatory policy. However we must be wary of falsely representing both the 'choice' open to the rule maker and the rationality of the process itself. Not all rule types are suitable for all purposes: trade-offs have to be made as the rule maker has to decide which to pursue. The rule making process is not one of complete comprehensive rationality. The rule is the culmination of a series of decisions of a process of amendment and reformulation as priorities alter or regulatory strategies, initially hazy and illformed, become increasingly defined. Finally, rule makers do not function in a political vacuum: politicians, consumer groups, the regulated and the media all criticise and make demands for the type of regulation which they think should be instituted, which demands are often competing. This is not a surprising occurrence. What is significant is the extent to which the choice of rule type is affected by them.

Which target? Tradeoffs in the choice of rule type

What emerges from an analysis of rule making by financial services regulators is that not all types of rule are suited to all purposes, and that the nature of rules imposes limitations on their effective use. The adoption of particular regulatory techniques may mean that only rules of a certain structure can be used. Conduct regulation—the requirement to act in a particular way, usually to exercise judgement—is a technique relied upon quite heavily by the regulators. It is a less onerous form of regulation than structural regulation (which structures the market in a particular way), permitting the person to carry on the business he or she would normally conduct, but requiring the exercise of higher standards of behaviour in so doing.[82] Firms are required to exercise standards of fair dealing, to give the best advice, or to recommend suitable products. The rule maker is essentially requiring the adviser to exercise judgement, relying on the adviser to exercise it well. The reliance may be misplaced, however. The suitability rule is one of the most widely breached rules in the regulatory system and the current concern over the sale of personal pensions stems from non-compliance with it. The regulators' answer, however, is not in a formalistic approach in which the rules are made more detailed. This is not because of the potential for creative compliance, but because of the inherent nature and limitations of rules themselves. The rule maker cannot formulate a more detailed rule: the circumstances of its application are too wide and varied and abstractions and generalisations cannot be made. However, neither can the rule maker create good judgement where none exists. The rule maker has reached the limits of rules as a regulatory instrument and has to rely on other forms of regulation to achieve the purpose: training and competence tests.

A rule such as the suitability rule thus relies on the skill of the regulated, and poses problems for enforcement. As noted earlier, enforcement studies indicate a preference among enforcement officials for detailed rules. Not only may detailed rules be incompatible with the regulatory technique chosen, however. As in the example above they may compromise the standard of behaviour attained. An example is the rule which requires brokers to deal at the best price available.[83] To aid enforcement the rule stipulated that this price is that shown on the automatic quotation screen at the time of the transaction. However very few deals are done at this price: most are done at better prices. Enforcement considerations therefore lead to a lower standard of behaviour being stipulated than that which the operators in the market expect to receive. Aiming the rule at enforcement thus compromises the achievement of high standards of conduct. It also has a deeper impact in that it affects the attitude the regulated have to the regulation to its relevance, and to the understanding they think the regulators have of the business which they are regulating.[84]

The context of rule making

Rules represent not one decision, but a series of decisions, each building on or modifying the last. Changes or developments may be prompted by changes in the market which occur while the rule is being formed, and will usually result in a lengthening of the rule to accommodate and provide for such changes. The reasons for increased length and precision may also be due to the need, as perceived by the rule makers, to accommodate the demands of those to be subject to the rule: rules are bargained over. The reasons for the complex and precise structure of the initial rules, for example, arose directly from the constant need to accommodate the demands of the regulated for certainty, or exclusion. As one official explained, a rule would initially consist of a provision which was quite wide in scope and fairly imprecise. Successive sub-clauses would then be added to accommodate the demands for precision, or to provide exclusions from the rules. It is a process which becomes self-perpetuating. Once exclusions start to be provided within the rule, its generality is undermined. This can reduce the flexibility available in its application. It can also reduce its certainty, as it becomes unclear whether, where specific exclusions are provided, everything not excluded must be included, or whether further exclusions may be implied. Finally, increasing the level of specificity and precision can paradoxically create more 'gaps' in the rules, which more detailed rules are used to fill.[85]

The process of rule making may also close off options. Rules cannot be seen as discrete decisions, made in isolation from other decisions on other rules in the rule system. When the rule making process is analysed from a holistic perspective, rather than simply looking at the formation of individual rules in isolation,[86] it becomes clear that decisions made as to the overall nature of the rule system affect and/or constrain the decisions made with respect to individual rules. This may determine the structure of the rule, or the sanction attaching to it, particularly where there is a hierarchical rule system. The choice of rule type is then constrained, and determined by the tier in which the rule should be, and not necessarily by what rule type is best suited to the particular substantive aim of the rule.

The knowledge and approach of the regulators themselves limits the rationality of the process and affects the type of rule formulated. The detailed structure of the initial rules, as we have seen, was due in part both to the DTI and to the provision of the right to civil action for their breach. It was also due to the lack of information and experience which everyone, regulated and rule maker, had of financial services regulation and uncertainty as to the new shape of the financial services markets. Regulation prior to the Act has been described as 'primitive',[87] and most aspects of the markets were being regulated for the first time.[88] The rule makers on the whole were civil servants and parliamentary draftsmen with little experience of financial services. The regulated were preoccupied with how the regulation would apply to their particular areas of business operation and were not in a position to review the rule books as a whole.

This inexperience was exacerbated by the speed with which the rules had to be made. There was no time for regulators to learn from experience in drafting their initial rules, nor was there even time to review the rules once drafting had finished to assess their overall coherence. The deregulation of the Stock Exchange, or Big Bang, had already occurred and the City was undergoing fundamental structural changes—changes which were giving rise to profound conflicts of interest in the way business was conducted.[89] The rules thus had to be formulated at considerable speed, and the regulators admit that the rules which emerged were essentially a first draft and in need of considerable revision.

Finally, rule making has been affected by the perceptions of politicians, the regulated and the media concerning the role regulation is meant to be playing in the market area in which it operates. In seeking legitimacy, the regulators seek to gain acceptance of these different constituencies, whose demands are not uniform. They affect the choice of rule type, however, as different types of rules are perceived as involving particular regulatory approaches, and as having different levels of efficacy. These assumptions have already been illustrated in the expectations surrounding the use of the Principles: the consequence of the more general rule would be to alter the perceptions of the regulation, increase its acceptability, and increase compliance. Detailed rules, on the other hand, are associated with greater control, and so have been equated in the rhetoric surrounding the regulation with tough regulation, with upholding investor protection.

The association of rule type with regulatory approach means that as the political climate varies in the demands it makes for the intensity of the regulation, so pressures for different types or rules are exerted. These vary with different groups: politicians, the regulated, the media; over time, with different constituencies making different demands in different periods; and between markets. The initial demand of the Government was for tough regulation. This then changed to requiring a more flexible regulatory approach. Detailed rules are perceived as necessary in market areas where private, individual investors are involved, where the objective of investor protection is seen to be of greater concern. However, in the wholesale market, where mainly professional investors are involved, investor protection concerns are outweighed by concerns for market efficiency, prompted by fears that 'over-regulation' will drive business from London. More substantive, purposive regulation which gives greater flexibility is thus demanded. Sir Kenneth Berrill, SIB's first chairman, identified the dilemma thus: 'in working out the rules, we need constantly to strike the balance between on the one hand stringent safeguards for investor protection and on the other market operation and development, including scope for flexibility and innovation'.[90]

Detailed rules have also been associated with effective regulation. In forming its initial rules SIB argued that detailed rules were necessary to ensure high standards of behaviour, given the competitive environment and actors who varied considerably in their competence and honesty.[91] In introducing the Principles, therefore, the regulators were concerned to convince those who were worried about regulatory laxity that the Principles would be meaningful, and enforceable. As one official stated, 'one of the aims was to sell the idea that rules did not have to be as specific as the old rule book to be effective'. Paradoxically, the Principles are now used by SIB to defend itself against charges of overregulation.[92]

Conclusion

The choice of rule type has been a significant part of the policy decisions made by financial services regulators. The determination of which type of rule to use has been a complex decision, in which rules have been used to induce compliance, affect the degree of control exerted and resolve the tension between certainty and flexibility. Underlying this use of rules have been deeper objectives which go beyond the substantive goals that one would expect of investor protection, market efficiency and international competitiveness. Rules have been used in an attempt to gain legitimacy for the regulation and to gain the acceptance of the regulatory and political community. Rules have been used as signals, to influence conduct, alter perceptions and communicate expectations. Rule type has been in a reflexive relationship with the function of the rule making body. It has defined, and been defined by, the position of that agency in the institutional structure.

There are trade-offs to be made in exercising the choice, and constraints which operate upon it. Other decisions made in relation to different rules limit the extent to which a choice is available: even if rule makers can choose which arrow to take from the quiver, they may not be able to determine what the quiver holds. Finally, the process is not wholly rational. There have been non-strategic reasons for the choice of rule type: lack of time, lack of knowledge, pressures for certainty and the self-perpetuating nature of detailed rules have led particularly to an increase in detail and complexity.

The reasoning of financial services regulators in the exercise of their rule making function is clearly limited at some levels to its particular context. However, this study identifies a number of uses of rule type which may be applicable in other contexts. Addressing issues of formalism and substantive rationality, of compliance and control, need not be context-specific; underlying aims to secure acceptance and legitimacy are also likely to be of general concern. Further studies on rule making are needed to determine whether these issues are of concern to other regulators, how they are addressed, and indeed whether rule type is itself a significant part of the rule making decisions of regulators operating in other fields. As rules abound and their effectiveness varies greatly, understanding the reasons why they were made is the first step in developing suggestions as to how they should be made.

Notes

1. I would like to thank Dr Rob Baldwin, Professor Denis Galligan, Dr Doreen McBarnet and Dr Chris McCrudden for comments on earlier drafts on this article. Responsibility for the views and errors contained remains my own.
2. A. M. Whittaker, 'Legal Technique in City Regulation' (1990) 43 C.L.P. 35, p. 48. (Mr Whittaker is an officer at SIB).
3. See, for example, G. Ganz, *Quasi-Legislation* (1987).
4. J. Beatson, 'Legislative Control of Administrative Rule Making: Lessons from the British Experience' (1979) 12 *Cornell Int' LJ* 199; M. Asimov, 'Delegated Legislation in the US and UK' (1983) 3 OJLS 253; R. Baldwin and J. Houghton, 'Circular Arguments: The Status and Legitimacy of Administrative Rules' [1986] P.L. 239.
5. J. C. McCrudden, 'Codes in a Cold Climate: Administrative Rule Making by the Commission for Racial Equality' (1988) 51 M.L.R. 409; R. Baldwin, 'Health and Safety at Work: Consensus and Self Regulation' and I. Ramsay, 'The Office of Fair Trading: Policing the Consumer Marketplace', both in R. Baldwin and J. C. McCrudden, *Regulation and Public Law*, (1987).
6. R. Baldwin, 'Why Rules Don't Work' (1990) 53 M.L.R. 321 and D. McBarnet and C. Whelan, 'The Elusive Spirit of the Law: Formalism and the Struggle for Legal Control' (1991) 54 M.L.R. 848.
7. R. Pound, *Jurisprudence* (1959) vol 1, p. 124. For a similar definition see J. L. Jowell, *Law and Bureaucracy* (1975), p. 135.
8. See, for example, Jowell (1975) p. 137, H. Hart and A. Sachs, *The Legal Process*, (1958), p. 157 and D. Kennedy, 'Form and Substance in Private Law Adjudication' (1976) 89 *Harvard LR* 1685.
9. W. Twining and D. Meirs, *How to do things with Rules* (1991), p. 131.
10. For other dimensional approaches to rules see C. S. Diver, 'The Optimal Precision of Administrative Rules' (1983) 93 *Yale LJ* 65, who identifies three: transparency, accessibility and congruence. (The first two are embodied here

in the dimension of structure, the last is a normative assessment: how far does the rule achieve its objectives); and R. Baldwin, (1990), *op. cit.* who identifies five: specificity, inclusiveness, accessibility (embodied here in the dimension of structure), status and force, and prescription or sanction.

11. For a fuller discussion of these dimensions, see J. M. Black, 'Regulators as Rule Makers: The Formation of the Conduct of Business Rules under the Financial Services Act 1986', unpublished D. Phil thesis, Oxford University, 1994 and *Regulators as Rule Makers* (forthcoming).
12. See further L. L. Fuller, *The Morality of Law* (1969), p. 64 and drawn on by Jowell (1975), p. 167.
13. C. S. Diver, 'Regulatory Precision', in K. Hawkins and J. A. Thomas, *Making Regulatory Policy* (1989), p. 199.
14. See particularly K. Hawkins, *Environment and Enforcement* (1984), 'Discretion in Making Legal Decisions' (1986) 43 *Washington and Lee LR* 1161, and K. Hawkins (ed.), *The Use of Discretion* (1992); D. J. Galligan, *Discretionary Powers* (1986) and R. Baldwin and K. Hawkins, 'Discretionary Justice, Davis Reconsidered' [1984] P.L. 570.
15. See K. Hawkins (1984), G. M. Richardson *et al.*, *Policing Pollution: A Study of Regulation and Enforcement* (1983) and B. Hutter, *The Reasonable Arm of the Law* (1988).
16. The use and manipulation by the regulated of specific rules to circumvent or undermine the purpose of regulation using an emphasis on legal form and literalism: see McBarnet and Whelan (1991).
17. Even in the U.S. where rule making by regulatory agencies has long been a focus of attention, this question has not been addressed. Rules have either been posited in opposition to adjudication or discretion in a debate on the relative merits of each or as to the models of justice or administration which they assume. Alternatively, the problems of rule making have been discussed in the particular context of U.S. statutory, judicial and executive constraints, which are prompting concerns at an 'ossification' of the process. On the need for less formal methods of rule making, see M. Asimov, 'Nonlegislative Rulemaking and Regulatory Reform' (1985) 34 *Duke LJ* 381, C. H. Koch and B. Martin, 'FTC Rulemaking through Negotiation' (1983) 61 *N Carolina L.R.* 275. For recent discussions of the issues see 'Symposium on Rulemaking' (1992) 41 *Duke L.J.*
18. I. Ehrlich and R. Posner, 'An Economic Analysis of Legal Rulemaking' (1974) 3 J.L.S. 257; R. A. Posner, *An Economic Analysis of Law* (1986), ch. 20; C. S. Diver (1983) and (1989), J. S. Johnston, 'Uncertainty, Chaos and the Torts Process: An Economic Analysis of Legal Form' (1991) 76 *Cornell L.R.* 341, and in the use of rules in the context of private ordering of transactions, see W. H. Hirsch, 'Reducing Law's Uncertainty and Complexity' (1974) U.C.L.A.L.R. 1233.
19. Diver (1983), pp. 73–4.
20. Diver (1989), pp. 233–5.

21. McCrudden (1988); Baldwin (1990).
22. McCrudden (1988).
23. Baldwin (1987), McBarnet and Whelan (1991).
24. Baldwin (1990); Hawkins and Thomas also argue that vague and imprecise rules impede the enforcement process, failing to recognise the limited ability of both officials and the regulated to cope with uncertainty: K. Hawkins and J. Thomas, 'Rule Making and Discretion: Implications for Designing Regulatory Policy' in K. Hawkins and J. M. Thomas (eds), *Making Regulatory Policy* (1989).
25. Baldwin (1990).
26. Whilst most enforcement strategies operated independently of the rules, enforcers did indicate that preferences for rule type varied with the enforcement strategy adopted, the type of employer and the nature of the risk: Baldwin (1990).
27. Although enforcement officials studied by Hutter were more equivocal as to their preferred type of rule, Hutter (1988).
28. McBarnet and Whelan (1991).
29. *Ibid.*, p. 849.
30. *Ibid.*, p. 851 and p. 854.
31. See, for example, R. Pound, *Introduction to the Philosophy of Law* (1954), P. S. Atiyah, *From Principles to Pragmatism* (1978).
32. M. Weber, *Economy and Society*, G. Roch and C. Wittich (eds) (1968).
33. Galligan (1986), pp. 68–72.
34. *Ibid.*, p. 71. See also R. Cotterrell, *The Sociology of Law: An Introduction* (1984), p. 319.
35. Galligan (1986), p. 71.
36. See Galligan (1986), ch. 4.
37. For a summary of the arguments see Jowell (1975).
38. Galligan (1986), p. 166.
39. For details see Black (1994) and forthcoming.
40. 'Investment' and 'investment business' are widely defined in Schedule 1 of the Act. For early proposals on the nature of the regulation see L. C. B. Gower, *Review of Investor Protection: A Discussion Document* (London, 1982), *Review of Investor Protection, Report, Part I*, Cmnd. 9125 (London, 1984) and *Report, Part II* (London, 1985). See also the White Paper: DTI, *Financial Services in the UK: A New Framework for Investor Protection*, Cmnd. 9432 (London, 1985).
41. Section 3 FSA (unless otherwise indicated, all section numbers refer to the Financial Services Act 1986).
42. Authorisation may be obtained through membership of the self-regulatory organisations (s. 7), the recognised professional bodies (s. 15), or directly from SIB (s. 25, s. 114). Automatic authorisation is conferred on insurance companies authorised under the Insurance Companies Act 1982 (s. 22) and friendly societies registered under the Friendly Societies Act 1974 (s. 23), and on persons authorised to carry on investment business in an E.C. member state (s. 31).

43. Under s. 114, Financial Services Act 1986 (Delegation) Order 1987.
44. SIB is also responsible for recognising professional bodies, clearing houses and investment exchanges.
45. These are the Securities and Futures Authority (SFA), the Investment Management Regulatory Organisation (IMRO), the Life Assurance and Unit Trust Organisation (Lautro), the Financial Intermediaries' Managers' and Brokers' Association (Fimbra) and the Personal Investor Authority (PIA). Lautro and Fimbra are due to cease operating in 1995, their members transferring to PIA. For the background to the formation of PIA see *Retail Regulation Review: Report of a study by Sir Kenneth Clucas on a new SRO for the Retail Sector*, SIB, March 1992.
46. Section 47A and B, inserted by s. 192 Companies Act 1989, s. 63A and B, inserted by s. 194 Companies Act 1989, and s. 63C, inserted by s. 195 Companies Act 1989.
47. Schedule 8.
48. Schedule 2, para 3(1), amended by s. 203(1) Companies Act 1989.
49. S. 47A(3), inserted by s. 193(1) Companies Act 1989, and s. 63A, inserted by s. 194 Companies Act 1989.
50. S. 62A, inserted by s. 193(1) Companies Act 1989. This right of action was initially given to all investors, and subsequently restricted to private investors only.
51. SIB is under a statutory duty to consult, Schedule 9, para. 12. The SROs consult with their members and representative bodies.
52. Sections 119–123.
53. SIB, *Statements of Principle*, March 1990; *Core Conduct of Business Rules*, January 1991. The Core Rules were designated to apply to members of SFA in December 1991 (SIB, *Core Conduct of Business Rules, Commencement for Members of SFA*, CP 55, December 1991) and to IMRO in August 1991 (SIB, *Core Conduct of Business Rules, Commencement for Members of IMRO*, CP 54, August 1991).
54. The main deregulatory element was the restriction in the application of some of the rules to private investors.
55. However, although the status of guidance reduces the level of detailed conduct prescribed by rules, in the realities of day to day business operation it is likely that the difference between conduct which is prescribed and that which is recommended is merely semantic.
56. The right of action provided to all investors for the breach of any SIB or SRO rule in s. 62 FSA 1986 was amended to make the right available to private investors only, and the Principles were exempted from its scope: s. 193(1) Companies Act 1989.
57. D. Walker, 'Financial Services: The Principles Initiative' (1989) BJIBFL 51.
58. *Ibid.*, p. 53.
59. See, for example, an article by IMRO's then chief executive, John Morgan, 'The Difference between Rules and Principles', *Financial Times* March 29, 1989.

60. The use of sanction in this way is akin to the use of an ouster clause in ousting the jurisdiction of the courts.
61. Although the potential for judicial review remains.
62. SIB, *Conduct of Business Rules: A New Approach*, November 1988, Commentary, para. 10.
63. See for example, *A New Approach*, Commentary, para. 10; D. Walker, quoted in the *Economist* July 30, 1988 M. Blair (chief legal officer at SIB), (1989) BJIBFL, and A. M. Whittaker (legal officer at SIB), 'Financial Services, Developing the Regulatory Structure' (1989) BJIFBL 5.
64. See R. Leigh Pemberton, 'The Markets, the City and the Economy' (1988) BEQB 59, and *Financial Times* February 27, 1988, February 29, 1988 and May 5, 1988.
65. For suggestions that participation in regulation may be a strategy of co-optation, see Baldwin and McCrudden (1987), p. 6.
66. This was recently re-iterated by SIB's current chairman, Andrew Large, *Financial Services Regulation: Making the System Work*, SIB, May 1993, para. 6.5.
67. This phrase was frequently used by officials when asked in interview what they thought 'self regulation' meant.
68. K. Hawkins, 'The Use of Legal Discretion: Perspectives from Law and Social Science', in K. Hawkins (ed.) (1992), p. 11.
69. See further *ibid.*, p. 36.
70. Baldwin and Hawkins (1984), p 582.
71. Empirical studies indicate that the ability of rules to control actions of officials is severely limited: Galligan (1986), Hawkins (1992), and Baldwin and Hawkins (1984).
72. Large Report (1993), para. 6.5.
73. For examples, see further Black, *op. cit.*, chs. 6 and 7.
74. SIB, *Regulation of the Conduct of Investment Business, A Proposal*, August 1989, paras. 21 and 24.
75. SIB, *Achieving and Judging Adequacy*, CP 39, April 1990.
76. See SIB, *Regulation of Conduct of Business, A Proposal*, August 1989; *Achieving and Judging Adequacy*, April 1990; *The Proposed Core Rules*, CP 42, October 1990.
77. The functional organisation of the regulation means that one firm could belong to more than one SRO. Divergencies in provisions were thus highlighted, and served to increase the complexity of the regulation for the regulated firm.
78. For elaboration see, for example, Gower's *Discussion Document and Report, Part I*, R. Leigh Pemberton, 'Changing Boundaries in Financial Services' (1984) BEQB 40, 'The UK Approach to Financial Regulation' (1986) BEQB 48, and the 1985 White Paper.
79. For a contemporary statement as to its view of the respective roles of SIB and the SROs and the aims behind the New Settlement see SIB, *A Forward Look*, October 1989.

80. R. Leigh Pemberton, 'The Markets, the City and the Economy' (1988) BEQB 59.
81. See, for example, a speech by the then chief executive of IMRO, John Morgan, to the National Association of Pension Funds, quoted in the *Financial Times* February 27, 1988.
82. An example of structural regulation was the old Stock Exchange rule that a person either had to sell shares for their own account (jobbers) or act as agents for another (brokers). When this was removed, so that brokers could also deal for their own account, it was replaced with conduct regulation, which imposed duties to deal at the best price etc.
83. Core Rule 22.
84. SFA's guidance now provides that in determining whether the firm has taken reasonable care to ascertain the best price available SFA will have regard to the conventions of the market: SFA rule 5–39.
85. See, for example, the debates over disclosure rules in retail regulation discussed in Black (1994), ch. 7: SIB, *Life Assurance and Unit Trust Disclosure*, CP 23, May 1989, *Life Assurance and Unit Trust Disclosure*, CP 27, August 1989, *Retail Regulation Review: Discussion Paper 3, Disclosure*, October 1991, and *Retail Regulation Review: Disclosure, Polarisation and Standards of Advice* CP 60, March 1992.
86. For the need to do this in relation to individualised decisions, see Hawkins (1992), pp. 27–35.
87. L. C. B. Gower, '"Big Bang" and City Regulation' (1988) 51 MLR 1, p. 6.
88. For details see Gower, *Discussion Document*, ch. 3.
89. For a summary of the nature of these conflicts, see the Law Commission, *Fiduciary Duties and Regulatory Rules*, CP No. 124 (1992), para. 2.4.12.
90. L. S. Gaz., February 12, 1986.
91. SIB, *The SIB Rulebook: An Overview*, October 1987, para. 10.
92. See, for example, the response of Mr Andrew Large to the Treasury and Civil Service Select Committee, Financial Services Regulation, Minutes of Evidence, 1992/3, H. C. Paper no. 733–i, para. 39.

References

BALDWIN, R. (1987) 'Health and Safety at Work: Consensus and Self Regulation' in R. Baldwin and J. C. McCrudden *Regulation and Public Law*.
—— (1990) 'Why Rules Don't Work' 53 *M.L.R.* 321.
—— and HAWKINS, K. (1984) 'Discretionary Justice. Davis Reconsidered' *P.L.* 570.
—— and McCRUDDEN, J. C. (1987) *Regulation and Public Law*.
BLACK, J. M. (1994) 'Regulators as Rule Makers: The Formation of the Conduct of Business Rules under the Financial Services Act 1986' unpublished D. Phil. thesis, Oxford University.

DEPARTMENT OF TRADE AND INDUSTRY (1985 White Paper) *Financial Services in the UK: A New Framework for Investor Protection*, Cmnd. 9432. London.

DRIVER, C. S. (1983) 'The Optimal Precision of Administrative Rules' 93 *Yale LJ* 65.

—— (1989) 'Regulatory Precision' in K. Hawkins and J. A. Thomas *Making Regulatory Policy* p. 199.

GALLIGAN, D. J. (1986) *Discretionary Powers*.

GOWER, L. B. C. (1982) *Review of Investor Protection: A Discussion Document*. London.

—— (1984) *Review of Investor Protection, Report, Part I*, Cmnd. 9125. London.

HAWKINS, K. (1984) *Environment and Enforcement*.

—— (ed.) (1992) *The Use of Discretion*.

HUTTER, B. (1988) *The Reasonable Arm of the Law*.

JOWELL, J. L. (1975) *Law and Bureaucracy*.

MCBARNET, D. and WHELAN, C. (1991) 'The Elusive Spirit of the Law: Formalism and the Struggle for Legal Control' 54 *M.L.R.* 848.

MCCRUDDEN, J. C. (1988) 'Codes in a Cold Climate: Administrative Rule Making by the Commission for Racial Equality' 51 *M.L.R.* 409.

Setting Standards

K. HAWKINS

I. Issues

The boundaries of regulatory deviance are drawn by administrative agencies: pollution, in other words, is an administrative creation. The broad legal mandate of the agencies about water pollution control is transformed into policy by senior officials and given practical expression in the setting of pollution standards.[1] Standards ('consents') are licences to discharge polluting matter. Pollution is in effect qualitatively and quantitatively controlled by the water authorities since standards are administratively negotiated. Not only, then, do the agencies possess power to enforce the law, they actually determine the reach of the law, for (in contrast with the police) they exercise a real legislative authority, enjoying broad discretion to define what makes 'pollution'. In this sense the water authorities create pollution, as Becker might say (1963:9), by making the rules whose infraction constitutes pollution. Definition and enforcement, a dual authority, are reciprocally related, a theme to be explored in this chapter.

For agency policy-makers the consent is an important tool expressing political and economic judgments about water quality. Its central purpose, as the head of one the agencies put it, is 'to produce a river which is suitable for the uses which are needed downstream—providing a potable water supply, for fisheries, just amenity. We might even decide we just want an effluent channel. . . . The consents are geared to produce the quality water you want in the river for the use you want downstream'. For the water authorities in general the practical significance of consents is that many rivers are now substantially comprised of effluent already discharged subject to consent. One of the southern authority's main potable supply rivers, for example, consists in dry weather of more than 50 per cent effluent; and another river flowing through a major city in the northern authority is over 90 per cent effluent.

Consents to pollute come in the form of emission or effluent standards which prescribe the temperature, amount, and kind of polluting matter which may be discharged from a particular source. Dischargers are thus permitted to pollute by the water authorities—but only (in theory) up to those levels set out in the discharger's consent document. A consent, however, is a movable threshold. Once fixed it may subsequently be reviewed and modified by the agency in consultation with the discharger. The pollution standards in a consent are defined locally by each water authority and are specific in application, with each consent negotiated on an *ad hoc* basis. Standards are expressed quantitatively and each one is formulated not to contemplate exceptional circumstances, instead addressing pollution without regard to mitigating features, such as accident. Simply by selecting those substances in impure water which are to be held potentially 'polluting' (pollution parameters) and the point at which such contamination is to be regarded as 'polluting' (pollution limits), together with temperature and volume restrictions, the agency establishes theoretically enforceable boundaries,[2] exercising, in other words, power to control the *potential level* of pollution. And since the water authorities are enforcement agencies they also control to a very great degree what may seem to be some sort of '*real level*' of pollution which comes to light in amount and kind—those events or incidents which are detected and processed, becoming statistics of non-compliance or even, in some rare cases, of prosecutions in annual reports.

In setting standards the water authorities, like other regulatory agencies, are confronted by two crucial problems. First, pollution control means cost. Since the cost of control usually returns few, if any, financial benefits to the discharger and has to be met by increased prices, dischargers regard pollution control costs as a burden to be avoided wherever possible. There is no simple relationship between control and costs. To make a significant impact upon a bad discharge is not necessarily very expensive. But to make even a minor improvement to effluent which already consists of relatively good quality water may involve the discharger in very heavy expenditure indeed (see Kneese, 1973). The power to define and enforce consents is ultimately a power to put people out of business, to deter the introduction of new industry or to drive away going concerns.

The second, related, problem in standard setting is one of the distribution of costs. A tension between officials' perceptions of equity and efficacy is, for them, a familiar dilemma. The conflict is between an essentially moral stance which prizes consistency and uniformity in the application of standards (similar discharges should be similarly controlled in degree and kind) and a utilitarian approach concerned with effect on water quality, regardless (again, in theory, at least) of the means of the discharger or the demands fortuitously being placed unequally on similar dischargers (see generally Ackerman et al, 1974). '[T]he load of polluting matter discharged should be controlled at all times to exactly the level that a river would accept without harming river uses or the environment' (unpubl. agency document). This is a recurrent dilemma in pollution control since efficacy insists that standards be tied to the setting of a discharge and settings vary enormously.

The weight to be given to the normally competing values of equity and efficacy is a persistent problem both in setting or reviewing standards and in enforcing them.[3] On one level the dilemma is between standards of general and of individualized application. On another the conflict is sometimes portrayed as between older, established ways of handling pollution control and newer, more 'scientific' approaches in which environmental impact in its widest sense is the primary concern. The utilitarianism of the scientist tends to view the principle of equity with a certain disdain because of its potential inefficiency: 'Equitability can be the biggest millstone,' said a senior officer with a special interest in consents. 'Often it goes against technical judgments. When it comes into play, some of the ultimate answers are not the best ones'. In terms of formal policy, a shift in recent years has seen equity ostensibly yield place to efficacy. In practice this should mean that the design of consents has been influenced less by conceptions about the similarity of discharges than by a scientific analysis of the impact a particular discharge may have on a particular watercourse and the polluting load that that watercourse can accept, according to agency plans for river quality.

Senior officials are firmly committed to the utilitarian view: 'treating discharges alike in terms of . . . the quality of the discharge, as distinct from the effect on the environment', said one, 'is absolute nonsense'. Yet senior staff and field men alike are aware of the fact that most dischargers are likely to be more sympathetic to an approach to standard setting and enforcement based on equity. Apart from the moral preference that like should be treated alike, dischargers' commercial instincts argue that no manufacturer or producer should be placed in a more favourable position than his competitors. The practical task, then, for a water authority setting or modifying a consent based on criteria relating to the particular watercourse becomes one of persuading a discharger disadvantaged by some criterion which makes sense in utilitarian—but not moral—terms of the force of the agency's position. In some cases this is apparently not as arduous a task as it may seem, simply because dischargers find it extremely difficult to portray themselves as 'similar' to their rivals on more than a very few criteria.

An individualized approach based on a concept of efficacy can be sold to dischargers because it is 'scientific'. Equity can be played down:

> I don't know if it should be that important if everything we do is based on good sound scientific principles, because if people perhaps in the same industry are situated in different places in the estuary, y'know, if one was to point the finger at the other and say 'But you allow him to discharge such and such and you only let us do this', then we should be able to turn round and say, 'Ah yes, but you're discharging in a different place and the river quality in this different place needs different treatment. . . . I think you can only treat them similarly if all other things are equal, if they're discharging into the same sort of watercourse in the same sort of position. Because obviously if you had one particular factory on the tideway and another producing an identical product and identical effluent in one of the tributaries, you couldn't treat them the same because they're patently different cases.

On this 'scientific' view of standard setting, the design of consents to discharge polluted water is to be based on a dispassionate consideration of the amount and kind of pollution load any watercourse can bear. The capacity of a watercourse and the tolerable limits of polluting effluent are calculated according to a mathematical model, which ideally should be applied routinely and objectively to all dischargers regardless of their means, the costs of treatment, their prior efforts, or the demands made of their competitors.

Field men, however, are constantly reminded that industrialists and farmers work on a principle of equity: they can readily discover the standards which their competitors must observe and may complain if they are being handicapped. But the agencies prefer to avoid complaints wherever possible, and this encourages an administrative inclination for equity of treatment, even though 'scientific' judgment may dictate otherwise. Negotiating about standards, especially when there is some disparity between apparently similar dischargers, though an infrequent event, can be one of the field man's trickiest tasks.

The picture of an agency which dispassionately administers scientifically-designed standards is blurred further by organizational practices. Later chapters [see *Environment and Enforcement*] will show that in pollution control work, standards are by no means treated as absolute proscriptions inexorably enforced. The agencies display a sometimes considerable flexibility both in the standards set and in the enforcement policy adopted, in recognition of the technical difficulties and costs of complying, the potential for error, and the stigmatizing effect of strict legal enforcement. Furthermore, the processes giving rise to pollution sometimes produce erratic discharges in which effluents may become heavily polluting, and river water is also inherently variable in quality. The result is that a degree of leeway is normally granted to dischargers, and a certain amount of pollution allowed to occur with impunity. Such leeway is the means by which the enforcement agency adapts to uncertainty. 'The present situation is that consent conditions are written as though they were absolute,' a senior official said:

The time it goes over—once in a lifetime—and you're a criminal. So the practice has been to regard them rather unofficially as having been satisfied if there is compliance on about . . . three occasions out of four, or four out of five, or two out of three—practices are variable from one authority to another. We've adopted a three out of four.'

Non-compliance with standards is thus organizationally sanctioned.[4]

Though the consent may not be strictly enforced, however, it is significant as a benchmark for both dischargers and enforcement authorities. It remains the criterion of pollution, despite its irregular enforcement; as such it creates a zone of officially tolerated 'pollution' which will vary as the standard and the setting of the discharge vary. And it remains the legally enforceable standard of pollution if customary enforcement practices are suspended.

In a substantial proportion of cases dischargers normally display little or no objection to the standards imposed by the agency, as a result of preparatory work in negotiations conducted by the field officer together with, in more important cases, his area supervisor. But where standards appear to a discharger to impose excessive demands he may exercise a right of appeal to the Secretary of State. Although hardly ever employed, this right of appeal in effect confers bargaining power upon the discharger who seeks to counter administrative extravagance, since the agencies attempt to avoid appeal at almost all costs:

It's certainly true that if people suggest they will go to the appeal procedure then we will do almost everything we can to avoid that, which means in some cases we're letting people get away with something that you wouldn't let others get away with, simply because they're being difficult—well, not being difficult, they would be just exercising their rights.

The desire to avoid appeal may partly be a reflection of the fact that standards have been selected by resort to the convention of established—but largely unexamined—practice which may be difficult to defend in a formal arena. Thus 'you've got to be very sure of your ground, because you've got to get up, not in a court, but in an appeal situation, and give evidence'. There is also a good organizational reason why agencies fear appeals: appeals introduce uncertainty (cf. Kaufman, 1960:154–5) and threaten their control over the design of consents. An appeal may establish a formal precedent which an agency might find undesirable, since standard setting is then taken out of its hands by people regarded as possessed of less expertise and possibly differing interests. A principle informing negotiations about consents is to avoid the risk of creating a precedent on appeal injurious to the agency's interests, a principle of wide applicability, for to appeal and win—or even to appeal and lose—publicizes the possibility of remedy which itself is a threat to the agency's whip hand.[5] Appeal gives status and recognition to protest.

Ironically, recourse to precedent is a useful tactic for the field man to employ when negotiating about standards. Reference can be made to standards usually set by the agency 'in cases of this kind', or attention drawn to practice in past cases. But the officer must maintain a certain flexibility and be able to deviate from usual practice where necessary, especially to avoid the risk that an administrative convention may be established which can subsequently be used by others with conflicting interests as a resource in their own negotiations. Obviously when the agency claims allegiance to an established precedent where its formal interests in pollution control or its organizational interests are served, it lends further support to the principle of equity. Indeed, the principle serves administrative interests well, for if standard conditions designed to be equitable as between apparently similar dischargers are routinely applied, the discharger who wishes to appeal against the standards imposed in his case is probably in a weak position if the agency is able to show the Minister that the standards were those typically in force. But in practical terms, the organizational imperative to avoid appeal means modifying demands in the course of negotiations if there is any suggestion that the agency might have to defend itself in an appeal: 'we try to see', said a senior officer, 'there are no grounds for reasonable objection'.

II. Procedure

Standard setting is not a large part of water authority activity because most dischargers have long since negotiated their consents, and a lack of industrial development has depressed the number of new consents applied for. More important in recent practice has been the review of consents already granted, a matter which has revived interest among agency staff in some of the major issues surrounding standard setting.

Consents tend not to vary greatly even between apparently very different effluents. Limits are normally imposed on organic pollution and solids per unit of volume and often on ammonia, while industrial discharges frequently have an acidity or alkilinity (pH) parameter. Certain kinds of manufacturing processes attract specific parameters such as various metals, cyanide, phenols, or temperature.[6] In most cases the choice of parameters and limits is made by the officer in the field in response to an application from the would-be discharger,[7] a choice which is routinely ratified by senior staff.[8] 'I just look at them,' said a senior official. 'I don't calculate them all'.

Once the field man and his area supervisor have settled upon the standards to be imposed, the consent is approved by the head of the agency and ratified by an advisory committee. Staff senior to the area supervisor play an active part in shaping the standard to be set only rarely, in cases involving what a supervisor described as 'very contentious issues: possibly major discharges from the Authority's own works or from industrial activity where there has been a history of wrangling and debate about standards from that particular section of the industry'. Those very large industries which produce a substantial volume of effluent will also have their consents individually negotiated in recognition of the greater potential impact on the watercourse. Otherwise, the usual approach is to apply the so-called 'Royal Commission' standards[9] of twenty parts per million (p.p.m.) biochemical oxygen demand (BOD)[10] and thirty p.p.m. suspended solids, with the addition of other parameters as appropriate. It is done intuitively. 'If we're talking about BOD or suspended solids, we have a feel for that . . . you make a "gut reaction" decision.' The choice relies heavily upon administratively established convention—'normal ways'—as a device for help in determining any case at hand. 'I think the first thing I tend to do', said an area supervisor, 'is . . . to look at the normal working standards normally applied and see whether they would be adequate to protect the river; and you then take into account the flow, the volume of the discharge or the rate of flow of the discharge and the volume.' In the majority of cases the process is unexceptional:

It starts at [field] officer level. The . . . officer will probably say 'Royal Commission', or whatever it should be, or something 'long these lines to some industrialist. The area man'll look at it and say 'You can knock them down'. 'There might be oil from this site, we're putting oil on as well.' Or, 'I think that one's a bit too strict.' Or he would perhaps modify it. And that's probably it.

There are also tacit constraints upon standard setting. Agency staff sometimes question themselves about a discharger's economic position and his capacity to comply with the parameters and limits which they would ideally like to impose. But predictive judgments about ability to comply are founded not upon some dispassionate analysis of economic facts and figures so much as upon characterizations of a discharger's know-how and willingness to comply, derived from his occupation, size, experience, and reputation. 'I think to myself "Well, has this company got the expertise to treat to this particular standard?"' said an area officer:

'Are they likely to be able to attain it and maintain it? Are they, will they have the expertise? Can they be trusted?' Y'know, I mean there are companies and companies. You expect that if you set a 10:10 standard[11] on a [large well-known company] for a, say a pickle liquor, or something, if you did that they would comply because they're big enough and they've got the expertise, and so on. If a company called Joe Bloggs and Company that were from a back street . . . came with a similar proposal and they obviously hadn't got two ha'pennies to rub together and certainly weren't gonna employ a chemist or anything like that, then . . . you would view that as a completely different proposition. You would probably say 'No. We won't issue you with a consent at all.'

These remarks emphasize the central judgment of reputability. The large, well-known company possesses 'expertise' and can be 'trusted', while the reference to employment of a chemist reveals that the small company lacking the attributes of wealth and competence is less able to comply, and implicitly untrustworthy. In the absence of constant monitoring, trust is of crucial importance to an enforcement agency's detection and compliance processes.[12]

Difficulties sometimes arise when the agency redefines pollution control policy for a stretch of water and wishes to tighten existing standards. Some of the issues typically raised during standard-setting negotiations surfaced in the following case from my field notes:

McDonald's[13] is typical of an old well-established firm which has been absorbed into a larger manufacturing group and modernized. The company makes a wide range of bathroom ware and discharges its effluent into a fishless urban stream. The authority's predecessor river board granted a consent some years ago with a number of parameters: 20 BOD, 100 p.p.m. suspended solids (relaxed from the usual 30), and various metals, including zinc, consented to the standard 0.5 p.p.m. Laboratory analysis for the presence of various metals was not routinely carried out, however, until a couple of years ago, when it was suddenly discovered that McDonald's effluent contained 256 p.pm. of zinc, a by-product of the glazing process. [Water heavily polluted with zinc paradoxically looks particularly clear.]

The agency demanded that McDonald's take remedial measures. Since then the firm has managed to bring the amount of zinc in the effluent down to about 20 p.p.m., but the agency has made it clear that it has to go below that. Dennis Blake, the area supervisor, wants to bring the zinc content much lower. He and his field officer, Brian King, are working to a timetable of compliance. After a slow start which prompted the field man to threaten to take a 'stat'[14] Blake now considers the firm to be abreast of the schedule he and King have set. Blake is satisfied that the threat of a formal sample was enough to put the wind up the company. McDonald's meanwhile went over Blake's head to complain to the director of the agency about the demands being made of them. The director, however, supported his staff; in fact, Blake said, he 'put the fear of God in them. You notice how they still turn white at the sound of his name'. [Later in the day during negotiations Blake will several times refer conspicuously to the director by name.]

As we drive to McDonald's, Blake explains that the company can probably only come down to about one p.p.m. zinc, and only then if it spends a lot of money. But it is technically possible. He predicts that McDonald's will ask for a consent of ten p.p.m. He says he will respond by asking for three, as a negotiating position, but is prepared to go to five. He has written a letter to his director of pollution control, suggesting two or three p.p.m., which has been returned with the annotation 'proceed as indicated'. Blake is adamant that he will not allow McDonald's to go above five p.p.m. To do so will effectively set a precedent for other dischargers who will soon discover that McDonald's is operating at an advantage. McDonald's belongs to a federation of companies in the same business and the area man takes it for granted that the firm's competitors will soon hear about the relaxed consent and apply to the agency for similar leniency. And if the agency subsequently has to treat other similar dischargers equally, Blake is concerned about a cumulative deterioration in water quality.

Blake has brought along a colleague who is an expert in the cleaning of various effluents to act as a consultant in the negotiations. He discusses with Blake the various treatment methods for zinc which the firm might try, raising the possibility of a sand filter. Blake is asked what a sand filter would cost. He says he doesn't know, then volunteers two thousand to five thousand pounds. [This is a fairly typical response of pollution control staff who prefer to err on the side of caution, in the form of wide ranges, when asked to estimate costs of pollution control remedies. Later that day, when involved in negotiation with the firm, the area man will adopt the lower figure for purposes of argument.] When I ask if the firm can afford it, he says it can, 'because of the size of the company', since it is part of a larger group. 'It's a lot of money to me personally. But it's nothing to a firm like McDonald's.'

We are met by one of the directors of the firm who takes us to a showroom for coffee. Conversation is soon transformed into preliminary negotiation. The director asks for ten. Blake says how pleased he is with the firm's improved performance, acknowledging the considerable efforts and progress it has made. But the legal requirement is 0.5. [A consent, once formulated, can be portrayed for purposes of negotiation as a 'legal requirement'.] He is, however, prepared to recognize that it is unreasonable to expect McDonald's to make one p.p.m. But he cannot go to ten. Blake talks of fairness to other manufacturers and observes that a discharge of ten will give McDonald's 'an unfair advantage'. [He will say later, 'If I'd relaxed the standard there I'd have had umpteen other people knocking on the door wanting to do precisely the same thing to save on trade effluent charges.'][15] The director nods in agreement and does not seek to argue the point. He has arranged for the agency staff to have lunch with the Managing Director and two experts in pollution control from the parent company who have travelled a considerable distance to attend. In the meantime, we are taken to inspect the treatment plant, where King takes a sample of the effluent. The engineer in charge reports that the firm has made great efforts recently to get some new equipment, but there has been a delay in delivery. [This is a story which pollution control staff hear regularly. In this case it seems not to be treated as a delaying tactic.]

We are then taken on a tour of the factory while waiting for lunch. Blake and King begin to complain to each other that they are now wasting their time, since matters are settled so far as they are concerned. After the tour we are ushered into the board-room where a magnificent buffet lunch is laid out. There is a splendid display of drinks. We are invited to help ourselves. Other directors join us, making a party of sixteen. There is talk over drinks until the Managing Director arrives. The pollution control men are not impressed with the lavish reception, King observing as he eats, 'They're trying to buy us off now'. Blake agrees: 'They're trying to soften us up. You enjoy it, but at the end of the day you have to ignore it'. He goes on to mention the occasional embarrassing moments which occur when firms try to pass backhanders.[16]

During lunch the Managing Director begins gently sounding Blake out, but the real business of the day does not begin until the table is cleared. The Managing Director makes a statement in which he adopts the position already put forward by his colleague, who meanwhile supports the various assertions with evidence. The statement culminates with a request for a consent to discharge ten p.p.m. of zinc. Blake repeats his earlier position, summoning the notion of equity between dischargers as a bargaining tactic to support his case, which seems to win the tacit approval of some of the directors present. For illustrative purposes, he uses the example of another similar manufacturer (whose identity, he reports, he must keep secret) who is able to conform to a one p.p.m. zinc requirement. After each side has declared its position, the negotiations are diverted to more technical questions about further remedies which might be tried, and then adjourned.

[The company, after persistent pressure from the water authority, was able, with improved treatment methods, to comply several months later with a standard of one p.p.m.]

III. Enforceability

Though the agencies do not regard standards as strict limits, their enforceability is important. From the field staff's point of view it is essential that the parameters in any consent should be clear and unambiguous. They should at least appear to be attainable, for standards which can be met, if at all, only after massive expenditure might tempt an unwilling discharger to question them by appeal. Unattainable standards, even though not appealed by a discharger, may still lead to practical problems of enforcement, since a continued failure to comply because of inefficient treatment plant or other 'good' reason which enforcement agents tolerate may suggest condonation by the agency of the continual breach of consent. Indeed in some cases in which there is a persistent failure to comply and no apparent harm to the watercourse, agency staff tend to assume that the standards may be too tight, as a senior official suggested:

> somebody who's not complying with consent conditions and yet is not having much effect on the river, it probably means the consent condition is not right to start with. . . . This is I think one of the reasons that we can tolerate some non-compliance because some of the consent conditions are a bit too strict.

In areas of regulatory control like pollution, where the law regards conformity as essentially a matter of scientific measurement, deviance is conspicuous. While clear, uncomplicated standards are useful for an enforcement agent negotiating to secure compliance, economy in the design of standards is also important. To be parsimonious in the use of parameters assists in the display of deviance; in a prosecution it aids a portrayal of guilt, as an agency head explained:

> I always try and have as few conditions as possible in a consent. . . . In a court it would be so easy if I was on the other side to get up and say that in respect of eight of the conditions 'We complied and we only contravened the suspended solids condition and we're very sorry it was 30,000 p.p.m. but you're not gonna cane us for that are you, Mr Magistrate? We've got eight of them right.' . . . So the simpler the conditions the better, the fewer conditions the better.

Yet the use of few parameters, if they are well chosen, can still be an efficient means of pollution control. To have one or two strict parameters often means that other pollutants, which can remain unspecified in the consent document, are brought under control. Or parameters which provoke the discharger's concern can be framed leniently, control being assured by the insertion of a much stricter parameter of more general application.

The issue of enforceability was given a poignant twist following passage of the Control of Pollution Act 1974 which made an inroad into the virtual monopoly of pollution control enjoyed by the water authorities. The Act (much of which is not yet in force) requires publication of sample results and permits members of the public to bring legal action against a discharger who fails to comply with consent.[17] The agencies will have to have available for public scrutiny a register giving details of consents granted and results of effluent samples.[18] A source of particular vulnerability for the agencies is that in its reorganization of the water industry the 1973 Water Act gives them responsibility for the management of the great majority of sewage treatment works, which in many areas are themselves significant—often principal—sources of pollution. The water authorities thus not only set and enforce pollution standards, they are major polluters themselves. Yet the pollution control sections of the agencies are ultimately impotent to secure the compliance of their colleagues in sewage treatment, lacking recourse to the legal sanction (even if it were organizationally possible to prosecute fellow-workers). Many sewage treatment works have been consented in the ritual process to the usual 20 BOD and 30 suspended solids standards, which they have regularly failed to meet. While the agencies maintained almost total control over the enforcement of standards, a control which could be preserved in the absence of publicity, this posed no problem: 'A lot of the standards were set by River Authorities; and the works weren't capable of achieving [them] anyway,' a senior official said. 'But the former River Authorities knew this and shut an eye to it.'

After passage of the 1974 Act, however, the authorities had to contemplate the prospect that they themselves might be prosecuted for the polluting effluents from their sewage works. The new-found public accountability has produced a quite dramatic protective response from the agencies, illustrating the familiar point that a previously unreachable ideal may be achieved if reduced within the compass of the practically attainable. The agencies have simply embarked on a large-scale programme to revise consents both of their sewage works and of other discharges. Where a discharger would consistently do better than his existing consent, standards have sometimes been tightened. But where a discharger has regularly been outside consent, standards have been relaxed to accord with the existing performance in 95 per cent of samples taken. Thus while standards were relaxed, this was accompanied by less tolerance of their breach, compliance being expected in 95 per cent, rather than 75 per cent, of samples: 'exposing ourselves five per cent of the time', as a senior man put it, thinking of his agency's sewage works. The review was described by an agency director as

> an attempt to span out 95 per cent of the quality of effluent at present discharged from the [sewage] works, taking out the fliers and the obviously bad discharges. . . . This means in practice that one out of every twenty samples will fail to comply with the standard and the Authority will be at risk from somebody, some member of the public, prosecuting.

This protective behaviour has occurred in reaction to the threat which publicity poses to the discretion with which the agencies would treat deviants—themselves and others. 'We've learned in the last go it's the public accountability thing,' said a senior official.

> The legal thing can go out the window. For the first time we've had to get our heads down and look at what we're doing. . . . We bend over backwards to do the thing [properly] . . . but what happens in practice—do we mean it? And in many cases we realize we don't. We find a factory who's not met its standards eight times in a year, and what do we do about it? Nothing! . . . So one reason [for reviewing consents] is it doesn't make any difference to the river system; another is that it means we're not prepared to follow the thing through on a legal basis. . . . It's going to be stupid when that booklet's published [giving details of sample results]—we'll get tons and tons of phone calls. . . . What we're now trying to do is say the river will stand a temperature of 30°C—if it'll stand it, don't let's bugger about. It takes some sort of public document for people to get you out of a rut.

Adaptive action by recasting the rules was necessary to preserve administrative control and avoid the public embarrassment of the authorities. Since financial, technical, and geographical constraints did not permit efforts to be made to bring their sewage works into compliance with the existing standards (even if time had been available), the only other means of demonstrably maintaining compliance was for the agencies to change the standards to fit the existing discharges.

For field staff, however, a review which led to the relaxation of a large number of consents appeared to be in direct contradiction to their efforts ostensibly to bring all dischargers into compliance—'trying to fiddle the books for when the Act comes into force' as a northern authority officer put it. For many, the task of policing relaxed consents was a source of considerable professional embarrassment following their attempts to secure compliance with the former, stricter, standards. For the pragmatic field officer, the exercise was purely cosmetic, not an effort to overhaul consents to place more rational demands (in terms of what a river could accept by way of pollution) on the discharger. 'To my way of thinking,' said an officer of wide experience, 'this is just paper—it's only figures on paper, this is—to prevent the general public being able to prosecute the authority if a sewage works has the odd bad sample.'

IV. Postscript

Pollution control standards are flexible markers of rule-breaking. How strict or lenient enforcement agents conceive them to be may well affect their enforcement behaviour. For example, working conceptions of cause may depend upon a judgment as to whether the standards breached were generally attainable at tolerable cost. If they were, a violation may prompt suspicions of deliberate wrong-doing in the interest of financial saving.

Until recently, pollution control standards were treated as guides rather than boundaries. They were a species of organizational rhetoric (now menaced by public access to the monopoly of enforcement), embodying compromise between conflicting values and recognition of the vagaries of the environment to be controlled. Standards which cannot be attained by negotiation or legal enforcement may at least be achieved if made less demanding. The threat of the 1974 Act was the challenge posed to agency control over the enforcement of regulatory deviance by publicity and a strict liability law. By redrawing the individual boundaries of deviance the agencies have sought to retain control. Their response may be the occasion for added impetus to the ostensible shift from equity to efficacy as the guiding principle in the design of water pollution standards.

Notes

1. See further Ackerman *et al.*, 1974; Davies, 1970; Richardson *et al.*, forthcoming.
2. It follows from this that the law may be irrelevant as a means of control or sanctioning unless the standard actually addresses 'pollution' specifically. What is visibly 'polluting' to a layman may legally be clean water. In one case, for example, a major water supply and fisheries river was contaminated by the escape of a large amount of chemical dye which ironically was normally used to trace the source of pollutions. As a result the water in the river for several miles turned conspicuously red. The chemical could not be treated, but simply passed through sewage works without effect. The dye did not affect the suspended solids, biochemical oxygen demand (see n. 10 below), or any of the other parameters laid down in the company's consent. The water authority in this case was theoretically powerless: there was no standard at the works where the pollution originated which addressed the colour of its effluent, therefore no pollution.

3. Consistent demands made of dischargers in the standards set, apart from expressing a deeply-held value of justice, also assist in efficient enforcement, Kagan (1978:81) argues, since rules which are uniformly and consistently enforced are more likely to meet with voluntary compliance.
4. An unpublished agency document reported that 'only about 50 per cent of samples of existing discharges conform with currently applicable quality conditions. . . .' Subsequent chapters [see *Environment and Enforcement*] discuss the reasons for such official tolerance.
5. There is a parallel here in the reluctance of regulatory agencies to pursue prosecutions where defendants are likely to contest the charge strenuously: see, e.g. Cranston, 1979:125.
6. Solids and ammoniacal nitrogen exert oxygen demand; metals may be toxic, cyanide in sufficient concentration lethal. Thermal pollution affects fish.
7. Applications for consent must address the nature and composition of the discharge, its maximum temperature and volume, and its highest rate of flow. A consent cannot be unreasonably withheld; it may be reviewed.
8. Field staff in the southern agency now play only a minor part in standard-setting.
9. These hark back to the Royal Commission on Sewage Disposal 1912.
10. BOD, an index of the biodegradable matter in water, is a measure of the amount of oxygen a given body of water will consume in five days at a temperature of 20°C.
11. i.e., 10 BOD, 10 suspended solids.
12. The suggestion that 'expertise' is linked with 'trust' perhaps indicates the extent of the belief that compliance is a substantially technical matter. In the USA, however, there seems to be a marked distrust of big business and it is unlikely that enforcement agents would take the same view.
13. Names of polluters and officials are fictitious throughout the book [see *Environment and Enforcement*].
14. The 'stat' is the formal sample necessary for purposes of prosecution. A number of different terms are used in both agencies: 'formal sample', 'legal sample', 'tripartite sample', 'sealed sample', 'official sample', 'prosecution sample', and 'statutory sample' (or, more commonly, 'stat'). They all refer to the same thing and are typically used interchangeably in conversation. The officers who have the most extensive experience of legal samples usually talk of 'taking a stat'; the process is described in ch. 8 [see *Environment and Enforcement*, Ch. 8].
15. The charges made to firms to discharge their liquid wastes into the foul sewer for treatment at the local sewage works: see Richardson *et al.*, forthcoming.
16. See also ch. 6 for a brief discussion of efforts by the regulated to co-opt enforcement agents by gifts or bribes.
17. Previously, an individual could only prosecute with the leave of the Attorney-General.
18. Control of Pollution Act 1974, s. 41.

References

ACKERMAN, B. A., ACKERMAN, S. R., SAWYER, J. W. and HENDERSON, D. W. (1974) *The Uncertain Search for Environmental Quality*. Free Press, New York.

BECKER, H. S. (1963) *Outsiders. Studies in the Sociology of Deviance*. Free Press, New York.

CRANSTON, F. (1979) *Regulating Business. Law and Consumer Agencies*. Macmillan, London.

DAVIES, J. C. (1970) *The Politics of Pollution*. Bobbs-Merrill, Indianapolis.

KAGAN, R. A. (1978) *Regulatory Justice: Implementing a Wage-Price Freeze*. Russell Sage, New York.

KAUFMANN, H. (1960) *The Forest Ranger. A Study in Administrative Behavior*. John Hopkins University Press, Baltimore.

KNEESE, A. J. (1973) 'Economics and the Quality of the Environment: Some Empirical Experiences' in Enthoven and Freeman (eds) *Pollution Resources and the Environment*. Norton, New York.

RICHARDSON, G. with OGUS, A. and BURROWS, P. (1982) *Policing Pollution: A Study of Regulation and Enforcement*. Oxford U.P.

Commissions and Codes: A Case Study in Law and Public Administration

M. CAVADINO*

The Mental Health Act Commission (MHAC) was established in 1983 as a 'watchdog' for patients compulsorily detained in hospital in England and Wales. Its history provides a useful illustration of the problems caused by the ambiguous approach taken to the creation of hybrid public bodies and rules in the United Kingdom.

The MHAC came into existence on September 1, 1983, as a consequence of the reforms in mental health legislation brought about by the Mental Health (Amendment) Act 1982 and subsequently incorporated into the consolidating Mental Health Act 1983. It is a multi-disciplinary body comprising around 90 part-time members drawn from the medical, legal, nursing, psychological, social work and other professions, with a 16-strong Central Policy Committee. Its statutory responsibilities include: keeping the 1983 Act under review; visiting and interviewing patients detained in hospital under the Act; investigating complaints about the treatment of detained patients and the operation of the Act; appointing psychiatrists to provide second opinions on compulsory medical treatments and on treatments giving rise to special concern (such as psychosurgery); and—the particular subject of this article—submitting proposals to the Secretary of State as to the content of the Code of Practice which supplements the Act.[1]

The genesis of the Mental Health Act Commission

The reform of mental health law contained in the Mental Health (Amendment) Act of 1982 had a long gestation period which may be

This paper is based on the findings of research which was financed by Research and Development for Psychiatry and carried out with the co-operation and valuable assistance of the Mental Health Act Commission. Particularly profuse thanks are due to the following individuals: Sir Louis Blom-Cooper, Viscount Colville, William Bingley, Elaine Murphy, John Finch, Mike Napier, Robert Bluglass, Eric Bromley, Maggie Hysel of Research and Development in Psychiatry, my colleagues Ian Harden and Cosmo Graham, and Lucille Cavadino.

dated from the publication of the Butler Report on Mentally Abnormal Offenders in 1975.[2] The Labour Government of 1974–1979 produced a Green Paper in 1976 and a White Paper in 1978[3] but no Bill. The Conservative Government which succeeded it produced a less extensive White Paper[4] simultaneously with its Bill in November 1981.

It was in this 1981 White Paper that the Government for the first time—seemingly quite suddenly—declared its intention to create a Mental Health Act Commission and a general Code of Practice covering all aspects of mental health law.[5] The Royal College of Psychiatrists had long been in favour of a Mental Health Commission, but the 1976 Green Paper made discouraging noises while the 1978 White Paper rejected the idea, at least partly on resource grounds.[6] Neither paper even mooted the possibility of a general Code of Practice. Yet the Conservatives' 1981 White Paper revived the idea of a Commission and introduced that of a Code. Why?

The two innovations were crucially linked to each other, and to a third issue: that of medical treatment for mental disorder. Strictly speaking, this third issue encompassed two matters: (a) procedures for authorising the compulsory treatment of detained patients, and (b) new safeguards surrounding the administration of 'special concern' treatments such as psychosurgery. Governments of both complexions had long favoured new procedures whereby a statutory second opinion should have to be required before either compulsory of 'special concern' treatment could be administered, rather than leaving the decision in the hands of a single consultant psychiatrist. But controversy had surrounded the question of what form the second opinion should take. MIND (the National Association for Mental Health, who were waging a vigorous campaign to introduce greater recognition for patients' rights into the law) argued that the final decision to administer compulsory or 'special concern' treatment should not rest solely with members of the medical profession, while the Royal College of Psychiatrists (R.C. Psych.) argued that it should.

The Labour Government of 1974–79 (whose Secretary of State for Health and Social Services, David Ennals, was a past Director of MIND) inclined more towards MIND's view. The 1978 White Paper favoured the creation of local multi-disciplinary panels (including lay members) who would have the final say on whether a compulsory or controversial treatment should go ahead. It noted that 'much, but not all, medical opinion was against the introduction of multi-disciplinary panels' but went on to express the hope 'that, in the light of the widespread support for them, the medical profession will feel able to go along with this proposal'. This hope was to prove vain, for the R.C. Psych. continued to campaign strongly against the introduction of multi-disciplinary panels.[8]

The situation was transformed, at least on this issue, by the election of a Conservative Government in 1979. The new Government announced its intention to introduce new mental health legislation, and in early 1980 confidential meetings took place between senior civil servants in the Department of Health and Social Security (DHSS) and representatives of the R.C. Psych., which centred on those areas of the 1978 White Paper about which the Royal College was most concerned.[9]

As previously stated, the R.C. Psych. had long favoured the introduction of a Mental Health Commission along the lines of the Scottish Mental Welfare Commission, a multi-disciplinary body with the general remit to exercise protective functions in respect of psychiatric patients, whether or not compulsorily detained in hospital. Parallel to this commitment to a Commission, the Royal College had from time to time made suggestions about introducing guidelines or a code or codes of practice relating to compulsory and/or 'special concern' treatments, which would, *inter alia*, specify circumstances when a second medical opinion should be obtained. Such ideas were part of the R.C. Psych.'s agenda in the 1980 meetings with the DHSS. At first, the Royal College envisaged that these codes could be produced by the College itself, and also mooted as a possibility that, where a second opinion was required, this would be provided by a psychiatrist member of a panel appointed by the College. However, in the course of discussion, the three ideas of the Commission, codes of practice and second opinions began to converge. It would be the Commission which drew up codes of practice and which appointed psychiatrists to provide second opinions.

As Larry Gostin (who was MIND's Legal Director and leading lobbyist at the time of the MHAC's creation) has said, 'it is highly likely that if a Commission had not been thought to be needed to resolve the intractable problems concerning consent to treatment . . . it would not have been established'.[10] For the Government and the DHSS, the idea of a Commission represented an opportunity to resolve the second opinion issue. The R.C. Psych. wanted a Commission and would be happy with a statutory second opinion system run by the Commission rather than itself, so long as the second opinion was to be furnished by a psychiatrist rather than a multi-disciplinary panel. On the other hand, the Commission was to be a multi-disciplinary body, so something remained of the multi-disciplinary principle. While this was a solution which clearly tilted towards the R.C.

Psych. position rather than that of MIND, it did contain an element of compromise.

Another advantage of the Commission concept was this. It allowed the Government to claim that it was taking effective steps to protect the rights of detained psychiatrist patients (whom history had shown to be a group vulnerable to abuse and neglect) but without taking the road favoured by MIND of specifying patients' legal rights more concretely in statutory form and extending the formal powers of the existing multi-disciplinary Mental Health Review Tribunals to cover decisions about patients' medical treatment as well as their detention. Pursuing MIND's course (dubbed 'legalism' by opponents, with pejorative intent) would have engendered both opposition from the R.C. Psych. and headaches about the exact drafting of patients' statutory rights. The approach taken—congenial to the R.C. Psych., less so to MIND—was along the lines of the philosophy that people (at any rate, professional people) are best ruled by other people rather than by fixed rules. By good people, that is—people of skill and status, who exercise power that is largely discretionary and informal, persuading rather than coercing professionals into improving their practices. Thus, the Health Minister, Kenneth Clarke, could later tell the Special Standing Committee on the Mental Health (Amendment) Bill:

We believe that legalism would be a mistake. It is essential for Parliament to set out the basic minimum rights of patients and to entrench the most vital of them, but the best protection for patients in practice is a powerful watchdog composed of professional people to take a continuing interest in cases and offer clear guidance to professionals...[11]

The idea of creating a Code of Practice which would guide rather than coerce fitted neatly into this strategy.

The DHSS remained concerned about the resource implications of creating a Commission and insisted that the Commission's statutory role would (unlike that of the Scottish Commission) be limited to the protection of detained patients and not extend to those psychiatric patients who enter or stay in hospitals informally (and who comprise 90 per cent. of patients in England and Wales). But as early as March 1980, DHSS civil servants told R.C. Psych. representatives in confidence that they agreed to the principle of a Commission. The Royal College was in effect invited to put up a detailed proposal for a Mental Health Commission. This it did in the form of a paper which was published in the College's *Bulletin* in July 1981.[12] These proposals were not accepted in full by the Government—

for example, the College suggested creating several regional Commissions rather than a single one, and still wanted the Commissions to have responsibility for the welfare of informal patients. However, some of the salient features of the R.C. Psych.'s paper—that the Commission should produce guidelines on the operation of mental health law, and that it should be the Commission which arranged for the provision of second opinions—found their way into the 1981 White Paper.

The concept of the Code of Practice was at this time vaguer than that of the Commission. The 1981 White Paper introduces the Code in the context of 'special concern' treatments.[13] This implies that the Code was at first primarily seen as an extra-statutory guide as to which treatments should be regarded as falling into this category, and hence its creation (like that of the MHAC itself) was originally an attempt to solve issues surrounding treatment. The idea of a more general Code of Practice was a later development.[14]

The passage of the 1982 Act

The proposed creation of the MHAC met with a mixed reception from MPs and Lords during the passage of the Mental Health (Amendment) Act 1982. Members who were close to MIND tended to be sceptical about the supposed benefits of the new 'quango,' as many of them called it.[15] Several members were unconvinced about the necessity for such a body, while some complained that the role and function of the MHAC were vague and sought to clarify them.[16] There were also calls, rejected by the Government at least for the time being, to widen its remit and to make it produce an annual rather than biennial official report.

However, the clauses in the Bill authorising the establishment of the MHAC and the drawing up of the Code of Practice proved to be repeatedly useful to Government ministers at various stages during the passage of the Bill through Parliament. Several times the Government successfully resisted attempts to write into the Bill specific provisions requiring professionals to act in particular ways by pointing to the future existence of a Commission and a Code which it was confidently assumed would provide an appropriate solution—indeed, one more appropriate than writing the proposed prescription into the primary legislation. For example, proposed clauses requiring professionals to consult patients' relatives or members of other disciplines at various stages were consistently opposed by arguing that this would be better dealt with in the Code, which would be drawn up by the MHAC. The following quotation from Lord Sandys is typical:

This is the sort of advice which the Mental Health Act Commission will, I have no doubt, include in the code of practice. It is difficult to make provision about the kind of consultation which would be involved in the precise terms required in a statute. I am sure noble Lords would agree with that . . . The code of practice will be able to give practical advice which will be much more helpful to those concerned than anything we, however well intentioned, attempt to enshrine in a statute, and I believe we should leave it to the commission. I am sure they will take note of this debate.[17]

Thus the MHAC began to assume more important dimensions. Already vaunted by the Secretary of State for Health and Social Security as 'the single most important innovation in the Bill,'[18] the MHAC began inflating into a potential panacea, capable of being all things to all MPs. Problems and competing demands were defused as they arose by delegating their resolution to the MHAC, usually by way of suggesting that the matter be dealt with in the Code. Kenneth Clarke's comment, 'No doubt they will meet all our wishes when they produce the code,'[19] referred only to one point, but could almost stand as a statement of the impossible task being remitted to the MHAC—to please everyone. It could be cast in this role because it would be (a) independent; (b) multi-disciplinary, and hence capable of reconciling inter-disciplinary differences; and (c) composed of respected professional people with a high degree of expertise and wisdom. And, perhaps most importantly of all, because it was not yet in existence and so had not had the opportunity to disappoint anyone yet.

The concept of the Code of Practice was also inflating. As we have noted, it was originally meant as a set of guidelines concerned only with compulsory and controversial treatments. By the time the Act was passed it had come to encompass guidance not only on all aspects of the operation of the Mental Health Act, but potentially all aspects of psychiatric treatment.[20]

The (variable) constitution of the MHAC

The MHAC is a 'quasi-governmental' body. It has often been referred to—for example in the Parliamentary debates on the 1982 Act[21]—as a 'quango' (a notoriously imprecise term which originally meant 'quasi-autonomous *non*-governmental organisation'). Formally, it is a 'special health authority' established by the Secretary of State under section 11 of the National Health Service Act 1977. Its members are appointed by (and may be dismissed by) the Secretary of State, who under section 13 of the NHS Act also has the power to give directions as to how the MHAC exercises its functions—all of which are functions delegated to the MHAC by and from the Secretary of State. This means that it is both the creation and ultimately—perhaps we should say formally and if needs be—the creature of the Secretary of State. And yet it was described as 'independent' by the Secretary of State in the Parliamentary debates on the 1982 Act,[22] and the impression fostered was that of a body which would operate at a good arm's length from the Government. The belief that this was so was further encouraged by the initial attitude of Government ministers, DHSS civil servants and the MHAC's first Chairman, Viscount Colville, all of whom maintained a comfortable distance between Government, Department and Commission. There was, therefore, from the beginning, a discrepancy between the formal (or *de jure*) constitution of the MHAC and the informal (or *de facto*) way in which people were encouraged to think of it.

A similar discrepancy between the *de jure* and supposedly *de facto* positions existed with regard to the relationship between the MHAC and the Code of Practice. Ministers repeatedly asserted that it would not be the Government but the MHAC which would be drawing up the Code, an assertion accepted without question even by critics of the Bill.[23] Yet the Bill at no stage said this. What eventually became section 118 of the Mental Health Act 1983 directed the Secretary of State to prepare the Code, and that Act contained no requirement for him to delegate this function to the MHAC. Subsequently, a statutory instrument required the MHAC, not to produce the Code of Practice, but merely to submit to the Secretary of State '*proposals* as to the content of the code of practice which he shall prepare'.[24] This discrepancy was to prove crucial in the years following.

It is a standard criticism of quasi-governmental and quasi-non-governmental arrangements in Britain that, imbued with a similarly ambiguous nature to that characterising the unwritten national constitution, they facilitate an unhealthy version of 'corporatism'.[25] By this is meant not only the delegation of governmental functions to agencies more or less removed from Government, but also the exercise of *disguised* and therefore *unaccountable* governmental control. Governments can insulate themselves from criticism by hiding behind bodies which are purportedly independent but which are in reality government-controlled, thus exerting power without responsibility—and the kind of constitutional ambiguity we have observed in the case of the MHAC facilitates such a process. On the other hand it is argued that quasi-government and quasi-non-government are appropriate for public functions which Government itself either takes comparatively little interest in or which, conversely, raise potential conflicts of interest for governments or parties of government. In either case it may be desirable for these functions to be exercised by a specialised body with a degree of genuine independence from government. Considerations of the latter type suggest that the creation of the MHAC could certainly be justified in principle; but fears of the former type—which were at times expressed about the MHAC during the Parliamentary debates on the 1982 Act[26]—could be said to be equally valid.

The Code of Practice

Codes of Practice have become an increasingly popular form of 'quasi-legislation'—popular with governments, that is. As Ganz says, they have 'been used almost as a panacea. Hardly a statute is passed without a provision for a code of practice or guidance'.[27] Baldwin and Houghton note that 'One view of such "rules" is that they offer a useful structuring of discretion; another is that they are often used cynically so as to make law without resort to Parliament'.[28] In the case of the Mental Health Act Code of Practice, a more telling criticism might be that it was used, not through a wish to slip any particular new rules in through the back door, but as a device to *avoid* or *defer* making rules when under conflicting pressures as to what kind of rules should be made. Partly, no doubt, this stemmed from legitimate desires to avoid overloading the main statute or hastily judging issues which required further consideration, but it could also be crudely described as a 'cop-out' by a government shirking its responsibility to arbitrate between the competing interest groups. A similar charge could be levelled regarding the discrepant messages conveyed as to whether the Code was to be essentially the creature of the Secretary of State or of the MHAC.

The MHAC interpreted its brief as not merely submitting proposals as to the Code's content to the Secretary of State (as the legislation formally stated), but as 'drafting the Code of Practice'. To this end, the MHAC adopted a cumbersome but highly participatory *modus operandi* whereby the Commission divided into 17 teams to discuss different topics and produce drafts for different sections of the Code. Subsequently, these drafts underwent successive stages of revision and editing. The process took from March 1984 until November 1985, when the MHAC's draft was submitted to the Secretary of State. The depth of confidence that then existed within the MHAC that its draft would be accepted without wholesale amendment is indicated by the fact that along with it was submitted a draft of a proposed introduction to the Code by the Minister. At the same time, the draft Code was distributed to other interested organisations and health authorities.

Although opposition to this draft from the Royal College of Psychiatrists was to be expected, the MHAC was unprepared for the degree of robustness that the draft evoked from the R.C. Psych. (and to a lesser extent from the British Medical Association). As Fennell says, the Royal College

> reacted with near outrage at the length of the Commission draft (it was over 200 pages), at its tone (which was described as overprescriptive), at the fact that it was being foisted on the psychiatric profession by an outside body, and generally that the code went too far into areas which were properly left to the clinical judgment of psychiatrists.[29]

On the substantive issues, there can be little doubt that the MHAC's draft was too long and in parts contained material which was over-theoretical for a document which was intended to provide practical guidance to busy professionals. Another major issue was whether power and responsibility should remain concentrated in the person of the psychiatrist or dispersed (as the Code suggested) by requiring a greater degree of multi-disciplinary consultation and co-operation. This had, of course, been a salient bone of contention at the time the 1982 Act was passed; clearly it had not been successfully resolved for all time by leaving it to the MHAC and expecting it to 'meet all our wishes when they produce the code'.[30]

The Royal College also objected to the procedure which had been followed by the MHAC, feeling that it should have engaged in full consultation with other interested bodies before submitting its draft to the Secretary of State. Formally and legally this was unnecessary: the MHAC's job was merely to submit proposals, the ultimate Code would be the Secretary of State's, and it was the latter who was required (by s. 118(3) of the 1983 Act) to engage in consultation. If, however, the MHAC's informal expectation that it had the right substantially to draft the Code were to be taken seriously then it should perhaps have consulted bodies such as the R.C. Psych. But—still on an informal track— MHAC members felt that the Royal College's complaints were misplaced since psychiatrist members of the MHAC (some of them highly positioned within the R.C. Psych.) were involved in the drafting process and the College had every opportunity to know what the draft was likely to include and to make representations to the MHAC at any stage. So the MHAC saw itself as having the role of reconciling differences of opinion between the disciplines by virtue of its own multidisciplinary composition. Thus the disagreement about the procedure adopted by the MHAC stemmed from the fact that no-one had made clear to the MHAC, the R.C. Psych. or anyone else what procedure *should* be expected regarding consultation over the drafting of the Code. This lack of clarity fuels what might be called 'the Babel problem': a situation in which different protagonists are—perhaps without realising it—not speaking the same language.

It was similarly unclear what kind of guidance the Code was supposed to contain. Should the Code (a) explain the law; (b) interpret the law; (c) explain the 'spirit of the law'; (d) encourage good practice; (e) require good practice? If (d) or (e), should this be confined to good practice regarding the operation of the Mental Health Act, or to wider issues of care and treatment of patients generally, including patients who are informal (not detained under the Act)?[31] How widely accepted should a good practice be before it would be right for the Code to recommend or require it? The R.C. Psych. felt that it was illegitimate for the Code to pursue aim (b), and that in pursuing (a) it had paraphrased the law in a manner which amounted to interpretation. The College accepted that the Code should guide on practice issues, but felt that it erred by prescribing practice which was not already sufficiently backed by professional consensus.[32]

Exacerbating the feeling behind these objections were the R.C. Psych.'s fears surrounding the general substantive issue of the status, power and responsibility of the medical profession. The Royal College had battled hard to retain for psychiatrists the ultimate power to decide on whether compulsory treatment should be administered, and although it had perhaps got the better of things in legislative terms it felt that it had been largely outlobbied by MIND and other interest groups.[33] Psychiatrists now feared that the MHAC had been 'captured'[34] by civil libertarians and would use the Code as a vehicle to reopen the attack on clinical responsibility. Such distrust concerning what was seen as an issue of fundamental importance could only be heightened by the 'Babel factor'.

The Government bowed to the R.C. Psych.'s hostility and shelved the MHAC's draft. Moreover, rather than remitting the Code to the MHAC for a rethink, the Government took it out of the MHAC's hands altogether, despite all the previous promises that it would be the MHAC and not the Government which drew up the Code. A second (much shorter) draft was drawn up within the DHSS and published in August 1987. The main authors of this draft were civil servants, but there had been some significant input by a representative of the R.C. Psych. As a result, the draft was inoffensive to psychiatrists, at the cost of saying very little apart from reiterating the provisions of the Act. Recognising that this draft had swung too far in the opposite direction, the Government shelved this one also. (It did the same with a later redraft, produced within the Department in early 1989 but never published.)

The deadlock was finally broken in 1989, when a small working group was set up under the auspices of the Department of Health to produce yet another draft. This group, the brainchild of civil servants within the Department, was chaired by an ex-Deputy Secretary in the DHSS and also included MIND's Legal Director who was seconded to the Department for the purpose, and an eminent psychiatrist who had become Vice-Chairman of the MHAC in early 1988. Compared with its predecessors, the draft produced by this group fairly sailed through: finalised in September 1989, it was laid before Parliament in December and duly came into force in January 1990.[35] How was this success achieved in the light of the previous failures?

Conclusions

The conditions for the success of the working group are instructive and may be of wider interest in situations where quasi-governmental bodies and procedures are deployed to arbitrate between conflicting interests. First, the Government finally took decisive action, letting it be known that it wanted the situation to be resolved and would if necessary use its power to impose a solution, although preferring a consensus outcome—almost 'knocking heads together'. Such decisiveness also had the effect of bringing to the scene a clarity which, as we have seen, had long been missing about where power and responsibility ultimately lay. On the other hand, the working group was by no means the Government's (or the Department's) poodle. On the contrary, it was given an almost totally free rein as regards its operations and the content of its draft. What made for a successful rather than unsuccessful venture in quasi-government was that a group was set up which could work independently with the unspoken assurance that the Government would support any reasonable forthcoming draft Code.

The group consisted of persons outside the Department who were also perceived as being independent of the different pressure groups although having good links with them. The group also avoided the worst problems entailed by formal consultation processes, proceeding by informal soundings out of individuals known to them personally rather than officially seeking the views of different bodies. Another feature of the group's *modus operandi* was that it moved from the concrete to the abstract, concluding rather than commencing with statements of principle. They started by identifying areas where practitioners were in need of practical guidance as to how to proceed, seeking to reach broadly acceptable solutions before finally distilling from these a few broad principles which were ultimately presented in Chapter 1 of the Code.

The result of this set-up was a group which worked well together and produced a draft—of middling prescriptiveness compared with the first and second drafts, and better written than either—with which most concerned bodies were on the whole pleased rather than displeased and which they were inclined to accept. This inclination was strengthened by a certain amount of battle-weariness among the contestants and a change of leadership within both the MHAC and the R.C. Psych., in both cases leading to a less confrontational approach.[36] Moreover, at the end of the day all sides wanted a Code to be enacted, and one containing some prescription—although the R.C. Psych. does not want too much intrusion on clinical responsibility, it nevertheless has an interest in controlling maverick members of its own profession of whose deviant practices it disapproves. The differences between the various pressure groups, although substantial and significant, are by no means as irreconcilable as the previous history had made them appear.

To some extent, the working group's success was facilitated by the failures which had come before. It capitalised on the widespread desire to resolve the prevailing situation; it had available to it the previous drafts and a mass of detailed comments and criticisms attached to them; and it could dispense with formal consultation because this had already taken place. But the main lesson of its success must be that quasi-government can be utilised in different ways, and the differences can have the effect of making a difficult task either impossible or feasible. The most important factor is whether government is prepared to shoulder its responsibilities and make clear its intentions and what it expects from quasi-governmental bodies, or whether it is content to allow the waters to be muddied for reasons of short-term expediency.

It seems inevitable that quasi-governmental and quasi-legislative devices will continue to be much (and in all probability increasingly) used; nor is this tendency to be deplored *per se*. But what is desirable is what Harden has called 'a constitution for quangos'[37]—along with a constitution for quasi-legislation[38]—to clarify where responsibility lies and in what ways such hybrid bodies and the makers of such rules are to consult interested parties and to be accountable to democratic processes. The temptation may be great for governments to avoid such clarification, but if they continue to do so in the future the travails of the Mental Health Act Commission concerning the genesis of the Mental Health Act Code of Practice could well be replicated repeatedly in other contexts.

Notes

1. Mental Health Act 1983, ss. 118 and 121; Mental Health Act Commission (Establishment and Constitution) Order 1983 No. 892); Department of Health and Welsh Office, *Code of Practice: Mental Health Act 1983* (1990).

    For an overview of mental health law see M. Cavadino, *Mental Health Law in Context: Doctors' Orders?* (1989); for an account of the passing of the 1982 and 1983 Acts see C. Unsworth, *The Politics of Mental Health Legislation* (1987).
2. *Report of the Committee on Mentally Abnormal Offenders*, Cmnd. 6244 (1975).
3. DHSS, *A Review of The Mental Health Act 1959* (1976); DHSS *et al.*, *Review of the Mental Health Act 1959*, Cmnd. 7320 (1978).
4. DHSS *et al.*, *Reform of Mental Health Legislation*, Cmnd. 8405 (1981).
5. *Ibid.*, paras. 29–42.
6. DHSS (1976), *loc. cit.* (see n. 3), paras. 8.38–8.42; DHSS *et al.* (1978), *loc. cit.* (see n. 3), paras. 6.33–6.35.
7. DHSS *et al.* (1978), *loc. cit.* (see n. 3), para. 6.28.
8. See Royal College of Psychiatrists, 'The White Paper on the Mental Health Act; the College's Comments', *Bulletin of the Royal College of Psychiatrists*, April 1979, 58–65.
9. R. Bluglass, 'The Origins of the Mental Health Act 1983; Doctors in the House', *Bulletin of the Royal College of Psychiatrists*, 1984, 127–134.
10. L. Gostin, *Mental Health Services—Law and Practice* (1986), para. 22.02.4.
11. H.C. Special Standing Committee 1981–2, Vol. XI, col. 797.
12. The College, 'Towards a New Mental Health Act: Mental Health Commissions for England and Wales', *Bulletin of the Royal College of Psychiatrists*, July 1981, 130–132. The proposals were introduced and commended by Robert Bluglass, who chaired the R.C. Psych.'s special committee concerned with the review of mental health law.

    The confidential meetings between the R.C. Psych. and DHSS civil servants are referred to in Bluglass, *op. cit.*, n. 9 above.
13. DHSS *et al.* (1981), *loc cit.* (see n. 4) paras. 38–39.
14. Cl. 40 of the original Bill was expressed in terms wider than those of the White Paper, stating that the Code would be 'for the guidance of medical practitioners and members of other professions concerned in the medical treatment of patients detained' under the Mental Health Act. This was even-

tually widened further to encompass 'the medical treatment of patients suffering from mental disorder' without limitation to detained patients (Mental Health Act 1983, s. 118(1)).
15. See e.g. H.C.Deb., Vol. 20, cols. 705, 711, 714 (March 22, 1982); Vol. 29, cols. 168–9 (October 18, 1982); H.C. Special Standing Committee 1981–2, Vol. XI, col. 582 (Sir Derek Walker-Smith, David Ennals, Charles Irving and Christopher Price MPs).
16. e.g. H.L.Deb., Vol. 427, cols. 1079–81 and 1097–8 (Lord Kilmarnock); ibid., col. 1085 (Lord Sandys) (February 25 1983).
17. H.L.Deb., Vol. 426, col. 849 (January 25, 1982). Cf. also H.C.Deb., Vol. 29, col. 114 (October 18, 1982) and H.C. Special Standing Committee 1981–2, Vol. XI, cols. 591–4 (Kenneth Clarke MP).
18. H.C.Deb., Vol. 20, col. 693 (March 22, 1982).
19. H.C. Special Standing Committee 1981–2, Vol. XI, col. 593.
20. In the first (November 1981) version of the Bill, the code was merely to be for the guidance of professionals 'concerned in the *medical treatment* of patients *detained* under' the Mental Health Acts (cl. 39(1)). By the time the Act was passed this had expanded to encompass guidance to professionals in relation to (a) compulsory *admission* to hospital, and (b) 'the medical treatment of patients suffering from mental disorder' without any limitation to detained patients (s. 53 of the 1982 Act; s. 118(1) of the 1983 Act).
21. See above, n. 15. For the term 'quango' and its ramifications, see e.g. A. Barker (ed.), *Quangos in Britain* (1982).
22. H.C.Deb., Vol. 20, col. 693 (March 22, 1982) (Norman Fowler MP).
23. e.g. H.C.Deb., Vol. 29, col. 114 (October 18, 1982) (Kenneth Clarke MP); H.C. Special Standing Committee 1981–2, Vol. XI, col. 591–4 (Kenneth Clarke MP) and col. 587 (Terry Davis MP).
24. S.I. 1983 No. 892, art. 3(2)(d).
25. See especially P. Birkinshaw, I. Harden and N. Lewis, *Government by Moonlight: The Hybrid Parts of the State* (1990).
26. See e.g. H.C. Special Standing Committee 1981–2, Vol. XI, col. 147 (Larry Gostin of MIND: 'We fear that Governments might hide behind the Commission.') and H.L.Deb., Vol. 426, cols. 1171, 1173 (February 1, 1982) (where Lord Kilmarnock expressed fears that the MHAC might become 'the Secretary of State's poodle').
27. G. Ganz, *Quasi-Legislation: Recent Developments in Secondary Legislation* (1987), p. 3.
28. R. Baldwin and J. Houghton, 'Circular Arguments: The Status and Legitimacy of Administrative Rules' [1986] P.L. 239–285, p. 239.
29. P. Fennell, 'The Mental Health Act Code of Practice' (1990) 53 M.L.R. 499–507, p. 500.
30. See n. 19 above and accompanying text.
31. Although this last point remained a controversial one (see the BMA's comments published in the *Bulletin of the Royal College of Psychiatrists*, February

1987, 63–66), it seems clear that s. 118 of the Act permits the Code to contain guidance on issues concerning 'the medical treatment of patients suffering from mental disorder' which are not directly connected with detention and compulsory treatment. Moreover, 'medical treatment' is defined widely in s. 145(1) of the 1983 Act to include nursing and care, habilitation and rehabilitation under medical supervision.

32. 'Comments of the Royal College of Psychiatrists: Mental Health Act 1982 Draft Code of Practice', 10 *Bulletin of the Royal College of Psychiatrists*, August 1986, 194–5.
33. See Bluglass, *op. cit.*, n. 9 above.
34. For the concept of 'capture,' see I. Harden and N. Lewis, *The Noble Lie* (1986). The R.C. Psych. had previously feared that the DHSS under David Ennals had likewise been 'captured' by MIND (Bluglass, *op. cit.*, n. 9 above).
35. Under s. 118(4) and (5) of the 1983 Act the Code was subject to the 'negative resolution procedure,' *i.e.* it had to be laid before Parliament and would come into force provided neither House voted against its adoption within 40 days. Although the National Schizophrenia Fellowship, concerned about the Code's provisions relating to consultation of relatives, persuaded some members to 'pray against' the Code, their objections were withdrawn after the Government provided reassurance about their concerns.
36. It was notable that the new President of the R.C. Psych., Dr. Birley, wrote to Parliamentarians on both sides of the House of Lords urging that Parliament should accept the Code (H.L.Deb., Vol. 515, cols. 81, 85).
37. I. Harden, 'A Constitution for Quangos?' [1987] P.L. pp. 27–35.
38. *Cf.* Baldwin and Houghton, *op. cit.*, n. 28 above.

Pluralism: UK

P. P. CRAIG

Participation: two arguments for the status quo

It might be contended that the general approach of the common law is correct in circumscribing the availability of such participatory rights. Two complementary arguments might be adduced towards this end. It could be argued that, if Parliament wishes to impose an obligation to consult, it could so specify in the enabling legislation as it has done in certain instances. There is, therefore, no warrant in the courts intervening where the legislature has declined to tread. This argument would be reinforced by a second, the essence of which is that there are existing checks on rule-making and that, in practice, the forces of pluralism will in any event ensure the adequacy of group consultation. Common-law intervention would, therefore, be not only unwarranted but unnecessary. These arguments will be addressed in turn.

Reasoning which derives its authority from general statements concerning legislative intent or parliamentary omission should be viewed with caution. A concealed premiss underlying the argument is that the parliamentary process operates in such a way that, on each occasion on which legislation is enacted, the question as to whether consultation in the rule-making process should be required will be carefully considered; that the advantages and disadvantages of the inclusion of such an obligation will be carefully weighed and the appropriate conclusion will be reached. There is little warrant for this assumption. The inclusion of such a requirement, or consideration of whether such a requirement should be included, is much more likely to reflect the existence of an articulate, relatively homogeneous interest group, which brings pressure to bear in order to ensure that any future rule-making will be dependent upon prior consultation with those it represents.[1]

The argument from legislative intent is, moreover, weakened in a less obvious manner. It is predicated upon the assumption that the particular institution would, in fact, be making rules, or acting in some fashion which could render consultation desirable. This assumption does not accord with the facts. In many instances a bureaucracy, particularly one which is dealing with large numbers of claims, will make rules as a mechanism for expediting its disposition of business. There may be no specific mandate in the legislation which enables it to do so, but it is a natural habit for a bureaucracy to conduct its affairs in this manner.[2] In these areas the legislature may not even have addressed its mind to the question of whether consultation should be required as an adjunct of the rule-making process.[3]

The argument derived from legislative intent may, as we have seen above, be reinforced by the contention that further judicial reinforcement of participatory rights is unnecessary. This contention itself has two limbs. One aspect of the argument is that existing parliamentary controls over rule-making are effective; the other is that *de facto* consultation between government and affected groups will, in any event, occur, and that this renders more formally structured participatory rights otiose. These arguments will be addressed in turn.

The weaknesses of legislative scrutiny cannot be developed here. They have been well documented by other writers.[4] Control on the floor of the House is extremely limited, and the committee system designed to supplement this control has been hampered by the restricted nature of its powers. Many rules will not even be reviewed by Parliament at all, because they do not come within the ambit of the legislation.[5] The Statutory Instruments Act 1946 is framed in formal terms. It covers powers to be exercised by Orders in Council, and powers conferred on a Minister of the Crown which are expressed to be exercisable by statutory instrument. Reliance upon a formal definition may well avoid awkward or intractable difficulties attendant upon a substantive definition. Such solutions always have a price. In this instance the price is the exclusion from the coverage of the Act of many rules which may in substance be legislative.

What then of the other aspect of the argument outlined above? Does not the literature on pluralist policy-making indicate forcefully that affected interests will be afforded participatory rights in practice? Do not the plethora of advisory boards, commissions, and the like reinforce this conclusion? A superficial understanding of pluralism or élite pluralism might suggest this conclusion. A more considered appreciation of pluralist democracy reveals the inadequacies of this argument.

Pluralism might well indicate the impact which non-elected groups can and do have upon the policy-making process. It might serve to dispel the illusion that Parliament is the sole repository of public power. It does not, without more, tell us whether this will occur in all areas of the administrative process. There is ample evidence for the continued existence of rules which are dispositive of a person's case, which may well be unpublished and the result of no consultation. The limitations of the empirical literature on pluralism should also be borne in mind. Even where this literature demonstrates that groups are consulted, this does not tell us whether the types of groups which are thereby consulted adequately cover the range of affected interests. Studies may demonstrate that this is so within a particular area. They may also, as we have seen, indicate that representative status is accorded to the select few.[6] The views of Sir Douglas Wass, who has had many years of experience within the central bureaucracy, are particularly apposite in this context.[7] He notes the tendency of departments to 'mark' certain pressure groups, to the exclusion of other groups, and of less organized interests.[8] Pluralism as an empirical phenomenon would lead one to expect group pressure upon the administrative process. The criticisms and qualifications of pluralism which were considered in this and the previous chapter [see *Public Law and Democracy*] would cause one to suspect that group pressure would be uneven. Whether it would now be more correct to depict decision-making as élite pluralist, or corporatist is more debatable.[9] The answer may vary depending upon the nature of the substantive area, and the attitude of the government in power. What is readily apparent is that arguments from the undeniable existence of participation, to the sufficiency thereof, are simplistic. Factual evidence that groups can influence policy is no basis for asserting that an adequate range of interests will be consulted. The early pluralists saw nothing inconsistent in recognising group power, *and* advocating the express acknowledgement of consultation rights.[10]

Participation, constitutional structure, and pluralism

The arguments adduced in support of the proposition that the law should not play any greater role than at present in fostering interest representation have been found wanting. This does not, in itself, make the case for a change in our legal rules. The status quo may rest on uncertain foundation, but, as with many areas of life, some more positive reasons are required before thoughts of reform will be taken seriously. A fact which is being increasingly perceived by administrative lawyers is that reasons for change in their own area may well be connected with broader issues of constitutional law and constitutional design.[11] The argument which will be presented within this section exemplifies this connection. It can be summarized in the following manner. Our approach to participation within administrative law continues to be coloured by a unitary conception of the constitution. It is only by revealing the inadequacies of this unitary view that we can begin to make a principled decision as to the role which participation should play within society.

The foundation of this unitary view is a concept which we have encountered before. It is the less-well-known face of sovereignty, that of legislative monopoly.[12] The presupposition is that Parliament is either the direct initiator or the controller of all that emerges formally as legislation. Ministers are accountable to, and controlled by, Parliament. Policy issues underlying legislation or delegated legislation will have been considered by Parliament.[13] Governmental agencies will be controlled by and accountable to Parliament through the relevant minister.[14] A natural correlative of this conception of legislative monopoly is that a duality in the meaning of the term 'legislative' is concealed. Norms which emerge with the formal seal of Act of Parliament or statutory instrument are the sole qualifiers as legislation. Parliament controls or initiates the passage of such norms and thus the fundamental tenets of representative democracy are preserved. The realization that there are rules of a legislative character which may be substantively indistinguishable from those actually passed by Parliament, but which are not subject to parliamentary control, cannot be squarely faced without undermining the very conception of legislative monopoly.

If Parliament really did still exercise the degree of control ascribed to it, if it did possess this legislative monopoly, initiating or controlling the passage of legislation, and if the only norms of a legislative character which regulated our lives were those emerging as a statute or delegated legislation, then all might be well. There would be little need for any further participation in this Illyria. The problem is immediately apparent: our constitutional structure does not conform to such an idyllic vision. Constitutional conventions should be founded upon some measure of empirical evidence. If our theoretical constructs depart too much from reality, they risk becoming at best empty vessels; at worst they serve as invalid premises for the development of more particular rules of conduct. Few can seriously maintain that the picture of power and legislative monopoly ascribed to Parliament accords with reality. A more accurate portrayal of our political system would highlight two themes, both of which have direct relevance to the issue of participation: the growth in the power of the executive, and the increasingly complex nature of public decision-making. The former undermines the ideal of parliamentary power, and thereby places the value of primary participation, in the form of the vote, in its true perspective. The latter challenges both the ideal of parliamentary power and the legislative monopoly of Parliament. In doing so, it thereby raises the question of whether other forms of participation are warranted.

The growth in the power of the executive has already been considered in a previous chapter.[15] By the early decades of this century the legislative process was already being both nationalized and centralized into the hands of the executive. Whether one now chooses to regard our political process as adversarial or oligarchic, the implications for participation are equally profound. At its most extreme, one commentator has described our system as elective autocracy.[16] The most basic act of participation is the vote, and elective autocracy provides this minimum form of participation for all. It is, however, only a minimum form of communication operating once every four or five years. The voters' own MP will have precious little input or control over the parliamentary process, which will be firmly in the hands of the executive. One need not necessarily go as far as this and yet still agree that the vote as a mechanism by which, through one's chosen MP, one participates in the process of government falls sadly short of an ideal contained in the 'golden age' of Parliament model. If the demise of real parliamentary power minimizes the participation resulting from the vote, the increasingly complex nature of government action both reinforces this and provides more positive support for other forms of participation. It is to this second theme that we now turn.

The last ten years have witnessed the emergence of a growing literature concerned with the problems attendant upon the existence of a large number of institutions which exercise governmental responsibility but which do not conform to the departmental norm. Such institutions are not new. They have been inhabitants of our governmental landscape for a considerable period of time.[17] In the nineteenth century, at least in the first half thereof, it was the norm to allocate areas of governmental responsibility to such bodies. Decline in the Board system resulted from the assertion of parliamentary supremacy. Parliament wished to exercise a degree of control over the activities of Boards which their institutional structure rendered difficult. The result was the absorption of a number of Boards into new Ministries.

While the number of such bodies declined, the general category has never disappeared. Governmental agencies of one form or another, which did not conform to the ordinary departmental norm, continued to exist. The post-war period has witnessed a rapid expansion in these institutions.

A plethora of such organizations now exist, with a bewildering variety of name, structure, and personnel. The rationale for their creation has been equally varied. Some have been established to allow greater expertise to be focused upon a particular task, others to foster greater public participation within the organization itself. The desire to set governmental functions apart from ordinary party-political pressures, or to ensure greater fairness in the allocation of resources in which the government has a direct interest, provides further reasons for their creation. Overlapping with these reasons is a more general desire to maintain some cohesion in the traditional Whitehall bureaucracy and to hive off certain specialized tasks to particular institutions.[18] The number of such organizations will increase as a result of the government's decision to restructure the Civil Service and to utilize agencies to implement major policy initiatives which have been decided upon by the executive and bureaucracy.[19]

Quangos or fringe organizations are not the only reason for the increasingly complex nature of governmental action in our society. Government contracting presents another facet of the same problem. Corporations which are, in formal legal terms, private may have a substantial input into the policy-making process.[20] While we can try to preserve strict constitutional theory by maintaining that such organizations are simply concerned with the means of fulfilling clearly articulated governmental policy, this may bear little relation to reality. Means and ends are not easily separated, and only the most sanguine observer would seriously suggest that the role of the private corporation is limited to the mechanism for effectuating goals which have already been defined by the traditional political processes.

What are the implications of these developments and what light do they shed on the question of participation.[21] The presence of such organizations challenges the ideal of the legislative monopoly of Parliament, the idea that Parliament controls or initiates all legislation, *and* that only those rules formally stamped as an Act of Parliament or delegated legislation are of concern in a representative democracy. The presence of fringe organizations questions this presupposition in two ways, both of which have direct implications for participation.

First, in relation to certain types of fringe organizations, the traditional devices of the 'golden-age' model do not work even in theory. Fringe organizations can be accommodated within our standard constitutional theory by accepting that they will be held accountable to Parliament either through the responsible minister, and/or through direct supervision by, for example, a select committee. What is apparent is that such premises run counter to the very reasons for the creation of certain fringe organizations. Some of these institutions will be established in order positively to secure a degree of independence from the normal party political processes. Decision-making divorced from the standard departmental norm is the aim.[22] In such areas it would be contradictory to insist upon the methods of accountability dictated by traditional constitutional doctrine.[23]

Secondly, while our traditional constitutional model encounters theoretical obstacles when applied to certain fringe organizations, the practical difficulties which lie in its path are no less severe. Ensuring effective ministerial accountability for those fringe organizations which are not intended to be independent in the sense articulated above is subject to all the normal difficulties attendant upon that doctrine. These are then exacerbated by the fact that the institution is not part of the normal departmental structure. Direct parliamentary supervision through one of the remodelled select committees may hold out some hope of more effective scrutiny. However, as Johnson has said,[24] the impact of such committees will probably be highly selective unless they are equipped with staff and resources on a scale not contemplated thus far.

While traditional forms of accountability may be subject to constraint in this area, this does not mean that other forms of control and accountability may not be available. There may be financial controls, and guidelines laid down in the enabling legislation. Monitoring the decisions reached by the organization is a further possibility.

Participation in the decision-making process can be another valuable device. We are accustomed to thinking of accountability in its traditional sense, from the 'top', from Parliament. The limitations, both theoretical and practical, of this technique have already been addressed. Accountability need not, however, be seen exclusively in this way; a different but important method of accountability may come from the 'bottom', from participation of those affected in the decision-making or rule-making processes of a particular institution. The existence of impediments to the traditional form of accountability from the 'top' does not necessarily preclude participation in the decision-making process from the 'bottom'. Thus, for example, the fact that we justifiably wish the allocation of media licences, or funds for the arts, to be dispensed free from party bias does not in any sense import the conclusion that interested parties should not be enabled to participate or air their views.

It is, of course, true that this type of accountability from the 'bottom' is different from the traditional concept. To accept that it is different is not, however, to suggest that one should be the sole bearer of the citizens' interest. The traditional concept rests ultimately on the idea that, in a representative democracy, decisions of a governmental nature should reside with the elected representative, directly or indirectly. This ideal has, as we have seen, come under increasing strain. The growth of executive power over the legislature has rendered the first limb of the 'golden-age' model, that Parliament wields real power, outdated. The consequence has been to reduce the importance of the citizens' primary participation, in the form of the vote, to a minimum.[25] This very growth in executive power, coupled with the increasingly complex nature of governmental institutions, challenges the second limb of that model, that of legislative monopoly. When we attempt to apply traditional notions of accountability, we encounter theoretical and practical difficulties of the type mentioned above. A greater regard for consultation in the process of rule-making can constitute a valuable form of secondary participation for the citizen in government. This applies as much to rule-making by the executive as to that by fringe organizations.[26]

Secondary participation in the sense adumbrated above should, of course, not be regarded as the panacea for all ills. The discussion of the conceptual and pragmatic difficulties which attend participatory rights in the United States is of relevance here also.[27] Problems of defining an interest, and of securing the adequate representation of affected interests, are endemic whenever such rights are accorded, even though their precise form may vary from country to country. The same is true concerning problems of cost and delay. However, if the genetic nature of the problems experienced in the United States is apposite, so too are the rejoinders and qualifications to those problems considered above. We shall return to this issue once again at a later stage. Before doing so, a separate, albeit connected, argument in favour of participation will be addressed.

Participation: voting, rule-making, and adjudication

Participation is obviously not a one dimensional concept, and political participation even less so. Although not free from difficulty, the meaning used by Parry will be adopted here. Political participation is the taking part in the formulation, passage, or implementation of public policies.[28] It is evident that political participation can take many forms. At one end of the spectrum there is the formalized vote. Voting[29] is obviously not the only mode of participation. Studies have dissected the decision-making process, indicating the varying political roles which are played between the inception of a project and the attainment of the final decision.[30] Formalized adjudication represents yet another avenue through which an individual can participate in the political process.

Viewing participation in the above manner prompts an interesting and important enquiry. Our legal rules at present protect both 'ends' of the spectrum, but treat the middle ground as no man's land. Formalized participation both through the vote or through adjudication, is legally protected.[31] Intermediate forms of participation, in, for example, rule-making by quasi-governmental agencies, are not afforded analogous treatment in the absence of specific statutory provisions. The remainder of this section will be devoted to considering the cogency of this position. Three reasons can be advanced to suggest that the present position is unsatisfactory.

First, the rationale for dividing the legally protected forms of participation into only two categories, those of litigant and of voter, is defective, being based upon an inadequate and simplistic theory. Rule application in an adjudicative context is held to warrant the requisites of procedural rectitude. Rule-making of a legislative nature is said to be different. No individual is a 'litigant', and participation is ensured, if at all, through the vote. Rule-making will be validated by the normal legislative process, and no further attention need be given to participation. The consequence is that the method by which the government chooses to deal with individuals becomes dispositive of their procedural rights. If it opts for an individualized form of adjudication, then the courts will activate the common-law procedural safeguards; if it chooses to treat citizens as a group, through the enactment of a rule, then the common law will abjure from supplying the omission of the legislature. The courts have been dormant, anaesthetized by the mention of the word legislative.[32] That the result may render procedural protection haphazard, and subject to fine distinctions between what is legislative and what is not, is itself objectionable. This point will be returned to below. What is equally important is that the reasoning fails to convince on a more theoretical level.

What *form* of participation an individual should be entitled to—the right to be heard as a litigant, the right to be counted as a voter, or some intermediate right to participate in the rule-making process—is, as Tribe has argued,[33] itself necessarily premised upon a substantive theory. The decision as to which kind of participation is warranted demands analysis, not only of the effectiveness of the different processes, but also of the importance of the interest which is being affected; and that in itself is dependent upon a substantive theory of values and of rights. Our law concerning process is perforce based upon a substantive theory, but it is one which is defective. The premiss is a rigid dichotomy between adjudication and rule-making in which due process is regarded as essential in the former in order, *inter alia*, to comply with the dictates of formal justice. No such obligation attends the latter; participation will be adequately ensured through the vote. What is more, the government's choice as to which avenue it should follow will be conclusive. It is clear that the concept of participation through the vote has become etiolated, and our law will remain seriously inadequate if it continues to be based upon a theoretical construct which so ill accords with reality. Important interests are dealt with through the promulgation of rules which the primary legislature cannot effectively control, and which demand a higher degree of participation than allowed by the vote. What is needed is a substantive theory which will recognize the possibility of a category intermediate between participation as litigant and participation as a voter. Only then will our law of process be equipped to meet the realities of modern government.[34]

A second reason why our present law may be regarded as unsatisfactory has been touched upon already. Distinctions between legislative and adjudicative action may be no easier to draw than those between, for example, judicial and administrative or administrative and executive action. Many cases may, although they arise as bipartite disputes, have broader multipartite implications.[35] The judicial focus may be directed towards, not just the resolution of a completed set of past events, but the modification of conduct in the future. Decisions reached by a public body through the medium of adjudication may entail broad issues of social and political choice which are indistinguishable from a legislative act. The distinction between legislative and adjudicative action is further blurred by the grant of broad discretionary powers to administrators. These may be concretized, either through the medium of individualized adjudication, or through the enactment of rules, or the one may follow the other. The former may have legislative effects every bit as real as the latter.[36]

Two consequences follow. On the one hand, it means that the problems of interest representation, of ensuring both that those interested in the suit do have a chance to make representations, and that the litigant presenting the case adequately represents future interests, may often be present even when the issue arises formally as adjudication.[37] We retain, however, a concept of adjudication in which the only interested parties are normally the actual litigants, a model which may well fit much private law adjudication, but is less suited to the subject-matter of public law. On the other hand, because we are still wedded to the idea that action of an overtly legislative nature generally carries no participatory rights, we refuse to countenance any interest representation when the general norm-creating nature of the action is made explicit. We seek to preserve our classical ideas of adjudication, and refuse to admit participatory rights in rule-making, partly at least because to do otherwise would be to admit the unthinkable, that public bodies external to Parliament make decisions of a broadly legislative nature. The force of our constitutional convention makes its presence felt once again.

The third and final reason why our present law may be open to question is that the very justifications for process rights in the context of adjudication bear some analogy to the reasons afforded for political participation in non-adjudicative contexts.

Justifications for due process in adjudication vary. One rationale emphasizes the connection between procedural due process and the substantive justice of the final outcome. Procedural deficiencies can undermine public confidence in the fairness of the proceedings, and render the substantive decision unreliable in the sense that an error is more likely to occur in the absence of procedural safeguards.[38]

Another justification often given for due process in an adjudicative context is that formal justice and the rule of law are thereby enhanced. The principles of natural justice encapsulated in the twin precepts of 'hear the other side' and 'let no man be a judge in his own cause' guarantee, as Hart has said, impartiality and objectivity.[39] Formal justice demands the regular and impartial administration of public rules, what Rawls terms 'justice as regularity'.[40] This contains a number of elements. The actions required by the law should be of a kind which men can reasonably be expected to do and to avoid. Like cases should be treated alike. Laws should be open, clear, and prospective.[41] Natural justice is a further such precept, designed to ensure that the legal order will be impartially and regularly maintained.[42]

A further rationale for due process views it as a condition for the moral acceptability of those institutions that allow some people to control the lives of others. Procedural due process is seen as playing a role in protecting human dignity by ensuring that the individual is told why he is being treated unfavourably, and by enabling him to take part in that decision. In serving these ends, the system is thereby responsive to demands for participation and revelation.[43] At its most fundamental, a right to be heard is regarded as integral to being a person. The very horror of the Kafkaesque world resides in the fact that the individual is caught up in a seemingly inexorable train of events over which he has no control and for which there is no explanation. The dehumanizing quality of the experience rapidly becomes evident.

Many reasons have been given for advocating political participation in contexts other than adjudication. Underlying these variants lie two more general theories, one of which has been termed instrumental, the other developmental.[44] Instrumental theories, as the nomenclature would suggest, regard political participation as a mechanism for achieving a particular goal, which may be to prevent tyranny or to counter excessive bureaucracy. Since the individual is regarded as the best judge of his own interests, this implies that he ought to have the chance to exercise his preferences. While practical necessity requires that the actual process of government should be carried out by a limited number, these should be responsive to the wider populace through consultation.[45]

There is a proximate connection between the instrumental theory of participation and the rationale for due process which is based upon a link between process rights and the substantive justice of the final outcome. Participation in non-adjudicative contexts, such as rule-making, may help both to prevent the use of rule-making power in a way which is substantively arbitrary, and also increase the likelihood that the chosen rule will effectuate its avowed purpose.[46] This is not to say that it will always be a success. The same may, however, be true of due process in an adjudicative context. The desired link between due process and the substantive justice of the final outcome may not be evident or may not operate in a particular area.

Developmental theories of participation have a different emphasis from those of an instrumental character. The benefit of participation does not reside wholly in its tendency to prevent arbitrary government. An ability to participate in decision-making is regarded as part of the broader development of the individual, both morally and politically. This theme has considerable pedigree.[47] The developmental theme can be found in the Aristotelian view of politics. The citizen finds his true realization through participation in political life. John Stuart Mill wrote in a similar vein.[48] He extolled the virtues and desirability of participation; only thereby would individuals develop their capacities to the full. The ability to organize one's life would be enhanced, and the very process of participation would be educative by forcing one to take into account the interests of others when deciding upon a course of action. Elements of the developmental approach can also be found in the work of more modern writers.[49] A connection between the developmental view of participation and the dignitarian justification for adjudicative due process can be perceived. Both ideas focus upon the importance of process rights to the individual as an individual. The dignitarian theme approaches the matter in 'defensive' terms: to judge me without hearing my case, or telling me of my alleged wrong, is to treat me as less than a complete human being. The developmental theme picks up the same idea and infuses it with more positive content: by allowing me to participate in public life, in additional to the skeletal right to vote, you will enable me to fulfil my own potential, and will facilitate my full membership of the community.

The two visions of pluralism and the limits of participation

The development of participatory process is not a panacea which will solve all the ills, real and imagined, which beset the exercise of public power in society. Difficult and taxing questions will need to be confronted if the law is to be expanded in this direction. Some of these problems have already been considered in the previous chapter, including the dysfunctional consequences which are said to attend the grant of participatory rights. Reference should be made to that discussion, since the generic nature of the problems there canvassed are relevant to any system which seeks to develop participatory rights.[50] They cannot be ignored in the United Kingdom. The object of this section is, however, to focus upon the way in which the choice between the two differing visions of pluralism outlined earlier constrains and shapes the development of participatory rights. This can be perceived in three distinctive, albeit related, ways.

First, both pluralist models place direct or indirect constraints upon the types of groups which can engage in participatory discourse with the central government. If certain groups within society wish to wield power in a manner regarded as antithetical to the organizing principle of the centre, their power will be curbed. The most obvious manifestation of this is the abolition of certain types of local authorities, and the placing of constraints upon the remainder, which has characterized Conservative policy in the 1980s.[51] Society may be pluralist, but groups which are seen as differing too radically from the philosophy propounded by the centre will be restrained, or even abolished. It should not, however, be assumed that this is a facet only of the more modern market-oriented pluralism. The earlier UK pluralists can be subjected to the same analysis, the difference residing in the types of group power which they would have to dominate in order for the policies of the centre to be fulfilled. The proximate connection perceived by the early UK pluralists between economic and political liberty, and the necessity for governmental intervention to secure the requisite economic foundations for democracy, led to constraints being placed upon a particularly potent form of group power, private property.[52] This could necessitate 'abolition' of the group power in certain areas, as in the context of nationalization; or it could produce constraints upon the way in which such power is used, as exemplified by the control over private property regimes necessary to effectuate ideas such as a right to work. Neither species of pluralism adheres to the unfettered idea that public policy emerges simply as a result of intergroup competition. Both require domination by the centre of power groups which threaten the prescriptive role for the state which the particular model erects.[53]

Secondly, a natural corollary of this theme is that, even where groups are not perceived as so threatening as to require direct control, the extent to which their views will actually shape or modify central policy will itself be constrained by the compatibility of those views with the fundamental policy dictates of the centre. The duty of a public body to give adequate consideration to the views of those who participate in the policy-making process will be considered immediately hereafter. However, neither of the pluralist models in the United Kingdom support the conclusions that any such duty of adequate consideration can be taken to include the *necessity* of modifying fundamental doctrinal positions on which the policy itself is premised. Both models assume that the realizable debating ground for policy alternatives will be bounded by the doctrinal assumptions of the model itself. This is just as true for the early pluralists as it is for their more market-oriented successors.

Thirdly, which of these models we choose to pursue will have significant implications for the specific area within which we espouse participation. The connection may be expressed as follows. Participation is one type of device which we have at our disposal for rendering public power accountable. There are many others, including financial constraints, internal bureaucratic control, judicial review, and political checks. Clearly questions can arise as to how well these different devices fit together in any particular area, and which of them should assume prominence. The 'package' of controls which we devise will be crucial, in part because certain of these tools may not sit easily together, in part because we may wish to know which should be regarded as of prime importance. The answer which we give will depend essentially upon the purpose which we believe regulation within a particular area serves. If, for example, we believe that the prime object of social welfare administration is bureaucratic rationality,[54] meaning the most efficient mechanisms for distinguishing true and false claims to welfare benefit, then we would concentrate upon matters such as internal agency controls, training of personnel, hierarchical checks within the organization, etc. Process rights would not be of central importance, and could hinder the realization of the basic aim. If, by way of contrast, we conceive that the principal objective of welfare administration is the settlement of 'entitlements' in a manner analogous to the settling of traditional common-law rights, then our perspective may well be very different. Process rights and the ordinary courts would assume much greater prominence.

The connection between this reasoning and the two differing views of pluralism becomes clear. Precisely what the background aim of regulation within a particular area is, and which areas should be directly regulated by the government at all, are both issues which are affected by the underlying political vision which one chooses to adopt. The appropriate role of participation within any single area can only be adequately determined when answers to these questions have been provided. Privatization is, for example, espoused partly because the market is felt to be more efficient in managing industries. Even a privatized industry may, of course, require regulation,[55] but the role of participation within such a regulatory framework will be seen as part of the general underlying aim of ensuring efficiency through the market. This will in itself shape who is entitled to be represented and the importance of participation when compared with other devices to ensure efficiency, such as management training, aggressive marketing, and financial oversight by the regulatory agency. Participatory rights will, when given, be accorded to 'consumers' of the activity and will be justified in these market terms. The objective of giving participatory rights was, for the early pluralists, broader. It was to enable the individual to participate in the process of government, and to foster the full development of the individual in society.

Notes

1. A public-choice theorist might argue that, as legislation is a bargain, the courts should not intervene to procure terms for one party which that party was not strong enough to secure for himself. This conclusion is dependent upon accepting the premises of this theory, and on agreeing that a particular statute should be viewed as a bargain rather than being designed to serve the public interest. See above, ch. 4 sec. 1 [see *Public Law and Democracy*, Ch. 4].
2. See, e.g., H. Parris, *Constitutional Bureaucracy: The Development of British Central Administration since the Eighteenth Century* (1969), 193; P.P. Craig, *Administrative Law* (2nd ed., 1989), pp. 188–95
3. A further argument against the view that the absence of a statutory requirement to consult should be taken to be conclusive is that it is not easy to

reconcile this with the general law on natural justice. One common explanation for the requirement of natural justice is that the court is supplying the omission of the legislature. If this is the justification for the imposition of procedural protection in cases concerned with adjudication, it becomes more difficult to argue that procedural due process should not apply to rule-making unless the legislature has explicitly so enacted. One might contend that there are essential differences between rule-making and adjudication which warrant procedural process in the one instance, but not the other. This is, however, a separate argument, and one which would have to be assessed in its own terms. See below, ch. 6 sec. 2d, for an exploration of this theme [see *Public Law and Democracy*, Ch. 6].

4. J. Beatson, 'Legislative Control of Administrative Rulemaking: Lessons from the British Experience' 12 *Corn. I.L.J.* 199 (1979); A. Beith, 'Prayers Unanswered: A Jaundiced View of the Parliamentary Scrutiny of Statutory Instruments' (1981) 34 *Parliamentary Affairs*, 165; Craig, *Administrative Law*, pp. 179–82; R. Baldwin and J. Houghton, 'Circular Arguments: The Status and Legitimacy of Administrative Rules' [1986] *P.L.* 239; G. Ganz, *Quasi-Legislation: Recent Developments in Secondary Legislation* (1987).

5. Craig, *Administrative Law* (n. 2 above), pp. 178–9.

6. See above, ch. 3 sec. 2c, ch. 4 sec. 7 [see *Public Law and Democracy*, Chs. 3, 4].

7. Sir D. Wass, *Government and the Governed* (1984), 104–5.

8. Wass believes that the political influence of such groups can only be checked by more, not less, popular involvement in decision-making; *ibid.* 106–8. See also J. Vickers and G. Yarrow, *Privatization: An Economic Analysis* (1988), for the influence of powerful firms on the privatization process.

9. See above, ch. 3 sec. 2c, ch. 5 secs. 2, 3 [see *Public Law and Democracy*, Chs. 3, 5].

10. e.g. H. J. Laski, *A Grammar of Politics* (4th edn., 1938), 80–2, 133, 246, 270, 324–6, 380–4; E. Barker, *Reflections on Government* (1942), 225–6, 250–1.

11. J. P. W. B. McAuslan and J. McEldowney (eds), *Law, Legitimacy and the Constitution* (1985); J. L. Jowell and D. Oliver (eds), *The Changing Constitution* (2nd edn., 1989); I. Harden and N. Lewis, *The Noble Lie: The British Constitution and the Rule of Law* (1986).

12. See above, ch. 2 sec. 3 [see *Public Law and Democracy*, Ch. 2].

13. e.g. objectors to new motorways who have attempted to question the need for these roads have been prevented from doing so on the hypothesis that policy questions will have been decided by Parliament even though the Commons may not have debated the policy; and see the unitary conception of contractual capacity contained in *Town Investments Ltd.* v. *Department of the Environment* [1978] A.C. 359. See also *Nottinghamshire County Council* v. *Secretary of State for the Environment* [1986] A.C. 40.

14. e.g. *Laker Airways Ltd.* v. *Department of Transport* [1977] Q.B. 643, and the criticism of the reasoning therein by R. Baldwin, 'A British Independent Regulatory Agency and the "Skytrain" Decision' [1978] *P.L.* 57. On a more

general level, the presupposition that Parliament either initiates or controls all that emerges as legislation goes some way to explain the absence of any developed legal conceptual structure within which to consider quangos; Craig, *Administrative Law* (n. 2 above), pp. 90, 92.

15. See above, ch. 2 sec. 4c [see *Public Law and Democracy*, Ch. 2]. For modern treatment, see, e.g., S. Walkland and M. Ryle (eds), *The Commons in the Seventies* (1977); N. Johnson, *In Search of the Constitution: Reflections on State and Society in Britain* (1977); S. Walkland (ed.), *The House of Commons in the Twentieth Century* (1979); M. Beloff and G. Peele, *The Government of the United Kingdom: Political Authority in a Changing Society* (2nd edn., 1985); H. Drucker, P. Dunleavy, A. Gamble, and G. Peele (eds), *Developments in British Politics 2* (1986).

16. J. Lucas, *Democracy and Participation* (1976).

17. D. Roberts, *Victorian Origins of the British Welfare State* (1960); Parris, *Constitutional Bureaucracy*; D. Fraser, *Evolution of the British Welfare State* (1973); N. Chester, *The English Administrative System (1780–1870)* (1981); H. Arthurs, *Without the Law: Administrative Justice and Legal Pluralism in Nineteenth Century England* (1985).

18. For general discussion, see *Report on Non-Departmental Public Bodies*, Cmnd. 7797 (1980); Hague, Mackenzie, and Barker (eds.), *Public Policy and Private Interests: The Institutions of Compromise*; A. Barker (ed.), *Quangos in Britain* (1982); R. Baldwin and C. McCrudden, *Regulation and Public Law* (1987). For a more radical explanation than those set out in the text, see, e.g., A. Cawson and P. Saunders, 'Corporatism, Competitive Politics and Class Struggle', in R. King (ed.), *Capital and Politics* (1983).

19. R. Baldwin, 'The Next Steps: Ministerial Responsibility and Government by Agency' (1988) 51 *M.L.R.* 622.

20. B. L. Smith and D. C. Hague (eds.), *The Dilemma of Accountability in Modern Government* (1971); C. Turpin, *Government Contracts* (1972); J. L. Jowell, 'Bargaining in Development Control' (1977) *J.P.L.* 414, and 'Limits of Law in Urban Planning' (1977) *C.L.P.* 63; T. C. Daintith, 'Regulation by Contract: The New Prerogative' (1979) *C.L.P.* 41.

21. Participation within such bodies can give rise to problems of patronage: see N. Johnson, 'Editorial: Quangos and the Structure of British Government' (1979) 57 *Pub. Adm.* 379; A. Davies, 'Patronage and Quasi-Government: Some Proposals for Reform', in A. Barker (ed.), *Quangos in Britain*, ch. 10.

22. Some control may be exercised e.g., through financial checks, and/or through the provision of guidelines in the empowering legislation. This does not affect the point being made in the text.

23. See Johnson, 'Editorial', and D. Keeling, 'Beyond Ministerial Departments: Mapping the Administrative Terrain: Quasi-Governmental Agencies' (1976) 54 *Pub. Adm.* 161, for a discussion of the differing meanings of control and accountability.

24. Johnson, 'Editorial'; N. Johnson, 'Accountability, Control and Complexity:

Moving beyond Ministerial Responsibility', in A. Barker (ed.), *Quangos in Britain*, ch. 12. For more general comment, see G. Drewry (ed.), *The New Select Committees: A Study of the 1979 Reforms* (1985).

25. A major alteration of our voting system, such as the introduction of proportional representation, could have some effect upon executive–legislative relationships. It would be likely to weaken the dominance of the single party in power because that party would often have to rely on the support of other smaller groupings to retain its majority.

26. The Rules Publication Act 1893, now repealed, is of interest in this respect. The Act imposed a duty on those making statutory rules (defined in s. 1(4)) to publish the rules in forty days before finalizing the rule. Public bodies, which were not defined by the legislation, were then able to make representations to the rule-making authority. The legislation primarily applied to those rules which were required to be laid before Parliament, but which were not actually subject to real parliamentary scrutiny, because they were operative immediately. Consultation with at least some interested parties was, therefore, seen to be a way of validating and controlling the enactment of such rules in the absence of parliamentary supervision.

27. See above, ch. 4 sec. 7b [see *Public Law and Democracy*, Ch. 4].

28. G. Parry, 'The Idea of Political Participation', in G. Parry (ed.), *Participation in Politics* (1972), ch. 1.

29. What voting system should be preferred is itself a complex question; see, e.g., S. E. Finer (ed.), *Adversary Politics and Electoral Reform* (1975); V. Bogdanor, *The People and the Party System* (1981); P. Dunleavy and C. T. Husbands, *British Democracy at the Crossroads: Voting and Party Competition in the 1980s* (1985).

30. See Parry, 'The Idea of Political Participation'; R. L. Nuttall, E. K. Scheuch, and C. Gordon, 'On the Structure of Influence', in T. N. Clark (ed.), *Community Structure and Decision Making: Comparative Analysis* (1968), 349–80.

31. For protection of participation through adjudication, see Craig, *Administrative Law* (n. 2 above), ch. 7; for protection through the vote, see H. Rawlings, *Law and the Electoral Process* (1988). This is not to claim that such protection is perfect.

32. This is subject to the qualifications contained in the more recent case law on legitimate expectations considered above, ch. 6 sec. 2a [see *Public Law and Democracy*, Ch. 6].

33. L. H. Tribe, 'The Puzzling Persistence of Process-Based Constitutional Theories', 89 *Yale L.J.* 1063 (1980). For discussion of this point in the US context, see above ch. 4 sec. 3b [see *Public Law and Democracy*, Ch. 4].

34. In general, participatory rights in rule-making in the United States are the creation of statute, see S. G. Breyer and R. B. Stewart, *Administrative Law and Regulatory Policy: Problems, Text and Cases* (2nd edn., 1985), ch. 5. The Constitution does not normally require due process in such cases: *Bi-Metallic Investment Co.* v. *State Board of Equalization* 239 U.S. 441 (1915), and compare to *Londoner* v. *Denver* 210 U.S. 373 (1908). However, some courts have read

certain regulatory statutes broadly, and have maximized the opportunity for hearing rights in legislative type cases: see, e.g., *The National Welfare Rights Organization v. Finch* 429 F. 2d 725 (D.C. Cir. 1970); *Environmental Defense Fund, Inc. v. Ruckelshaus* 439 F. 2d 584 (D.C. Cir. 1971). The reasoning in such cases comes close to the courts supplying the omission of the legislature; see C. R. Sunstein, 'Interest Groups in American Public Law', 38 *Stan. L. Rev.* 29, 67–8 (1985). For more general discussion, see above, ch. 4 sec. 7a [see *Public Law and Democracy*, Ch. 4].

35. A. Chayes, 'The Role of the Judge in Public Law Litigation', 89 *Harv. L. Rev.* 1281 (1976). The multipartite implications of adjudicatory decisions can emerge, directly or indirectly, within varying stages of administrative law; see, e.g., the way in which they arise in the choice between rule-making and adjudication as a method of advancing agency policy; Breyer and Stewart, *Administrative Law and Regulatory Policy* (n. 34 above), pp. 397–415, 466–90.

36. See, generally, D. J. Galligan, *Discretionary Powers: A Legal Study of Official Discretion* (1986).

37. Craig, *Administrative Law* (n. 2 above), pp. 373–4; J. Vining, *Legal Identity* (1978), 20–5.

38. See, e.g., D. Resnick, 'Due Process and Procedural Justice', in J. R. Pennock and J. W. Chapman (eds.), *Due Process: Nomos 18* (1977), 217; Resnick accepts that there are non-instrumental justifications for due process, pp. 217–18.

39. H. L. A. Hart, *The Concept of Law* (1961), 156, 202.

40. J. Rawls, *A Theory of Justice* (Oxford, 1973), 235.

41. J. Lucas, *The Principles of Politics* (1966), 106–23; J. Raz, 'The Rule of Law and its Virtue' (1977) 93 *L.Q.R.* 195.

42. Rawls, *A Theory of Justice* (n. 40 above), pp. 238–9.

43. F. I. Michelman, 'Formal and Associational Aims in Procedural Due Process', in Pennock and Chapman (eds. , *Due Process: Nomos 18* (n. 38 above), ch. 4; R. B. Saphire, 'Specifying Due Process Values: Toward a More Responsive Approach to Procedural Protection', 127 *U. Pa. L. Rev.* 111 (1978); J. L. Mashaw, *Due Process in the Administrative State* (1985), chs. 4–7.

44. Parry, 'The Idea of Political Participation' (n. 28 above).

45. An instrumental perspective does not, however, dictate one particular form of political society; *ibid.* 22–6.

46. Even behaviouralists can advocate participation as a method of ensuring that élites remain responsive to a wider populace; e.g. L. W. Milbrath, *Political Participation: How and Why do People get Involved in Politics?* (1965), 144.

47. See below, ch. 10 [see *Public Law and Democracy*, Ch. 10].

48. J. S. Mill, *Utilitarianism, On Liberty, and Considerations on Representative Government*, ed. H. B. Acton (1972), 208–18.

49. See the discussion of, e.g., Bachrach, Pateman, Macpherson, and Bottomore, above, ch. 3 sec. 2c, d [see *Public Law and Democracy*, Ch. 3]. See also Lucas, *Principles of Politics* (n. 41 above), pp. 268–9.

50. See above, ch. 4 sec. 7b [see *Public Law and Democracy*, Ch. 4].

51. See, e.g., M. Loughlin, M. D. Gelfand, and K. Young (eds.), *Half a Century of Municipal Decline, 1935–1985* (1985); M. Loughlin, *Local Government in the Modern State* (1986).
52. e.g., the treatment of property rights by Laski; see below, ch. 6 sec. 3 [see *Public Law and Democracy*, Ch. 6].
53. Whether constraints placed upon local democracy are 'worse' than those placed upon private property is itself dependent upon a background political theory.
54. The example is drawn from J. L. Mashaw, *Bureaucratic Justice* (1983). The point is, however, generalizable to other areas; see P. P. Craig, 'Discretionary Power in Modern Administration', in M. Bullinger (ed.), *Verwaltungsermessen im modernen Staat* (1986), 79–111.
55. See, e.g., T. Prosser, *Nationalised Industries and Public Control* (1986); Vickers and Yarrow, *Privatization* (n. 8 above). See, more generally, S. G. Breyer, *Regulation and its Reform* (1982).

INDIVIDUALIZED DECISIONS AND PROCESSES

Using Legal Discretion

K. HAWKINS

Understanding decision-making

Social scientists have not generally been concerned with decision-making in specifically legal contexts until quite recently. Much of the work on decision-making in general has been conducted from the perspectives of economics or psychology and concerned with problems in business, management, and public policy. Organization theory has produced much of the best-known writing; this has focused on decision-making within organizations as a form of administrative behaviour (H. A. Simon 1947). In the last twenty-five years, however, important work concerned specifically with legal decision-making has appeared. Some of it has been carried out by criminologists, some by sociologists or (in the United States in particular) by political scientists whose work is heavily influenced by sociology. Much of the sociological research has attempted to advance an understanding particularly of the negotiated decisions that are frequently made in various parts of the legal system outside the courtroom in settings which lend themselves to ethnographic work in the anthropological tradition.[1] There now seems to be a growing interest in the collective or organizational aspects of discretion, which are emphasized in Part II of this book.

While various approaches to the social scientific study of decision-making are evident in the literature, it is possible to discern two broad contrasts. One distinctive approach tends to be normative and starts from the premiss that decision-making is a fundamentally rational matter. The other is descriptive and concerned with understanding the natural processes of decision-making.[2]

Rational decision theory and its variants

Much early social science work on decision-making was marked by various assumptions which led to a particular emphasis on the notion of rationality. Rational-choice theories of decision are concerned with the effective attainment of a particular objective or set of objectives and give rise to explanations that have an individualistic focus and presuppose a very orderly view of human behaviour. The approach has been extremely influential in areas as diverse as industrial and governmental decision-making, on the one hand (C. R. Miller 1990), and analysis of decision-making in criminal justice, on the other (see, e.g., Gottfredson and Gottfredson 1980).

Rational decision theory, derived originally from work in microeconomics or statistical decision theory, is premised on the fundamental assumption that decisions are purposive choices made by informed, disinterested, and calculating actors working with a clear set of individual or organizational goals. Decision-making, in this view, is intentional and consequential activity carried out by a rational individual. Another of the guiding assumptions of rational-choice models is that decision-makers survey the likely outcomes of various courses of action that are possible and under consideration. Furthermore, consequences are assumed to be capable of being fully anticipated, and the decision-maker is then assumed to choose the alternative that promises most closely to attain the objectives of the decision. Once objectives are known (or reasonably inferred), such a model permits predictions to be made of the ways in which other decision-makers will choose. Decision-making is reduced here to a 'technical calculus' (C. R. Miller 1990: 165). In short, decision-making is seen as

intentional, consequential, and optimizing. That is, [it is assumed] that decisions are based on preferences (e.g., wants, needs, values, goals, interests, subjective utilities) and expectations about outcomes associated with different alternative actions. And [it is assumed] that the best possible alternative (in terms of its consequences for a decision-maker's preferences) is chosen. (March, 1998: 1–2)

Choice implies a set of alternatives. Under classical theories of rational choice it was assumed that the decision process involved a costless search among alternatives, irrespective of the limits on the human capacity to process large amounts of information. In his celebrated book *Administrative Behavior*, however, Herbert A. Simon (1947) modified the idea of rational choice, suggesting that rationality was in fact bounded, that people do not act in a purely rational way in making decisions. Bounded rationality recognizes limits on decision-makers' knowledge and ability: people do not have unlimited resources and capacity to search for and process information, but are assumed to work with only a limited number of logically possible alternatives. As a result, Simon argued, decision-makers would typically search among alternative decisions only until a satisfactory alternative, likely to be influenced by past decisions and practices, presented itself. The decision-maker 'satisfices', that is, takes the first satisfactory solution encountered. Bounded rationality, however, remains firmly rooted in the tradition of classical rationality, since 'it is simply a "rational" (that is efficient) adaptation to a set of empirical limitations' (C. R. Miller 1990: 167).

Rational decision theory has been influential in a number of the social science analyses of legal decision-making, especially those by criminologists concerned with criminal justice decisions (e.g. Gottfredson and Gottfredson 1980). For example, Leslie Wilkins defines as a rational decision 'that decision among those possible for the decision-maker which, in the light of the information available, maximizes the probability of the achievement of the purpose of the decision-maker in that specific and particular case' (1975: 70). In research terms the 'decision' is treated as a choice made by an individual based upon unproblematic 'factors' or 'criteria' which may be deduced from the creation of a putative relationship between observed input (various kinds of information) to a decision-maker and observed output (decision outcome). It seems to be assumed that an item of information revealed to be statistically related to output represents a matter taken into account by the decision-maker, and it becomes a 'factor' held to 'explain' the decision. The validity of this method resides in its capacity to predict the outcomes of further decisions, but it assumes an empirical association between a fact and an outcome and the decision-maker's cognitive awareness of such a fact. Furthermore, an ability to predict an outcome need not explain how that outcome was reached: there is no explanation or understanding of how input is translated into output.

The conceptual and methodological stances of this work reveal various preconceptions about the behaviour under study which are empirically questionable. Official objectives of the particular part of the legal system under study, or what decision-makers claim they are trying to do, are taken as given. Though rational-choice approaches tend to look for their explanations for legal decisions in formal policy, rules, criteria, or procedures, these cannot automatically be assumed to have guided the decision-maker to a particular outcome. It is often difficult, indeed, to conceive of a legal decision or a legal institution as having a single or clear set of purposes; purposes are often confused, unknown, or disputed (Hawkins 1986: 1181 ff.). Furthermore, decisions are seen as the work of autonomous individuals, exercising a discretion unconstrained by colleagues or context. They are treated as simple, discrete and unproblematic, and not relatively complex, subtle, and part of, or the culmination of, a process, in which external constraints, such as organizational and occupational rules, norms, procedures, and resources also operate. Indeed, decisions may well appear to be simple, discrete matters because the structure of the legal process requires them to be presented, described . . . [or] sent forward for consideration in that form. (*ibid.*: 1187).

Rationalist work, with its emphasis on outcome rather than process, draws attention to discrepancy rather than regularity. And discrepancy, if made apparent, is regarded as pejorative, rather than as something intrinsically interesting.

Another difficulty particularly relevant to rational-choice explanations (though it bedevils other approaches also) is what might be termed the problem of recognition. How can we recognize *that* a decision has been made? And how can we know *how* a decision has been made (Manning 1992; 1986; Hawkins and Manning, forthcoming)? Peter Manning points out that observed or assumed decision outcomes may be produced in any number of ways, but the process of choice, and the nature of that choice, have to be inferred following what is taken to be a decision. If we want to know how the power to choose was exercised, we are accordingly reliant upon the accounts or the justifications of the decision-makers (Manning 1986). A real 'decision' must be distinguished, for instance, from the ratification of an earlier choice, or a conscious accounting for or rationalization of the decision in terms regarded as more legitimate by the decision-maker.

The usefulness of rational-choice theory in understanding legal discretion varies. Some individualistic decision-making behaviour by legal actors may be explicable in rational-choice terms, as when, for example, lawyers or claimants have to decide whether to settle out of court or litigate in personal injury claims (Ross 1970; Genn 1987). In this case the decision focus is particularly narrow and the lawyer's or claimant's objectives are presumably clear (getting the highest figure by way of compensation); nor is there much doubt also about the commercial objective of the defendant's insurance company, namely to incur as little cost as possible (see, generally, Phillips and Hawkins 1976). Other kinds of legal decision, however, tend to show up the explanatory limitations of rational choice models of decision-making. The clear set of decision goals and the well-informed actor they presuppose are conditions that probably obtain in relatively few settings.

Moreover, as Part II of this book shows [see *The Uses of Discretion*], rational-choice theories are premised upon an individualistic conception of decision-making which does not often accord with reality. Discretion is often a collective enterprise. But legal actors may not have a clear set of objectives, and these may not be consistently held by disparate individuals. Decision-makers do not necessarily make the choice that is most likely to attain the formal—ostensible—objectives of the decision; they may frequently have regard to the possibly conflicting interests of others. Within legal bureaucracies, for example, it is not safe to assume that all members of an organization consistently hold the same conception of what their legal mandate is (even assuming their conception is clear), or consistently seek to attain the same goals. The logic of rational decision theory is to assume that decision-makers' conception of their interests are congruent with the interests of their organization, and that their decisions are not informed by matters of self-interest, economics, politics, and the like.[3]

A naturalist perspective

Rational decision theory has been challenged by studies which have stressed the natural processes by which decisions are made. Naturalism relies on descriptive analysis, and seeks to understand how decisions are actually arrived at unencumbered by any particular assumptions about what the nature of the behaviour is, or should be. This view is associated in the socio-legal field with sociologists in the traditions of interactionism or ethnomethodology such as Aaron Cicourel (1968) and Robert Emerson (1969, 1983). Such writers have enhanced our understanding of discretion in both an exterior sense, by showing how an appreciation of context and pattern is valuable in extending the focus beyond the individual case, and also in an interior sense, by exploring the significance of meaning to individual legal actors who must choose.

Naturalism questions the goal-directed conception of classical rationality. On the contrary, decision goals are seen as the consequences of efforts to explain or rationalize action which may be the product of routine or the influence of other actors, rather than as some overarching set of objectives which first informed the choice that came to be made. Naturalism suggests that actions are not necessarily the product of intention, or of conscious choice or planning, even though decision outcomes may to some extent be predictable. Actions are not seen as necessarily the result of the exercise of conscious choice. Instead, the notions of context and meaning are central to naturalist views of decision-making.

The approach stresses the *ad hoc*, particularistic nature of legal decision-making, the decision routines employed, and the relatively haphazard way in which information may be used in the process. Work in the naturalist tradition draws attention to the moral and symbolic content of decisions and the systems of belief and meaning held by decision-makers. This sort of work is concerned with such matters as the ways in which decision-makers make sense of their decision task, or the possible consequences of various courses of action, their adaptive behaviour, and the contexts in which they choose. It addresses how information may acquire meaning and relevance depending on the way it is framed and made sense of, an approach characterized by March's dictum that 'Interpretation, not choice, is what is distinctively human'. (1988: 15)

The use of the notion of frame which, when applied to interpretation, refers to the structure of knowledge, experience, values, and meanings which a decision-maker uses to make sense of the decision problem and the available information (see Hawkins 1986; Manning and Hawkins 1990) is particularly relevant here. Framing is a way of connecting criteria with outcome.[4] Naturalism is not an individualistic conception; rather it emphasizes a holistic view of discretion and decision-making as a collective process. In this it contrasts with the normative and instrumentalist tendencies in rationalist writing. Where writers in the rational tradition (e.g. Gottfredson and Gottfredson 1980) tend to be concerned with the substance of decisions ('factors' or 'criteria'), naturalism attends to the processes by which decisions are made and their contexts: organizational, social, political, or economic.[5] At the same time, naturalism also draws attention to an individual's or organization's need for survival. An organization is seen as responsive and adaptive to its environment, while its members are viewed as people with their own values, needs, expectations, and agendas, rather than as dispassionate individuals all working coherently together to achieve a set of formal organizational goals. Decisions are made, but do not necessarily reflect some conception of the formal aims of an organization so much as the interests of individuals in maintaining their own position. People both anticipate and adapt. They follow rules, but they also create rules, norms, patterns of behaving. They make decisions in ways that are situatedly rational, that is, rational in a particular context. On this view, there are not necessarily any broad, clear, taken-for-granted organizational or other goals whose attainment is sought through choice. Instead, decisions are seen very much as embedded in their own particular contexts, as the response of a decision-maker to a particular set of circumstances. In general, decisions are taken and action occurs, but these things do not happen as a result of conscious planning or choice (see, generally, Weick 1979). The remaining discussion in this chapter is written from the point of view of naturalism.

Elements of a holistic perspective

Case and policy decisions

If we wish to understand the nature of legal discretion empirically, it is important to take a systemic or holistic view of it. The individual case is usually taken as the unit of analysis in most social science studies of legal decision-making, though what is usually more appropriate, as the following section argues, is a broader focus which, *inter alia*, looks at the careers of cases within a decision-making system, and at interactions between cases or groups of cases. The concern for the individual case is not surprising, since legal actors and legal bureaucracies typically assemble problems for decision into individual 'cases', each relating to a particular and concrete matter and consisting of physically segregated units by which they may be handled.[6]

Even so, individual case determinations are themselves more complex than they might seem, even within an ostensible single decision point. As Lempert shows (1992), what may on the surface appear to be one simple discretionary decision quite often involves a rather more complex series of decisions. A legal decision may require a judgment first as to the nature of the problem for decision (questions concerned with 'what actually happened', or 'what the present position is', and so on), then whether the problem or event is addressed by, or constitutes a breach of, a rule. If a breach is found, or a rule applies, further decisions may be needed as to what action should be taken, and, if so, what precise action is required. Subsequent decisions are thus contingent upon earlier ones, suggesting that analytically it is possible to distinguish 'core' and subsequent 'contingent' discretion.

Discretion is also often regarded as a feature of decision-making by individuals. This, too, is not surprising, since individuals or panels of individuals are often allocated formal authority to make legal decisions. Decision-making in law is, however, to a greater extent than is apparent from much of the literature, a collective enterprise. Indeed, it is hard in reality to sustain the idea of the individual actor exercising discretion according to legal rules or standards alone, unencumbered by the decisions or influences of others. Some legal decisions are explicitly designed to involve groups, like boards, tribunals, and juries. Where a discretionary outcome is the product of a number of decision-makers acting together, discretion is exercised, as it were, in parallel. In the event that differences between individual decision-makers cannot be reflected in a majority vote, they have to be negotiated into an outcome that can be presented as the group's decision. In resolving individual differences, expertise, experience, status, and personal charisma are important matters shaping discretion, since they confer an interpersonal authority to have cases decided in particular ways (see, further, Hawkins and Manning, forthcoming).

A focus on the individual case and the activities of individual decision-makers often does not portray the reality of legal discretion in another way. A good deal of decision-making in legal bureaucracies is concerned with matters of policy—deciding in general how to decide in specific cases. Policy-making involves making decisions about the objectives and meaning of the law, and about how these ideas are to be shaped into strategies to permit their implementation. This is 'the very heart of the discretionary process' (Galligan 1986a: 110). Policy is the means by which discretion is at once shaped and transferred down through an organizational hierarchy. Policy decisions speak abstractly to the future in varying degrees of generality and exist in the form of a series of statements often incorporating matters such as the objectives of decisions, criteria to be taken into account, information to be used, and procedures to be followed. Policy, like legal rules, acts therefore as one of the constraints in the context or field within which individual decisions have to be made (see Manning 1992).

A serial view of discretion

An argument was made above for a view of discretion as part of a sequence of decisions and occurring as part of a network of relationships in the legal system (see Emerson 1983; Emerson and Paley (1992)). Substantial power is wielded by those making earlier determinations in the handling of cases, for discretion is exercised not only in parallel in legal systems, but also in series. A decision made at one point in the system may profoundly affect the way in which a subsequent decision is made, owing to the structural position of the individual at the point at which prior discretion is exercised (see Lempert (1992)). It makes sense to see many legal decisions as comprised of a number of discretionary determinations following in sequence. Cases are processed over time by means of a referral system: the creation of any legal case and its subsequent career are shaped by decisions made in a dynamic, unfolding process. Once created, individual cases in the legal system are typically handed on from one decision-maker to another until they are resolved, discarded, or otherwise disposed of.

A serial perspective also draws attention to the fact that effective power to decide is frequently assumed by actors other than the person allocated formal authority to exercise discretion. What is described as a 'decision' reached is sometimes nothing more than a ratification of an earlier decision made in the handling of a case, even though that prior decision may appear in the guise of an opinion or a recommendation. The nature of a discretionary determination may change or be changed depending on where in the legal system authority to decide is located. In such circumstances, matters such as the flow of information from one point in the system to another become particularly important (see Emerson and Paley (1992), and Manning (1992)). Since discretion is diffused among those supplying information, evaluations, and recommendations to the proximate or ultimate decision-makers, it is important to distinguish the real exercise of discretion from mere ratification. Some people who supply information or assessment may have such an enormous influence on the subsequent handling of a case that it becomes difficult to conceive of the visible, official point of decision being the place at which real discretion was exercised.

The diffusion of discretion means that decision-making power is dispersed in legal systems.[7] Power resides, *inter alia*, in the capacity of decision-makers to drop or divert cases. The primary concerns of decision-makers are often shaped by a concern to handle and manage a stream of cases seen in organizational context (Emerson 1983). Earlier decisions may serve to close off the scope of discretion afforded to subsequent decision-makers entirely or partially, either by excluding a subsequent decision-maker (by discarding the case) or by narrowing that decision-maker's range of choice. For instance, legal actors handling cases prior to trial take advantage of their structural position in various ways. The low visibility of many legal decisions (like arrest of disrespectful teenagers by the police), and the degree of credibility accorded the source of information used in formal decisions, are two illustrations (see Lempert (1992). Officials have their own views on the merits of particular cases and are often able to dispose of them informally in ways that accord with their sense of justice, or that allow them to make or honour bargains over other cases. This is largely how plea-bargaining works in the United States (e.g. Heumann 1978; Feeley 1979). Again, in trials or hearings, advocates can exert considerable control over what is and (equally importantly) what is not put before the adjudicator. On the other hand, Lempert (1992) shows how decisions can be made contrary to a legal mandate because no opportunity is provided for review in the legal structure.

Another form of power resides in opportunities afforded by the legal system to those who create, assemble, or supply material relevant to a decision to formal decision-makers. These people are able, artfully or unwittingly, to frame the contents of reports or other information to give prominence to a particular point of view (see Hawkins 1983 for examples). The use of language in documentary reports may often reveal where the effective source of power or influence in the making of a decision actually resides. Officials may frame how discretion may subsequently be exercised, not only by describing or presenting the case in a particular fashion, but also by making related decisions in a certain way. Thus, for instance, the penalty imposed by the authorities for prison misconduct may well profoundly influence a subsequent determination about the prisoner's release by the parole board (see Hawkins 1986: 1196 ff.). What has happened earlier in the processing of a case, or similar cases, has powerful indications for the present decision, as well as for future ones. Furthermore, from the point of view of the earlier decision-maker, one decision is not made independently, but in a way that takes account of the implications of other cases for the present one and vice versa (see Emerson 1983: 425; Emerson and Paley (1992)).

The matters attended to by decision-makers and the nature of the constraints to which they feel subject may change as time passes and cases move in the sequence of handling decisions. It follows that decision-makers at different points in the handling system might be expected to have different priorities; indeed, there may be not only different sets of resource constraints operating, but also quite different value systems. When discretion is viewed in serial perspective, the nature of the links between different parts of the decision-making system, and how they are bound together, become apparent. The history of a case is especially important because the legal method compels the selective social reconstruction of the past. In recreating history, the method of law is to pare down to remove the uncertainties of the real world 'in the course of successive transformations over time from the original event or act to final adjudication' (Cicourel 1968: 28). What happens, to quote Cicourel's description of the juvenile justice system, is that

Each encounter or written report affects the juvenile and events considered illegal in such a way that the contingencies in which the participants interpret what is going on, the thinking or 'theorizing' employed, are progressively altered or eliminated or reified as the case is reviewed at different levels of the legal process and reaches a hearing or trial stage. (*ibid.*: p. xiv)

One implication for policy here is that, because discretion in effect exists in legal systems in a certain equilibrium which may be disturbed if a rule is changed at one point in an effort to limit or extend discretion, discretionary play in one part of the system may be transposed elsewhere. This suggests that those who would change rules, policies, or procedures must adopt a holistic approach in appraising the legal system and seek to anticipate more effectively the precise impact of different structures and forms of rules on discretionary behaviour.

Adjudication and negotiation

Decisions are produced in legal systems in a number of ways (see, generally, Galanter 1986), of which adjudication and negotiation are two core forms in three-party or two-party settings.[8] Those who are interested in what courts do tend, accordingly, to think about discretion in terms of adjudicative decision-making, while those who are interested in how law works are more concerned with the negotiation that characterizes most of the decision-making outside the courtroom, much of which serves to divert cases being litigated from ultimate adjudication. The central position accorded to adjudication and the work of the courts in legal scholarship about the uses of discretion is understandable, given that courts provide authoritative statements of what the law regards as legitimate conduct (see Bell (1992)).

Adjudication is of particular interest to lawyers because, in settings where decisions are made by adjudication, such as courts, rules are the basis by which disputes are resolved. That is, it is only possible to resolve a matter by adjudication on the basis of a specific rule or standard that speaks generally to similar cases (Aubert 1984). Adjudicated decisions are based 'on those aspects identified as salient by applicable legal categories' (Galanter 1986: 158). Adjudication grants to an authoritative, independent individual or group possessed of special forms of knowledge (ibid.) the power to make a decision binding upon the parties to a dispute. It imposes a solution to a problem defined in adversarial terms, that is, in contrasting and partisan ways; it therefore implies argument. Adjudication also implies participation in decision-making by the affected parties through the supply of information and adversarial debate, but such participation does not extend to the exercise of discretion itself. The outcomes of adjudicated decisions tend to be cast in uncompromising binary terms (win-all or lose-all; guilty or not guilty). This, allied with the logic of the surrounding structure of rules, also helps to give adjudicated outcomes some measure or predictability.

Another influence on outcome is the set of values that people bring to the decision-making task (Lempert (1992); see also Asquith 1983), and in this connection who adjudicates is of central importance to the kinds of decision that are made. One way of controlling how discretion is exercised, of course, is in the choice of who decides (a matter well appreciated by connoisseurs of the political skirmishing which inevitably attends proposed appointments to the US Supreme Court).[9] Lempert (1992) shows that changes in the appointments to the housing board he studied took place (with corresponding changes in the decisions reached) when those with power to appoint concluded that some decision-makers were too pro-tenant. The law itself may state some preferences in terms of its formal requirements for the sort of people who are to be appointed to adjudicate (see Schneider (1992)).[10]

The focus of concern in almost all of the chapters in the present volume is not, however, the judge, the usual object of attention in many legal analyses of discretion, but those many other people who do not work in the courts. Matters are more complex outside the courtroom. Decisions are typically made by means of negotiation by officials who create, handle, and process cases, who often have a rather unspecific legal mandate, and whose activities are relatively loosely circumscribed by rules. Indeed, rules which act outside the courts tend to be framed in such a way as to foster the use of discretion to attain policy objectives (see Bell (1992)). In bureaucracies, furthermore, discretion tends to be squeezed out to, or effectively assumed by, the periphery, where it may be exercised largely invisibly and immune from organizational control.

A very large proportion of legal or potentially legal cases are disposed of by negotiation. In many others negotiation precedes adjudication. When outcomes are reached by negotiation, discretion assumes a different character (see Handler (1992)). While the rule structure may be much looser, people usually negotiate nevertheless within a broad framework of rules, or with a mandate conferred on them by law (see, generally, Hawkins 1984). Negotiation, however, is particularistic decision-making in the sense that what is usually at issue is the substantive problem in an individual case. The application of a rule in an adjudicated decision, in contrast, does not necessarily speak to the substance of the matter at hand. Negotiation relies upon bargaining, and to that extent any agreed outcome is tolerable to both sides in a dispute (Gulliver 1969: 17). The decision reached is usually a compromise, jointly arrived at, rather than an all-or-nothing adjudicated verdict. Negotiation therefore achieves at least a measure of mutual commitment to the decision reached. The incentive to arrive at a decision is not provided by the compulsion of the adjudicator (though the threat of an adjudicated decision being imposed on the problem is often enough to coerce the parties to negotiate), but the self-interest of both parties in reaching a quick, preferably amicable, and cheap resolution of their problem. Negotiation also implies participation in decision-making, but it does not involve parties arguing before a dispassionate third-party adjudicator who reaches a decision and then imposes it upon the parties. Instead, negotiation takes place in informal settings, to that extent less visible and less suited to the promotion of predictable outcomes, given the less apparent place of legal rules in the context of decision, and the emphasis upon reciprocity. Negotiation is pervasive in both private-law and public-law settings, though in some respects its character may differ as between them. For example, the bargaining that takes place in settlement of a claim (Genn 1987) may differ from the bargaining that may be involved where an official may be surrendering power formally to enforce the law in return for an offer of compliance (Hawkins 1984).

None of this is to suggest, however, that legal rules exert little or no influence upon decisions. Organizations and their members may not always be constrained by rules, but they also use rules as defences, as resources, or as matters about which to negotiate. To the extent that bargaining achieves its ends, it works partly because of constraints provided by cultural and occupational norms, and the like. But it works also because of the existence of legal rules that give rise to stable expectations about how the ultimate decision-makers are likely to act (if a dispute proceeds that far)[11], and the formal machinery that may ultimately be recruited to adjudicate on a dispute. The relationship between rules and discretion is, however, complex.

Discretion in using rules

In reality it is impossible to treat rules and discretion as discrete or opposing entities. Discretion suffuses the interpretation of rules, as well as their application. In thinking about the relationship between rules and discretion, it is important to distinguish between fact-finding and fact-defining decisions, on the one hand, and decisions about action, on the other. Rules themselves have to be defined as to their meaning and relevance. Even where the meaning of a rule seems clear, the facts upon which the application of a rule may depend have always to be interpreted. To claim that one is dispassionately following a rule is to take for granted the interpretative work—the choices—surrounding fact-finding, and to assume that the 'facts' assembled are relevant to the application of a particular rule.

Interpreting a rule involves, at the minimum, discovering its meaning, characterizing the present problem, and judging whether that problem is addressed by the rule.[12] And, even where a rule is granted meaning, there will still be scope for the further exercise of discretion by officials, not only as to its applicability, but also as to the accuracy or genuineness of information relevant to the exercise of discretion. The facts, writes Galligan,

> can be ascertained only by imperfect means, relying on imperfect procedures—the evidence of others, one's own perceptions and understandings, and the classification of those perceptions; also, there are limits to the time that may be spent in the quest for factual accuracy . . . any decision requires assessment and judgment, both in fixing the methods for eliciting the facts and in deciding how much evidence is sufficient. Understood in this special sense, there is some justification for talking of discretion in settling the facts. Similarly, in applying a standard to the facts, the decision-maker has to settle both the meaning of the standard and the characterization of the facts in terms of that meaning. (1986a: 34–5)

The form and complexity of a rule have important implications for the degree of discretion created. Schneider (1992) suggests that the simpler the rule the more likely it is that the principle embodied in it will be adhered to, while the more complex the rule the greater the discretion available to individual decision-makers in its interpretation and application. Similarly, complex systems of rules, though highly specific, may also have the effect of creating greater discretion in practice, as Long's (1981) study of US tax legislation suggests. This recalls Damaska's comment that 'there is a point beyond which increased complexity of law, especially in loosely-ordered normative systems, objectively increases rather than decreases the decision-maker's freedom' (1975: 528). On the other hand, a broad legal mandate, such as that typically granted to regulatory bureaucracies, will give rise to huge areas of administrative discretion. In such circumstances, as Bell (1992) suggests, rules are reference points about which a legal actor may organize the exercise of discretion. Pollution standards operate in this way, for example, when an inspector has to decide whether to act and what action to take when confronted with a discharge in excess of the consented amount (Hawkins 1984).

The particular way in which discretion will be exercised may not be predictable from the particular forms of rule. For instance, where the form of a rule or set of rules is devised to circumscribe or channel discretion, the objective may not actually be achieved. Sometimes the opposite effect will happen. Since discretion is adaptive in character, rules may serve to displace discretion to other sites for decision-making within a legal system, and thereby possibly to enlarge it, or create the conditions for its exercise in more private, less accountable, settings. A telling example of this effect is to be found in the efforts to curtail discretion selectively to release prisoners on parole in California by use of legislatively fixed, presumptive sentences which served to push effective power to dispose of serious criminal cases into the hands of those who engage in pretrial bargaining (Hawkins 1980).

Furthermore, the form in which legal rules are cast may create discretion in such a way as sometimes to divert, or even to subvert, their broad purpose. This latter point can be illustrated with an example from health-and-safety regulation. Factory inspectors often have a choice about which safety law to prosecute. One set of rules may be drafted in specific terms and organized around relatively absolute duties, while others embody general principles and duties. An example of the former is s. 14(i) of the Factories Act 1961, which imposes an absolute duty upon employers in its demand that dangerous machinery be guarded. In contrast is s. 2(i) of the Health and Safety at Work Act 1974, which, for some factory inspectors, has a daunting indeterminacy in the duty it imposes upon employers, 'to ensure, so far as is reasonably practicable . . . health, safety and welfare at work . . .' The qualifying phrase requires a calculation by the inspector of the risks, costs, and technical implications which may affect possible compliance, a matter that will ultimately fall to the court to decide if a prosecution is pursued and the charge is defended. As a result, inspectors are encouraged to prosecute violations of absolute rather than general duties by the easier proof required to show breach of an absolute duty and the correspondingly greater chance of a guilty plea and a successful outcome to the case. This systematic bias, in turn, has important policy implications in terms of the sorts of legal breaches, accidents, or risks that are treated most seriously and are publicly addressed by the courts: in this example the result is that safety matters are given disproportionate attention at the expense of occupational health problems, where deaths may in fact be far more numerous (Hawkins 1989).

Thinking about the relationships of rules and discretion draws attention to tensions, ironies, and contradictions. Rules are valuable to legal actors, not simply because they can offer secure guidance, but because any ambiguity, factual or normative, surrounding them gives leeway for the exercise of discretion, which grants flexibility in their application. Similarly, rules are important resources for legal actors, allowing, for instance, justification for decisions after the fact (Bittner 1967a). Indeed, discretion in general carries with it various functional benefits for legal systems, such as allowing the handling of the gap between rhetoric and reality in the legal system (McBarnet 1981), obscuring lack of consensus or ambiguities in policy (Prosser 1981), or foreclosing the use of costly formal procedures in the law (see Lacey (1992)). Rules devised to attain some general purpose may give rise to conspicuous lack of justice when applied in a particular, concrete case, demanding a decision to mitigate or even to avoid their effects. Yet one of the consequences of a concern for discretion as a problem in socio-legal studies has been to view discretion critically as the reason for the lack of fit between the values and rules of the written law and the practices of legal actors (the 'gap' problem). Legal actors, however, do not necessarily behave in random or even inconsistent ways, but often predictably, though to say this does not deny that apparent inconsistency in decision-making also exists within the system.

Rules in using discretion

To explore the use of discretion empirically reveals an orderly process at work, since discretionary decisions are rarely as unconstrained as they might appear.[13] It is precisely these social constraints that lead to highly patterned outcomes of discretionary decisions in the aggregate and may prompt some to conclude that little or no effective discretion really exists (see Baumgartner (1992)). Patterns in discretionary outcomes provide a marked contrast to the characteristic legal view of the use of discretion as individualized decision-making that is potentially capricious. While lawyers may conceive of a part of a legal system without rules as one of 'absolute discretion' (Raz 1979), it does not make sense from a social scientific point of view to speak of 'absolute' or 'unfettered' discretion, since to do so is to imply that discretion in the real world may be constrained only by legal rules, and to overlook the fact that it is also shaped by political, economic, social, and organizational forces outside the legal structure. It is important not to strip away these contexts since they exert considerable influence. Implicit social rules constrain discretion powerfully.

There are also, of course, many explicit social or organizational rules to assist in the exercise of discretion. Organizations, for instance, have their own routines (see Feldman (1992), and Manning, (1992)). Furthermore, organizations may well shape discretion in ways not necessarily anticipated by law by imposing their own constraints on the way in which their members exercise discretion. If, for example, they introduce decision-making procedures designed to make their staff more accountable, this can lead to adaptive behaviour by individuals to minimize criticism from superiors. The effect of such controls can be to produce decisions which emphasize conformity to regular or expected organizational practices at the expense of an attempt to advance the bureaucracy's broad legal mandate. If organizations seek to make their staff more productive (as with the use of arrest or prosecution quotas by enforcement agencies, for example), individuals may spend their time generating the appropriate indices of output for organizational reward and neglecting other important activities.

Nevertheless, many advantages accrue for individuals from making decisions in compliance with a set of rules, legal or otherwise. In administrative agencies, for example, bureaucratic rules guide officials in making decisions or in providing protection from criticism in difficult cases. Because rules offer guidance, they are an important ingredient in the efficiency with which decisions can be made, since they permit ready repetition of the process of deciding without the need to treat all potentially relevant matters in a new case afresh. This aspect of rule-guided behaviour also leads to the emergence of other, informal, rules of thumb, where legal actors have to deal with a stream of cases presenting similar features. The result is that organizational actors often decide, not on some conception of the merits of a particular case, but according to general, established, rule-governed procedures (see below). Where legal decision-makers impose other constraints and forms of guidance on their own exercise of discretion in circumstances where the law grants them discretion, one consequence may be to persuade them that, from their own internal perspective, they possess less discretion than may, on an external view, be the case (Galligan 1986a: 12–13; see also Lempert (1992)). The following discussion touches on a few of the more prominent occasions for rule-governed behaviour.

Routine and repetitive decisions

Discretion is often exercised in routine and repetitive ways (see Feldman (1992), Emerson and Paley (1992), and Manning, (1992)), and the frequency with which legal actors make decisions has important implications for the nature of their discretion. Those legal actors, however, who make decisions relatively infrequently are likely to approach the matter in a more complex way, taking more time and considering more information.

The experience of having to make regular and repetitive decisions (such as those made by police officers or regulatory inspectors, social security officials, probation officers, or members of tribunals or panels) leads to the development of shorthand ways of classifying cases and appraising each one by focusing on the extent to which it presents typical features. The adoption of a simplified categorical approach supplants individualized consideration; instead, simple rules of thumb dictate what information is relevant and how each case should be dealt with. In the context of legal decision-making, it is clear that discretion is heavily influenced by conceptions of precedent, by understandings of the 'normal ways' of acting and deciding when confronted with certain kinds of problem or case or in certain kinds of situations (see, further, Sudnow 1965; Lloyd-Bostock 1991). Only those cases not regarded as fulfilling simple criteria—those designated as not 'normal'—prompt any special consideration (Sudnow 1965). For example, Emerson and Paley (1992) show how decision-makers often seek to cope with complexity by moulding cases into a binary type: 'good' or 'bad' (or 'serious' or 'light' Mather 1979: 27). When a case has been so categorized, the decision about disposal tends then to be straightforward: a 'good' case is routinely handled in a particular way. However, what Emerson and Paley show is how this apparently simple decision-making process is actually reached, and how complex the processes leading up to this simple categorization can actually be.

Patterns are easily discerned in the use of discretion by legal actors, and are the more readily visible where decisions are made routinely (see Baumgartner (1992)). To the extent that decisions do not involve simplified routines for assessing information and reaching decisions, however, it is possible that they may be less predictable. Shorthand decision methods are employed partly because familiarity with the broad features in typical cases gives decision-makers confidence in being able to see similarities in other cases, and partly in the interest of saving the time and anxiety otherwise involved in addressing each new case afresh. Typification elides legal rules. It is a means of imposing order on a potentially disorderly process that opens the way to the application of organizational and other norms beyond, or instead of, legal norms (matters sometimes regarded as 'extra legal'), to permit ready disposal of a case. Hence typification reaches to a wider range of matters deemed relevant. As decision-makers gain experience, this process of typification becomes simpler and more routine (Rubinstein 1973; see, generally, Rock 1973). Indeed, it seems clear that certain kinds of highly repetitive decisions are made virtually automatically (Lloyd-Bostock 1991).

Repetitive decision-making builds up in the participants a stock of knowledge about the decision-making proclivities of others. Where discretion is exercised by those who know the decision-making behaviour of other actors well, such familiarity allows a high degree of mutual predictability of decision-making. This can make it possible for decision-makers to penetrate or look behind earlier decisions to understand the 'real meaning' of a decision, much as people sometimes 'read between the lines' to understand what is really being said. It also makes practices like plea-bargaining or settling out of court possible, enabling legal actors to make decisions in anticipation of what 'the other side' will do, or decisions in anticipation of what others to whom cases will be handed on will do (see Emerson and Paley (1992)). The repetitive decision-maker who deals with others for whom decision-making about a matter with legal implications is an infrequent or even unique occurrence, however (the 'repeat-player' faced with a 'one-shotter' (Galanter 1974)), is at a considerable advantage, in terms of having decisions made in ways deemed desirable (see Lempert (1992)).[14] The absence of a repeat-player from one side in an adversarial hearing will lead to considerable imbalance in the distribution of power.[15]

Time and precedent

Legal cases have both a history and a future. Discretion may be backward-looking, or reflective, and have a responsive or a reactive character. In other cases, however, it may have a strong anticipatory or predictive character. In fact, it is only because of a rich knowledge of the past that people may predict. Predictive decisions can be pre-emptive or prospective, made in anticipation of what another person or organization is likely to do in response to one's own actions. Emerson and Paley (1992) provide a way of thinking about the organizational handling of cases with their use of the concept of horizon, which draws attention to the different contexts in which cases must be processed in the legal handling system. Their chapter is concerned with discretion as a phenomenon that looks both forward and back, showing how people decide on a course of action by reference both to past events as well as anticipated future events. Decision-makers contemplating court action, for example, have to anticipate questions that might arise concerning the nature or sufficiency of available evidence and decide on strategy accordingly. Similarly, pre-trial decisions about charge or plea are usually informed by knowledge of the sentencing proclivities of judges. Sometimes anticipatory behaviour may be used in an artful way by a decision-maker to attain a particular kind of preferred outcome, as when a clerk allocates certain kinds of case for trial to a judge whose sentencing practices are well known so as to ensure that the case is likely to receive what he or she would regard as the appropriate punishment (Lovegrove 1984).[16]

The past is particularly instructive, not only for judges in commonlaw systems, but also for officials in legal bureaucracies where regular working practices crystallize into organizational precedent. Precedent serves important functions. It grants access to a repertoire of accustomed ways of handling problems, and is another device to make the task of decision-making quicker and easier. Sainsbury (1992) quotes an example of an official who said that, after thirty years, 'you have a pretty good built-in computer telling you what assessments should be'. Where initial cases may be regarded as a marker for the handling of later cases, precedent reveals the relevance of a decision-making sequence for organizations. One decision usually carries implications for others, since decisions create expectations. As Martha Feldman (1992) points out, when people work in organizations, they do so under the weight of a whole series of expectations which others have of how they will make decisions, a matter contributing greatly to the regularities of discretion. Equally, legal officials may use a decision to inform an interested audience (for an example, see Hawkins 1986: 1196 ff.). Precedent also serves to instruct other decision-makers. As Lempert (1992) shows, officials tend to develop their own sense of precedent for their decision practices, a sense which hardens with the passage of time (Rubinstein 1973) and may culminate in suggesting to later decision-makers in effect that, for all practical purposes, they themselves possess relatively little discretion in particular instances.[17] The practical consequence of this behaviour is that organizationally- or subjectively-created precedents can acquire the same binding force as legal rules. Furthermore, as Lempert points out, the forces contributing to the creation of precedent are not necessarily the same as those that keep it in effect. Finally, but not least, precedent acts as a refuge when the exercise of discretion is questioned.

Moral evaluation

Another set of rules has a strong moral component. Indeed, it is clear from a large number of studies that assessments of moral character made by legal decision-makers are one of the most pervasive and persistent features in shaping the exercise of discretion. For instance, moral disreputability tends to prompt more punitive responses by law enforcers (e.g. Emerson 1969; Hawkins 1984); and in the handling of cases, both criminal and civil, much turns on the credibility of the individuals caught up in the case, whether as complainants, defendants, or witnesses.[18] Credibility itself is a matter bound up with assessments of moral character, social class, educational attainment, and so on. The identity, status, or character of a person offering information or evaluation to a decision-maker may be extremely influential and may prompt a series of common-sense assessments about the trustworthiness, competence, or sincerity of the person concerned. For example, lawyers will do their best to make their own witnesses appear credible, by attributing to them, or giving them the opportunity to display, qualities which are socially valued. Therefore they will do their best in a trial to make their witnesses seem respectable, sincere, honest, and so on. Similarly, the lawyer on the other side will try to convey precisely the opposite impression (practising lawyers have all sorts of practical insights into the nature of legal discretion).

Important implications follow from this. Joel Handler shows how social workers appraise needy clients and how, in handling the uncertainties of the work and the demands placed upon them by clients, social workers develop 'practice ideologies'. These are based on the social construction of moral character and serve to screen out incompatible information and resist change, thereby allowing the easier and more efficient exercise of discretion. Moral concerns tend to predominate in such screening decisions, and the kind of moral character ascribed to an individual determines the agency's obligations to the client. These processes are means by which an agency is able to admit clients who are important to its success as an organization and reject undesirables. They operate systematically and establish yet another set of regularities in discretionary outcomes. They are also another aspect of the power implicated in the exercise of discretion.

Notes

1. There is now a large literature, particularly in the field of criminal justice, where there are many studies of policing, plea-bargaining (in the United States especially), and the work of supervisory officials such as probation officers.
2. The rest of this section is not intended to be exhaustive but rather illustrative. It does not deal, for example, with political perspectives on decision-making. The politics of discretion have not for the most part been closely studied by socio-legal scholars (but see, e.g., Wilson 1980), though, given the amount of legal decision-making which involves organizational activity and the extent to which attempts may be made to influence how decisions are made where uncertainty or conflicts of value exist, it is clear that this is an important area

for more work involving the study of power in the exercise of discretion. A perspective from politics on the use of power to advance, protect, or preserve some conception of the interests of groups or individuals contrasts with rational or bureaucratic conceptions of the use of discretion, where there is no place for political activity because their organizing assumptions are that decisions are made to attain the organization's goals using the best available information, or that decisions are made in accordance with legitimate organizational rules and procedures.

3. In explanations of decision-making that emphasize bureaucratic features, the substantive rationality of rational-choice theory is replaced by a conception of procedural rationality (H. A. Simon 1979). On this view of decision-making, decisions are seen not so much as deliberate choices, but more as outputs from bureaucracies which follow well-established procedures and patterns of behaviour (Allison 1971: 67). Many decisions made by organizational actors, however, tend not to be guided by calculation in an individual case, but are made merely by following some procedure already laid down by existing rules or by some routine established by existing practice. On this view, decision-making is less the exercise of individual choice, more an output of a bureaucratic process. Decisions are viewed in a more practical light, in contrast with classical rationality; 'satisficing' behaviour leads to adoption of a course of action that is satisfactory; outcomes are not fully explored; conflicts between different positions or alternatives are not fully resolved. Nevertheless, decision-making is still very much an orderly process. Objectives are attended to in sequence, which leads to a conception of organizations as adaptive mechanisms that develop operating procedures deemed appropriate for certain problems or situations which guide decision-making. Decision-makers are assumed to operate with less calculation than in classical rationality; their choices are influenced instead by regular policies and practices. The operation of self-interest in decision-making is defeated, in this view, through the operation of control devices such as rewards based on performance, or seniority, or rules ensuring fair treatment, or the possibility of a career advancement. Here, decisions are by no means consciously worked through in the way posited by the rational-choice approach, but instead tend to be relatively quickly done on the basis of existing practice, and with relatively little need for, or use of, information. To follow a precedent is not necessarily to select the most rational means of attaining a particular goal (see, generally, Pfeffer 1981*b*).

4. Manning, for example, explains the complex behaviour that actually takes place by which legal actors produce decisions as a result of the processes of framing within a decision field.

5. Emerson and Paley (1992), and Manning (1992) (in particular), are firmly rooted in this approach. Some have complained that much naturalist work is too concerned with the banal, but Manning's chapter is especially interesting for moving the focus from routine and repetitive decisions to decision-making in conditions of crisis.

6. The domination of individualistic approaches to legal decision-making and their limitations is a theme of Emerson and Paley (1992), and Manning (1992); see also Emerson 1983.
7. A central problem in the use of discretion is how power and authority are exercised and to what constraints they are subject. If power in decision-making is a capacity to affect outcomes, to write rules can be to exercise power, but equally (as is clear from a number of the chapters [see *The Uses of Discretion*]) to exercise legal discretion is also to exercise power. Power has a structural aspect, arising in particular from the way in which the legal system is organized, the division of labour and hierarchical arrangements within legal organizations. It also has an interactional aspect: it seems to be generally accepted that power is a characteristic of social relationships, whether among individuals or organizations, and that it varies according to particular relationships or social contexts as well as over time.

Where power is exercised legitimately, it becomes an exercise of authority (Weber 1947; see also Bell (1992)). Power in institutions may be transformed into authority when values and practices which are accepted and expected in a particular social context are regarded as legitimate within that context. So far as decision-making by legal actors is concerned, legitimacy is important, not least because it encourages compliance on the part of those who are the objects of decision. With time, the distribution of power within a social setting may be accepted and thereby legitimated, creating stable expectations of patterns of influence.

Acquisition and use of power are also matters of politics. To the extent that decisions are made in legal bureaucracies which involve the resolution of uncertainties or differences of position, such decision-making becomes a matter of organizational politics. The tie between politics and power involves a particular view of decision-making, and one which contrasts with theories of decision-making that are premised on rational conceptions or bureaucratic procedures. In both rational-choice and bureaucratic models of choice, writes Pfeffer,

> there is no place for and no presumed effect of political activity. Decisions are made to best achieve the organization's goals, either by relying on the best information and options that have been uncovered, or by using rules and procedures which have evolved in the organization. Political activity, by contrast, implies the conscious effort to muster and use force to overcome opposition in a choice situation. (1981b: 7)

Handler (1992) explores conceptions of power, using Lukes's work (1974) as a starting point. This discussion is important for drawing attention to the fact that power may not only be exercised upon those taking part in decision-making but beyond, by excluding people or issues from decision-making altogether. An analysis of the problem of discretion must, on this view, therefore take into account the fact that some people may not be granted the opportunity of access to discretionary determinations. Handler goes on to point out,

however, that, according to Lukes, this view fails to account for the fact that power may be exercised in such a way as to shape whether and how matters for decision are conceived of in the first place. Disputes here are pre-empted.
8. There are, of course, many deviations from the pure forms that are discussed here; for a comprehensive survey, see Galanter 1986.
9. Another way of controlling the exercise of discretion, whatever the decision-making form employed, is, of course, by training and socialization, which serve to inculcate appropriate decision-making values. Training involves the acquisition of norms and values central to the occupational task, and acts to frame the exercise of discretion in desired ways. Training serves, therefore, as an important means by which to advance or change bureaucratic or legal policy (see Schneider (1992)).
10. Requirements for certain sorts of training or experience may be set out, as in the case of the membership rules for the Parole Board for England and Wales in the Criminal Justice Act 1967.
11. This is not to suggest that bargaining leads to decision outcomes that are not predictable in an aggregate sense (Baumgartner (1992)) or cannot be predicted by someone with an inside knowledge of how particular legal actors routinely exercise their discretion.
12. On the use of rules generally, see Twining and Miers 1982.
13. This is a theme of many of the book's chapters. See, in particular, those by Schneider (1992), Baumgartner (1992), Feldman, (1992), Lempert (1992), Emerson and Paley (1992), and Manning (1992).
14. Galanter (1974) also observes that it pays a recurrent litigant to expend resources to influence the making of relevant rules by lobbying, and so on.
15. Handler (1992) is concerned with the distribution of advantage in bargaining arrangements leading to the making of legal decisions.
16. While the behaviour of known individuals may be rather predictable, that of outsiders to the legal system, like potential witnesses, tends to be rather unpredictable (Hawkins 1989).
17. Lempert (1992) also provides evidence, however, which indicates that this is not necessarily a general effect. This suggests that the development of such precedent may be contingent on other matters, such as the relative frequency with which cases are encountered by decision-makers.
18. Similarly, the credibility of the source of information used in decisions is very important: for examples, see Hawkins 1986; Manning 1988*d*.

References

ALLISON, G. T. (1971) *The Essence of Decision: Explaining the Cuban Missile Crisis.* Little Brown, Boston.
ASQUITH, S. (1983) *Children and Justice: Decision-Making in Children's Hearings and Juvenile Courts.* Edinburgh University Press, Edinburgh.

AUBERT, W. (1984) *In Search of Law*. Martin Robertson, Oxford.
BAUMGARTNER, M. P. (1992) 'The Myth of Discretion' in K. Hawkins (ed.) *The Uses of Discretion*. Clarendon, Oxford.
BELL, J. (1992) 'Discretionary Decision-Making: A Jurisprudential View' in K. Hawkins (ed.) *The Uses of Discretion*. Clarendon, Oxford.
BITTNER, E. (1967) 'The Police on Skid Row: A Study of Peace Keeping' in *American Sociological Review* 32.
CICOUREL, A. (1968) *The Social Organization of Juvenile Justice*. John Wiley & Sons, New York.
DAMASKA, M. R. (1975) 'Structures of Authority and Comparative Criminal Procedure' in *Yale Law Journal* 480–544.
EMERSON, R. M. (1969) *Judging Delinquents: Context and Process in the Juvenile Court*. Aldine, Chicago.
—— (1983) 'Holistic Effects in Social Control Decision-Making' in *Law and Society Review* 17.
—— and PALEY, B. (1992) 'Organizational Horizons and Complaint-Filing' in K. Hawkins (ed.) *The Uses of Discretion*. Clarendon, Oxford.
FEELEY, M. M. (1979) *The Process is the Punishment: Handling Cases in a Lower Criminal Court*. Russell Sage, New York.
FELDMAN, M. (1992) 'Social Limits to Discretion: An Organizational Perspective' in K. Hawkins (ed.) *The Uses of Discretion*. Clarendon, Oxford.
GALANTER, M. (1974) 'Why the "Haves" Come Out Ahead: Speculations on the Limits of Legal Change' in *Law and Society Review* 9.
—— (1986) 'Adjudication, Litigation and Related Phenomena' in L. Lipset and S. Wheeler (eds.) *Law and the Social Sciences*. Russell Sage, New York.
GALLIGAN, D. J. (1986) *Discretionary Powers: A Legal Study of Official Discretion*. Clarendon, Oxford.
GENN, H. (1987) *Hard Bargaining: Out of Court Settlement in Personal Injury Actions*. Clarendon, Oxford.
GOTTFREDSON, D. and GOTTREDSON, M. (eds.) (1980) *Decision Making in Criminal Justice*. Plenum, New York.
GULLIVER, P. (1969) 'Introduction: Case Studies of Law in Non-Western Societies' in L. Nader (ed.) *Law in Culture and Society*. Aldine, Chicago.
HANDLER, J. (1992) 'Discretion: Power, Quiescence, and Trust' in K. Hawkins (ed.) *The Uses of Discretion*. Clarendon, Oxford.
HAWKINS, K. (1980) 'On Fixing Time: Reflections on Recent American Attempts to Control Discretion in Sentencing Parole' unpublished paper presented to Howard League Seminar Series 'The Future of Parole', London School of Economics, October 1980.
—— (1983) 'Assessing Evil: Decision Behaviour and Parole Board Justice' in *British Journal of Criminology* 23.
—— (1984) *Environment and Enforcement: Regulation and The Social Definition of Pollution*. Clarendon, Oxford.
—— (1986) 'On Legal Decision-Making' in *Washington and Lee Law Review* 42(4).

HAWKINS, K. (1989) '"FATCATS" and Prosecution in a Regulatory Agency: A Footnote on the Social Construction of Risk' in *Law and Policy* 11(3).

HEUMANN, M. (1978) *Plea Bargaining: The Experiences of Prosecutors, Judges and Defense Attorneys*. University of Chicago Press, Chicago.

LACEY, N. (1992) 'The Jurisprudence of Discretion: Escaping the Legal Paradigm' in K. Hawkins (ed.) *The Uses of Discretion*. Clarendon, Oxford.

LEMPERT, R. (1992) 'Discretion in a Behavioral Perspective' in K. Hawkins (ed.) *The Uses of Discretion*. Clarendon, Oxford.

LLOYD-BOSTOCK, S. (1991) 'The Psychology of Routine Discretion' unpublished paper, Centre for Socio-Legal Studies, Oxford.

LONG, S. (1981) 'Social Control in the Civil Law: The Case of Income Tax Enforcement' in H. L. Ross (ed.) *Law and Deviance*. Sage, Beverly Hills.

LOVEGROVE, A. (1984) 'The History of Criminal Cases in the Crown Court as an Administrative Discretion' in *Criminal Law Review* December.

LUKES, S. (1974) *Power: A Radical View*. Macmillan, London.

MANNING, P. K. (1986) 'The Social Reality and Social Organization of Natural Decision Making' in *Washington and Lee Law Review* 43(4).

—— (1988) *Symbolic Communication: Signifying Calls and the Police Response*. M.I.T. Press, Cambridge, Mass.

—— (1992) '"Big Bang" Decisions: Notes on a Naturalistic Approach' in K. Hawkins (ed.) *The Uses of Discretion*. Clarendon, Oxford.

—— and HAWKINS, K. (1990) 'Legal Decisions: A Frame Analytic Perspective' in S. Riggins (ed.) *Beyond Goffman*. Aldine DeGruyter, Berlin.

MARCH, J. G. (1988) 'Introduction: a Chronicle of Speculations about Decision-Making in Organizations' in *Decisions and Organizations*. Blackwell, Oxford.

MATHER, L. M. (1979) *Plea Bargaining or Trial? The Process of Criminal Case Disposition*. D. C. Heath, Lexington.

McBARNET, D. (1981) *Conviction: Law, the State and the Construction of Justice*. Macmillan, London.

MILLER, C. R. (1990) 'The Rhetoric of Decision Science, or Herbert A. Simon says' in H. W. Simons (ed.) *The Rhetorical Turn: Invention and Persuasion in the Conduct of Inquiry*. University of Chicago Press, Chicago.

PFEFFER, J. (1981) *Power in Organizations*. Ballinger, Cambridge, Mass.

PHILLIPS, J. and HAWKINS, K. (1976) 'Some Economic Aspects of the Settlement Process: A Study of Personal Injury Claims' in *Modern Law Review* 39(5).

PROSSER, T. (1981) 'The Politics of Discretion' in M. Adler and S. Asquith (eds.) *Discretion and Welfare*. Heinemann, London.

RAZ, J. (1979) *The Authority of Law*. Clarendon, Oxford.

ROCK, P. (1973) *Making People Pay*. Routledge and Kegan Paul, London.

ROSS, H. L. (1970) *Settled Out of Court: The Social Process of Insurance Claims Adjustment*. Aldine, Chicago.

RUBINSTEIN, J. (1973) *City Police*. Ballantine Books, New York.

SAINSBURY, R. (1992) 'Administrative Justice: Discretion and Procedure in Social

Security Decision-Making' in K. Hawkins (ed.) *The Uses of Discretion*. Clarendon, Oxford.

SCHNEIDER, C. E. (1992) 'Discretion and Rules: A Lawyer's View' in K. Hawkins (ed.) *The Uses of Discretion*. Clarendon, Oxford.

SIMON, H. A. (1947) *Administrative Behavior*. Macmillan, New York.

—— (1979) 'Rational Decision-Making in Business Organizations' in *American Economic Review* 69.

SUDNOW, D. (1965) 'Normal Crimes: Sociological Features of the Penal Code in a Public Defender Office' in *Social Problems* 12.

TWINING, W. and MIERS, D. (1982) *How to do Things with Rules* (2nd edn.). Weidenfeld and Nicolson, London.

WEBER, M. (1947) *The Theory of Social and Economic Organization*. Free Press, New York.

WEICK, K. (1979) *The Social Psychology of Organizing* (2nd edn.). Addison Wesley, Reading, Mass.

WILKINS, L. (1975) 'Perspectives on Court Decision-Making' in D. Gottfredson (ed.) *Decision Making in the Criminal Justice System: Reviews and Essays*. Government Printing Office, Washington D.C.

WILSON, J. Q. (1980) *The Politics of Regulation*. Basic Books, New York.

Discretionary Powers in the Legal Order
The Exercise of Discretionary Powers

D. J. GALLIGAN

Discretionary powers in the modern state

We can now consider how these points of tension around the ideal type of formal rational authority are manifested in the modern state. It has become commonplace that a notable characteristic of the modern legal system is the prevalence of discretionary powers vested in a wide variety of officials and authorities. A glance through the statute book shows how wide-ranging are the activities of the state in matters of social welfare, public order, land use and resources planning, economic affairs, and licensing. It is not just that the state has increased its regulation of these matters, but also that the method of doing so involves heavy reliance on delegating powers to officials to be exercised at their discretion. Similarly, within areas that have long been the subject of state control, like criminal justice and the penal system, there are signs of a trend towards greater discretion. Of course, discretionary powers are not an innovation of the modern state since any system of authority relies on discretion in varying ways and degrees; the argument is only that in recent decades the quantity of discretion has increased to make it a more significant facet of state authority. This expansion can be seen not only in powers expressly delegated, but also in the levels of unauthorized discretion, in the sense that officials assume to themselves the power to depart from change, or selectively enforce authoritative legal standards.[1]

The extended reliance upon discretionary powers has accompanied and to a large extent been a product of the growth of state regulation.[2] There is ample evidence of that expansion in the later part of the nineteenth century and the major part of the twentieth as the state has extended its activities into a wide range of social and economic affairs. An adequate explanation of this extension would be a complex undertaking; it can clearly be linked to changing ideas about the nature of society, and about the proper role of the state in achieving ideals of social justice and welfare; it can also be related to the extended franchise. It can be linked to the increased necessity for state intervention and regulation of the economy, partly in order to achieve those social ideals, and partly in order to maintain and protect the basic structure of capitalist economies.[3]

The more immediate issue is why increased state activity should lead to a wider delegation of discretionary authority to subordinate officials. Here we must distinguish two sets of questions: one set requires a distinction between the factors which may influence the legislature or other authority in delegating powers in a discretionary form, and the extent to which those powers in the hands of the subordinate authority remain discretionary in the way they are exercised. There are always these two aspects to an enquiry into discretion: why power took a discretionary form when delegated, and how it is exercised by officials in practice. More will be made of this distinction in chapter 3 [see *Discretionary Powers*, Ch. 3], when we consider the strategies that an authority might adopt in exercising its powers; in the following discussion, I shall allow this distinction to be blurred by considering in a general way the factors that contribute to the growth of discretionary powers. A second set of questions requires a distinction to be made between factors that have as a matter of history contributed to the discretionary nature of authority, and the possibility that there is some more fundamental, conceptual link between the nature of modern societies and the role of the state, and the rise of discretionary powers.

To begin, it is possible to identify some of the general historical factors contributing to the increase of discretionary powers. In order to effect comprehensive controls in areas of welfare, the economy, and the environment, it has been necessary to vest in a variety of officials and agencies a wide range of powers. At one level of explanation, this diffusion of powers itself provides part of the explanation for the increase in administrative discretion; the very magnitude of the task of regulation has encouraged legislatures to delegate authority to subordinate bodies, and to allow more specialized policies and strategies to be devised by them. This is not to suggest that all delegations are made in discretionary forms, but only that, with growth in the size, diversity, and complexity of regulation, it is inevitable that the legislature, itself having limited capacities and resources, should delegate specific problems and undertakings to subordinate and specially created authorities.

A second and closely related factor in the expansion of discretionary powers has been the belief that many regulatory undertakings are to be approached as technical or scientific matters to be settled by specialist authorities. Once problems are characterized in this way, there are good reasons for leaving them to the discretion of the expert agencies with only minimum guidance from the legislature. Here we encounter the shades of the Weberian idea that administration is a matter of logical reasoning and scientific application. There is, however, no clear line between matters which can be resolved by the application of technical skills, and matters which require decisions of policy and value. Subordinate authorities, even those with specialist skills, are often required not only to make judgements based upon technical expertise, but also to make decisions about the allocation of resources, about the goals to be achieved, and about the distribution of burdens and benefits among groups and interests.[4] A striking example of both the inseparability of matters of expertise and matters of policy, and the way that issues may be characterized in terms of the former, thereby obscuring the latter, is the treatment and control of offenders within the penal system. For a large part of the present century questions of rehabilitation and deterrence have been seen as matters to be resolved in accordance with the social and medical sciences; on this basis, extensive controls over offenders have been taken in order to allow scope for the diagnosis and treatment of each according to his own peculiarities.[5] In this and in many other areas of social order and welfare (mental health, child care, juvenile crime), the claims of science and the readiness to classify problems as scientific have justified an increase in the discretionary powers of agencies and experts. Only in recent years has it come to be realized that each claim must be examined closely, and that, within areas like the penal system, discretionary powers must be reduced and restrained according to firm legal standards.[6] Similar concern has begun slowly to penetrate other areas where discretionary assessments over matters of the most value-based kind are made in the name of science.[7]

A third matter which signals a tendency away from formal to substantive authority, and thus to greater reliance on discretion, relates to the nature of the tasks undertaken by the state. Most of the activities of the state can be grouped in two broad categories: those related to individual and social welfare, and those related to social order. These categories naturally overlap, but within the first are matters like the alleviation of poverty, the provision of sound housing, education, medical care, and a clean environment; the second includes regulation of the economy, control of business practices, regulation of interest groups, law and order, and the reduction of crime. The tasks undertaken by the state within each category are the products partly of political ideals, and partly of practical necessity; but typically in each area broad policy goals are set—an unpolluted environment, alleviation of poverty, a balanced plan for urban development, industrial harmony, fair competition, reduction in crime rates—and powers are delegated to achieve them. In doing so, there are two kinds of factors which have an important bearing on the movement away from formal rules to a more substantive rationality. In the first place, decision-making is purposive in the sense that, however abstract the goals may be, the whole rationale for conferring powers is to achieve those goals to the highest degree possible. It is understandable that there should be among officials a concern to relate individual decisions directly to those goals, and a natural impatience with formal, complex, and possibly unsatisfactory, mediating rules: the mother in need must be helped, the offender must be rehabilitated, the factory polluting the river must be stopped, and, if an oversupply of public houses is contributing to local disorder, the number of licences must be reduced. The test of success, and to a large extent the legitimacy of official decisions, is how well they realize specified goals, not whether they have been made by the impartial application of fixed rules. Of course, there are likely to be pressures towards formalism, so that the end result will reflect an accommodation of the two; the point of importance, however, is the natural tendency for purposive, goal-based decisions to veer towards substantive rationality, and therefore towards discretionary authority.

Another type of constraint which adds to the substantive nature of decision-making has two components: one is the complexity of the matter, the other the variability of the problems encountered. With regard to the first, complexity is a function of the nature and variety of the interests involved or affected, the degree of real or potential conflict between them, the range of solutions that might be open, and the consequences of any particular solution for other areas of social or economic activity. In many areas of state regulation the level of complexity in achieving a particular social goal is likely to be extremely high. To take an example, the control of pollution levels may seem a relatively simple matter of setting firm standards and then ensuring their enforcement. But soon the complexities appear: what, after all, is an acceptable level of pollution in a world heavily reliant on industrial enterprise; what trade-offs are to be made between reduced pollution and more expensive goods; to what extent are the consequences to be taken into account of pollution controls upon the capacities of the manufacturer, in terms of expense, reduced output, and perhaps redundancies; how much, if at all, do the interests of anglers and swimmers in having clean rivers and streams count? Factors like those give some indication of the complexities that may arise in achieving broadly defined social objectives. Indeed, as we shall see, the concept of complexity itself has become important in understanding modern legal systems; the point for the moment is simply that any resolution of complexity might best be attained by a broad grant of power to an administrative authority, allowing it considerable discretionary freedom in settling upon courses of action.

Considerations of variability are closely related; in achieving objectives broadly defined, there may be such differences from one situation to another that it is hard to find a sufficiently common basis for comparisons and generalizations. Each instance may present its own special and peculiar problems, and so provide little guidance for the future. An example commonly used is the notion of personal need in social welfare; but examples abound in other areas—the potential of an offender to be rehabilitated, the settlement of a particular industrial dispute, a specific conflict between the use of resources and the interest of ecology, or the construction of a development plan for one city rather than another. The relevant factors are likely to be closely locked into that specific case in a way which may not recur in quite the same way in any other case. Variability has the effect, therefore, of directing both the form in which power is conferred, and the methods by which it is exercised, towards the discretionary end of the scale.

These three sets of factors emphasize different aspects of the overall problems of pervasive state regulation, and taken together they help to explain the discretionary trend both in the allocation and the exercise of authority. The concern of the state to attain through its instrumentalities specific social goals results in a strong tendency to substantive rationality. This does not mean, however, that substantive rationality provides a complete or adequate picture of authority; on the contrary, the very tensions identified at the theoretical level can be seen in play at the practical level; the currents flowing towards substantive rationality are opposed by a concern for the formality of clear and general rules mediating between the general goal and decision-making in specific cases. Just how these conflicts are to be resolved can only be determined in the context of specific powers, where the various considerations can be assessed in settling upon a suitable strategy for making decisions.

There is a fourth, important factor contributing to the discretionary nature of authority, and this can be related to shifts in the power structure of modern states. The extension of state activities is to some extent a product of and at the same time contributes to the formation of interest groups—employers, employees, corporations, consumers, conservationists, professions, etc.—who are able to exert influence upon the legislative and administrative processes.[8] It is characteristic of interest groups of this kind that they are highly organized, usually with a core of permanent professional staff, able to command a high level of loyalty from their members, often organized on democratic principles, and sometimes financially powerful. The protection and advancement of their members' interests is their primary aim, although the interest group as an entity might acquire substantial independence of its rank-and-file members, and become powerful in its own right.

It is sometimes said that interest groups and corporate bodies of this kind constitute the really important centre of power in modern societies; whether or not there is sufficient evidence for that claim, it is clear that they have had considerable impact on the exercise of political power. One sign of this can be seen in the fact that matters of political importance are sometimes settled within those institutions themselves; another sign is the shift away from the chambers of the legislature as the most important focus of power. Corporate groups can pursue their aims in the ministry and the department, the party room, the boardroom, and in the administrative agencies. Parliament remains juridically sovereign, but many problems are in practice settled elsewhere, within the administrative organizations for example, with the result that parliament's role may be merely to give its formal approval, and to provide the necessary legal authority. The result is a greater diffusion of powers amongst subordinate authorities, and politicization of the administrative process. Where problems are left to be resolved and the strategies settled in the exercise of administrative discretion, it is inevitable that groups should try to bring pressure to bear in order to achieve results satisfactory to themselves. In these circumstances, it is more difficult to formulate general decision rules; and in any event the reasons for doing so are weakened. It may be that there is no clear sense of the public interest, so that decision-making becomes a matter of seeking compromises and adjustments amongst the groups and interests that will be affected. If we add to these considerations the growing complexity and variability of the matters sought to be regulated, then the pull of general rules is further weakened.

Interest group politics are an old phenomenon and, provided that there are controls on their internal organization and their activities, they are a legitimately democratic form. But while interest groups and corporate structures are important in understanding the political organization of modern societies, this does not mean that the state has become a kind of neutral umpire in the contests between groups. Indeed, the existence of such groups may increase the need for the state to exert direction over them, and thus create a tendency towards a highly regulated, state corporatism. An illustration of this occurs in regulation of the economy.[9] Western capitalist economies are organized primarily on principles of private ownership and rely heavily on market forces; but, because of economic crises on the one hand and the state's commitment to social goals such as welfare and increased prosperity on the other, it is necessary for it both to support private economic organization and at the same time to direct it towards public goals.[10] Economic activity becomes mingled with political and social activity, the public and the private spheres become substantially merged; in order to maintain this 'corporate' character, whereby the state 'directs and controls predominantly privately owned businesses towards four goals: unity, order, nationalism and success',[11] the state may need to be able to act quickly, to change policy abruptly as circumstances change, and to use economic incentives as a way of achieving social goals.[12] It is important in this environment that the legal structure be sufficiently open and malleable to accommodate the state's needs. This encourages a weakening of the concern for general legal rules and for a division of institutions; it also creates a greater reliance on discretion. What occurs in economic matters may also be detected in other areas of state regulation.

Each of the four factors considered, when viewed against the background of Weber's modes of authority, goes some way towards explaining why legal authority is so often of a discretionary kind. It is sometimes suggested, however, that there are more fundamental connections between the activities of the modern state in providing social welfare and maintaining social order, and the discretionary mode of authority employed. R. M. Unger has argued that welfare and corporatist tendencies have substantially eroded the idea of legality, meaning something like a mixture between formal rational authority and the rule of law, in favour of a pervasive substantive rationality based on notions of equity and communal solidarity.[13] Unger's account of types of authority in many respects parallels Weber's, but while Weber, at the time he wrote, detected only limited elements of substantive rationality in modern law, Unger sees these elements as dominant. He appears undecided, however, whether this is simply a normal cycle in the life of a society, or the signs of a more fundamental transformation of the nature of legal authority.[14] P. S. Atiyah has identified a similar shift from formal authority to substantive, or as he prefers to put it, from principles to pragmatism. He has found evidence for these trends in various areas of law, and in contract law in particular, and has offered an explanation which also relies heavily on the interventionist nature of the modern state.[15] His view seems to be in general to deprecate the consequential erosion of the certainty and formality which he considers central to the notion of law. Similar views appear in the writings of E. Kamenka and A. E. S. Tay who have identified three legal paradigms: *gemeinschaft*, *gesellschaft*, and bureaucratic-administrative.[16] Modern societies contain elements of each; but just as capitalist, liberal societies signified a move from the organic, communal society of *gemeinschaft*, to the individualistic, exchange relationships of *gesellschaft*, so there is in modern societies a shift to a system in which the individual right-bearer is no longer at the centre of the stage. Here the dominant mode is bureaucratic rationality, which is the mode of authority most suited to achieving the goals of public policy.

Whether there has been a qualitative change in the nature of legal authority is rather hard to assess. It is clear that the activities of the state and the political environment are conducive to substantial reliance on discretionary authority; it is equally clear, however, that in respect of many matters, state power proceeds according to the tenets of formal rationality. Any claim that there is a predominant tendency towards discretionary authority must depend on careful examination of the many areas of state regulation, and this has not yet been done in a systematic way. It is sufficient to recognize that authority centres around the focal points of formal, substantive, and reflexive rationality, and that the patterns that emerge are not constant but ever grouping and regrouping. And in any event, the interesting task is to see what place law has in respect of any particular pattern.

There is, however, a related argument which needs to be considered in more detail. Here the ideas of complexity and variability are taken up, and given greater importance in understanding the nature of authority. A recurring idea in the views of those who see a turning from formal authority is that the objects, which the state is trying to achieve through regulation, are so complex (and here complexity may be taken to include variability) that they cannot be realized effectively by a system of authority based on the formulation and application of general rules. Any attempt, then, either to understand the structure of the state in terms of that model, or to use it as a practical guide in allocating and exercising powers, is likely to result in distortion of the way power is exercised in practice, and to hinder the achievement of its objectives. It is important to

Unger's account that the modern state, in its concern to regulate the economy, to control the use of resources, and to achieve ideals of social welfare and justice, has to deal with matters which are too complex to allow for the generalizations characteristic of rules. The best that can be done is to provide broad policy goals to be achieved through the discretionary choices of officials.[17]

This approach has been pushed to heights of great sophistication and abstraction by other social theorists. One of the most interesting of these is Niklas Luhmann who has made complexity a key concept in his analysis of modern societies.[18] Luhmann has developed the theory that modern societies are exceptionally complex; this results from the great diversity of interests, groups, and values within them, from the possibilities that are made available by science and technology, and from the nature and interrelatedness of economic systems. Societies are unable to assimilate the full range of possibilities that are opened up by such complexity, and so the main function of social activity is to reduce complexity; this is done by selecting aspects of the environment and thereby simplifying it. Subsystems within society develop in this way; they simplify or reduce the complexity of the total system, and yet at the same time they match or correspond to the total environment. Thus, modern society becomes characterized by the proliferation of subsystems, whether economic or political, cultural or educational, each of which performs certain functions within society. The proliferation of subsystems has the effect of reducing complexity and making social life possible, but, at the same time, the 'functional differentiation', or splitting of society into a mass of subsystems, also increases complexity. And modern society, argues Luhmann, is characterized by functional differentiation.[19]

This has important implications for law. One of the effects of the positivization of law, and the putting aside of natural law constraints, is that the law is whatever the lawmaker decrees. This has enabled legal regulation to be extended over any area of activity for whatever reasons seem fit. However, the increasing complexity of other social subsystems, and their differentiation from the total environment and from each other, make it correspondingly difficult for legal systems to formulate abstract norms capable of accommodating the complexity and variability of the situations sought to be regulated. This affects not only the model of formal but also substantive rationality; since, even if the goals and values are stated, there is no guarantee, because of functional differentiation, that they can be achieved simply by the decrees or decisions of one set of officials. Legal regulation is then at risk either of distorting reality by imposing formal rules, or of being forced to rely on extremely abstract statements of purposes. In either case, instead of being an integrating force, law may add to the conflicts amongst subsystems; this may hamper efficient regulation and undermine social stability.

The question, then, is whether there are avenues open for the development of legal regulation in modern societies. Luhmann makes clear that no one centralized system or structure—whether legal, moral, or economic—is capable of providing overall social integration and regulation. It is necessary, therefore, to concentrate on the relationships between sub-systems, and to create mechanisms which make possible a certain compatibility or harmony amongst them. Here the idea of reflexion is important: each subsystem must develop the capacity to limit its own actions, so that it may co-exist with other subsystems. This process of reflexion becomes a primary method in modern societies for integrating subsystems.[20] Within the legal context, this means that with respect to certain issues, perhaps industrial relations or economic regulation, the capacity for comprehensive legal rules is limited; instead, an assessment of the subsystems affected and the conflicts between them may show that other techniques are suitable, perhaps mediatory procedures for the first, and greater reliance on market forces for the second. Put more simply, the legal system must approach each problem with a view to solving it in the best way possible, recognizing that techniques other than simply formulating general norms may be required. Thus law becomes concerned with providing institutions and procedures within which conflicts between subsystems can be resolved, rather than attempting to provide comprehensive social control.[21] So Jurgen Habermas, who, from a different perspective, has been concerned with similar issues, has concluded that 'since ultimate grounds can no longer be made plausible, the formal conditions of justification themselves obtain legitimating force.[22] Connections can be seen between these theoretical conclusions and the popular call for greater representation and participation in public decisions.

It has been possible to give only the barest outline of theories which are highly suggestive in demonstrating the scope for legal regulation. They advance a number of hypotheses which can be verified only by extensive empirical analysis of different areas of legal control. Nevertheless, a number of implications for discretionary authority may be suggested. In the first place, the complexity and functional differentiation of society provides an important part of the explanation for the expansion of discretionary authority. Differentiation generates conflict between subsystems, and it falls to the state to be involved in its resolution. This often takes the form of delegation to specialized bodies on the understanding that they may have the capacity to develop suitable and innovatory techniques of dispute resolution. Secondly, however, there is always the question of the most suitable balance between precise normative direction by way of legal standards, and the delegation of discretionary authority with its emphasis on flexibility and capacity to adjust. Legal regulation is undertaken to achieve goals which are considered desirable for moral and political reasons, or out of the necessities of preserving order; the difficult question always is to know how far regulation may go in achieving those goals without creating intolerable levels of conflicts or consequences which undermine the very point of the regulation. Thirdly, if one of the consequences of complexity is increased reliance on reflexive techniques, then issues about decision strategies and participatory procedures become increasingly important. Greater attention should then be devoted to developing different techniques suited to different tasks, to understanding the implications of those techniques in terms of legal and political values, and in developing an adequate framework of accountability and legal regulation.

* * * * *

Constraints on the exercise of discretion: (a) Those of a more practical kind

Following this account of administrative theories and their relationship with issues that arise in exercising and regulating discretionary powers, we may now consider, again briefly, some of the main influences and considerations that bear on both the substantive outcomes and the decision strategies adopted. This is of interest not only in providing illustrations of the preceding discussion of administrative theories, but also in understanding better the exercise of discretion.

Efficiency and effectiveness

Efficiency is a concept which is most at home in the mechanical sciences, and which can be transplanted only with difficulty to government administration.[23] Within business organizations, the concept of efficiency refers to the relationship between the resources used and the returns that result; the mediating notion is profit, and a business is operating most efficiently when the greatest profit is produced by the least use of resources. Although simple to state, this relationship might be highly complex in practice since the firm must take into account matters like capital expansion, re-investment, and worker satisfaction. Nevertheless, there is within business a reasonably clear framework within which to make assessments of efficiency

The position is different in an administrative agency. Here the primary object is not to profit from some marketable commodity, but to achieve stated goals—to relieve poverty, to provide a tolerable environment, or to regulate public houses. The overall test of success in such cases is the extent to which the respective goals have been achieved. It is useful then to distinguish between efficiency and effectiveness; efficiency pertains to the relationship between resources and outcomes, while effectiveness refers to realization of goals.[24] Effectiveness is of vital importance in the exercise of discretionary powers, but is not itself an easy test to apply. The goals may themselves be unclear, or they may be interconnected so closely with other programmes and with wider notions of social welfare that judgments of effectiveness are hard to make. In any event, rationality and purposiveness would seem to require that the discretionary authority sets for itself, within the terms of its authority, some goals against which effectiveness may then be tested.

Within the general concern to achieve goals effectively, efficiency has an important role. Administrative decisions involve costs and resources, and it is possible to make some appraisal of the relationship between expenditure and results. Appraisals can be made at different levels; it may be a matter of technical assessment, whether mechanical equipment might perform some of the tasks that are otherwise labour intensive, or whether re-organization of an institution would enable more work to be produced, and perhaps bolster staff morale. The concern in such cases is to determine whether the same or better results could be achieved, in terms of goals, by a lesser dissipation of resources; for this purpose, efficiency is a natural and useful concept to employ.

Difficulties may arise, however, when assessments of this kind extend beyond reasonably clear technical matters, and attempts are made to calculate the costs and benefits of broader policy goals. A noticeable aspect of the current scepticism towards regulation is the demand that administrative programmes be subjected to rigorous analysis, to make clear the goals being sought, the wider consequences, the costs and benefits, and in some cases the net benefits to society. In America, President Reagan has issued executive orders requiring agencies to prepare, with respect to all major rules they adopt, an analysis of their impact, including the costs and benefits, and in particular the net gains to society.[25] There are undoubted virtues in generally raising the rational basis of discretionary decisions, in being clear about alternative courses of action and their respective costs, and in seeking the maximum benefits from official actions. There are considerable dangers, however, in extending the techniques of cost-benefit analysis into the assessment of the overall social advantages of particular programmes. Regulatory action may be prompted by a range of values and intangible considerations which cannot easily be assessed in economic terms; moreover, it is often difficult to separate the costs and benefits of any one programme from a wider assessment of government action.

The notion of effectiveness in realizing given goals is a basic concept, both in the exercise of discretion and in its legal regulation. It is important for this reason to modify the suggestion that might derive from an hierarchical and instrumental view of authority, that effectiveness is achieved simply by finding the most suitable means to a given goal. In the first place, the goals may themselves be various and conflicting so that the advancement of one goal may be to the detriment of another. The assessment of effectiveness then depends on how the goals are ranked in relation to each other. In the second place, effectiveness may also depend partly on satisfying certain subsidiary goals, or recognizing certain values in the process of reaching a final outcome. In complex matters, for example, effectiveness may be judged according to how well disputes amongst interest groups are resolved and participatory procedures applied.

Apart from these general considerations regarding efficiency and effectiveness, an authority with discretion may have to decide the more specific issue as to what strategies of decision-making are to be adopted. Various factors, as we have noted, will be influential; but the different decision strategies may themselves contribute in different ways to effectiveness. The range available is between comprehensive planning, with a clear statement of goals and heavy reliance on clear standards, and incrementalism, with a loose framework of standards to be developed in a piecemeal manner. Rules which are formulated carefully to achieve stated goals might be expected to attain a high level of effectiveness; they have the obvious advantages of guiding officials in making decisions, and enabling citizens to know how decisions are likely to be made. Rules also generally simplify decision-making by representing a distillation of knowledge and experience gained in the past, and thus limiting the range of considerations to be taken into account in dealing with specific circumstances.[26]

Rules have costs, however, and these may be grouped in three categories. Firstly, because of complexity and variability, rules are not always the most rational means for achieving purposes. The cost of rules in such circumstances is to impose a selective and possibly oversimplified view of purpose; as a result, the gains in certainty might not be sufficient to compensate for the costs in reduced effectiveness. Secondly, there is the problem of overinclusion and underinclusion; Ehrlich and Posner have noted a 'necessarily imperfect fit between the convergence of a rule and the conduct sought to be regulated';[27] sometimes a rule will have results which are not the best in terms of purpose, while other situations which should come within the rule may be excluded. The extent to which either of these situations occurs is a function of the precision of rules on the one hand, and complexity and variability on the other hand, or, as Ehrlich and Posner put it, heterogeneity. The difficulties of heterogeneity can be overcome to some extent by allowing the decision-maker to depart from the rules by invoking broader purposive standards of interpretation, and by relating particular circumstances directly to purposes. But this has obvious costs, since the higher the incidence of departure from the rules, the less useful they are as guides to outcomes, and the more opportunity there is for officials to make mistakes. Decision-making then becomes increasingly incremental, until a point may be reached at which rules give little indication of how decisions are in fact being made. Thirdly, in any assessment of effectiveness two different standpoints should be considered; that of the officials regulating, and that of the citizens being regulated. From the citizen's standpoint, efficiency in regulation is likely to be closely connected to precision in rules; but from the regulator's point of view, the precision of rules may seem an insupportable obstacle in a more precise achievement of purposes. Discretionary powers are not concerned always with regulating behaviour, but, where they are, a compromise may have to be sought between the importance of rules and the virtues they engender, and the need for close correlation between action and purpose. Finally, there is a wide variety of surrounding transaction costs which relate to any decision strategy: for example, the relative costs of setting up procedures and institutions to formulate rules, compared to the equivalent costs of incremental processes. Costs of this kind must be assessed and balanced in relation to particular tasks.

Political considerations

Although the primary concern of a discretionary authority is to achieve given objects, this requires policy choices and assessments which are subject to a range of influences. These may flow from superior officials or from interest groups pursuant to formal participatory procedures; alternatively, such influences may derive from a multitude of informal sources, and produce effects of a subtle, variable, and intractable kind. Also of interest are the perceptions of the authority itself as to its place in the political system, and the values and goals that warrant its support.[28]

Influences stemming from the political system in this way can affect not only substantive outcomes, but also the formal strategies that are employed. An example can be seen in the workings of the parole system: it is characteristic of parole that in both Britain and elsewhere legislators, executives, and parole boards are reluctant to stipulate with any precision the criteria by which parole decisions are to be made. One reason for this is the belief that the early release of offenders is such a sensitive political issue that the emphasis must be on exercising great caution to ensure release of only the best risks, and to prevent any suggestion that parole is a matter of right rather than privilege. Decision-making tends then to be incremental, with emphasis on the unique features of each case, and with little effort being made, at least publicly, to develop patterns of consistency.[29] Similarly, within the welfare system efforts are made in certain areas to retain powers in a highly discretionary form in order to avoid or conceal the necessity for difficult policy choices about the distribution of resources. For example, a discretionary format may allow reduction in expenditure without appearing to make changes to the system; alternatively, a discretionary structure facilitates inaction on the part of the authority, and this may itself be a useful way of satisfying political pressures.[30]

Organizational factors

We have seen at some length the importance of the organizational structure of administrative authorities in affecting the substance and strategies of decision-making, so my remarks here may be short. The general point is that the way an authority is organized, in terms of the authority structure and so the degree of power which one official has over another, the extent to which it is hierarchical, the degree of autonomy particular officials have to act as they think best, the position regarding promotion—each of these factors, together with a range of others, has some effect on the way discretionary powers are exercised. To give just one example, Keith Hawkins in his study of the enforcement of standards regulating the pollution of water has noted the conflicts that occur between the organizational structure and the control exercised over field officers on the one hand, and the need for the field officer to have substantial discretion in negotiating and bargaining with polluters on the other. Compliance with anti-pollution standards is gained most effectively by these tactics rather than the immediate threat of prosecution; but the field officer may experience considerable conflict between the need for such flexibility and the control deriving from the superior officials within the water authority.[31]

According to the Weberian approach, decision-making within large organizations tends towards the ideal of formal rational authority; but there are other factors as well, some of which may pull in other directions. The level at which decisions are made, the competence of staff, the complexity of the tasks, the extent to which powers are diffused and decentralized, the effects of outside influences: these are some of the factors that may bear upon the approach taken within any organization towards decision-making.[32] To take another example that might occur within a welfare agency: the relatively low competence of officials at the front desk deciding individual cases is likely to influence officials at higher levels to formulate rules of a fairly strict kind. Yet, ironically, it is often at the front desk that an incrementalist approach is most appropriate in dealing with individual cases. In practice, the result is likely to be a compromise between the strict application of rules, and the freedom to depart from them in specific cases, subject to various mechanisms at higher levels for monitoring anomalies.[33] By contrast, the fact that the members of parole boards do not form part of a highly structured organization, but are laymen who meet to decide specific cases, and that each comes with some professional expertise, adds to the incremental nature of parole decision-making. Similarly, where authority is diffused amongst small units, the members of which are highly motivated and who hold shared values, decisions may be made most effectively in the relative absence of detailed rules.[34]

Economic factors

Administrative agencies are concerned unavoidably with issues of resource use and allocation. All agencies must be concerned to use their resources efficiently and effectively in achieving goals; some agencies also are concerned with the direct allocation of resources to others. The resources available to any agency are limited, and, consequently, it is hardly necessary to note that economic considerations affect both the substantive decisions made, and the form which decision-making follows.[35]

As to substance, the limitation on resources is a significant factor in deciding the courses of action to be followed. It is usual for an authority to have to make choices as to which amongst possible goals is to be sought, or the methods to be employed in doing so, and in this way the discretionary nature of decisions is increased. We have noticed before the selectivity used in arresting, prosecuting, and trying offenders, and how limited resources may lead to the non-enforcement of legal duties. In a different context, the inevitable selectivity required in deciding to advance one goal rather than another, because of limited resources, has been described as 'emphatic' discretion, the suggestion being that this is the most highly discretionary aspect of decisions.[36] Emphatic discretion in the use and allocation of resources gives the agency an increased control over parties whose interests are affected, by making the grant of concessions or benefits depend upon compliance with certain standards or conditions. This position of control in the allocation of resources may even enable the agency to achieve goals which are not strictly within its powers.[37]

The level of effectiveness in achieving goals may itself be affected significantly by scarcity of resources. In one study of water pollution control, it was found that this factor resulted in paucity of research and a backward state of scientific knowledge; this meant in turn that the standards of control imposed were somewhat arbitrary.[38] The same studies have emphasized the effects which limited resources have on the methods employed to secure compliance with standards. Because the water authorities could not maintain a system of strict detection and enforcement through prosecution and the threat of it, they are forced to negotiate, bargain, and strike compromises with polluting firms.[39] Thus, there appears to be a direct connection between the resources and capacities of an authority, and the tactics it employs in achieving its goals. It is also interesting to note the significance of the economic position of the polluting firm in the field officer's decision whether any action, and if so what, should be taken against breach of the standards. Indeed, this factor was an important component in the formation of an officer's moral evaluation of particular cases of pollution.[40]

Economic factors may also influence the selection of a position between comprehensive planning and incrementalism, but there is no set correlation between the two. Where limited resources are being distributed, comprehensive planning with rules governing allocation may seem the most effective way of accommodating economic limitations. However, rules can become rigid and give rise to expectations which, it may turn out, cannot be met. A more incremental approach has the virtue of allowing flexibility in accommodating decisions within economic constraints, especially if the latter are themselves likely to change;

the resulting costs are a relative lack of planning and co-ordination in the way resources are allocated.[41]

The nature of the task

Attention has been drawn to the connection between the growth in discretionary powers and the kinds of tasks that the modern state has undertaken; we saw in particular how the ideas of complexity and variability help to explain both the initial delegation of discretionary powers, and the strategies adopted by the delegate in exercising them. It may be helpful now to see further how these concepts affect the decision strategies. Where to site a new airport, for example, is clearly a matter of considerable complexity, involving many interests, and so is suited to an incremental approach which allows consideration of various possibilities, and ensures the maximum representation of interests. Where, however, decisions of a highly complex kind are made regularly, there are countervailing forces pulling towards more comprehensive planning. For example, the grant of licences to conduct public houses might be considered complex in terms of the interests and factors involved; but decisions on these matters have to be made regularly, and thus tend to be governed by relatively settled, if oversimplified standards. In such cases, complexity is compromised in order to gain in efficiency and consistency. On the other hand, being responsive to complexity and variability may be of primary concern, especially in economic regulation. It has been remarked that regulation of economic affairs requires quick and informal adaptability, and for this reason is often dealt with most suitably towards the incremental end of the scale.[42] Similarly, where there is a high level of variability from one case to another, a relatively loose framework of guiding standards may lead to more satisfactory solutions. It has been found, for example, that the provision of financial assistance to families experiencing a period of temporary difficulty was made in a most effective and responsive manner when the emphasis was on a case-by-case approach with few controlling rules.[43] However, its success was partly due to other factors; the organization was small, it was not overworked, and its members shared a common set of values and ideals; without these factors, the position might have been different.[44] So, while complexity and variability have natural affinities with incrementalism, they may have to compete with other factors which point to greater reliance on advance planning and general standards.

Constraints on the exercise of discretion: (b) Those of a more value-based kind

It is not only practical matters of the kind considered that affect both the substance and form of discretionary decisions. Other considerations are of a more clearly value-based kind, and include not only the personal moral views of officials, but also the background of moral and political values. These divide into two broad but connected categories, the first being the political values that are linked to rational decision-making, the second being the implications drawn from moral principles, in particular from ideas of fairness. It can now be shown how these matters affect discretionary decisions by taking a number of specific topics: the moral attitudes of officials, the implications of rational action, the concept of arbitrariness and the principles of consistency, considerations of fairness, and the importance of guidance.

The moral attitudes of officials

It has been shown in earlier discussion how the moral attitudes of officials are likely to be important factors in assuming discretion to depart from, modify, or selectively enforce binding legal rules. These same moral attitudes are of significance in the exercise of discretion generally. They are likely to affect the official's perceptions of the statutory purposes, and to influence the interpretation put on them; such moral attitudes will also influence the formation of subsidiary and instrumental goals that are set in achieving overall purposes; and, finally, those attitudes will be of importance in shaping the methods of enforcement. This is not to suggest that an official will simply apply his own moral policy; the position is more complex. It has been shown that sentencing judges have ample scope for giving effect to their own penal philosophy, but they may have to modify that philosophy in order to comply with clear legislative directions or rulings from higher courts.[45] Similarly, a judge's philosophy may have to be modified and compromised in order to fit into existing institutional structures and practices. These various factors may in turn cause the sentencing judge to adjust his penal outlook.

There is clearly a close and complex relationship between an official's moral views, his practical actions, and the environment in which he works. The image, then, is not of an official with clear and rigid views which he brings to bear in performing his duties; it is rather of an official with a range of views, some more firmly held than others, some more important than others, which influence the tasks he is required to perform, but which also are influenced by these same tasks. There is also a close interrelation between the views an official has and the wider community morality. Indeed, the official's perceptions of the community morality may be a major component of his own views. It has been shown in studies of regulatory enforcement that officials are highly sensitive to community morality, and that their actions and interpretations are much affected by it.[46] The police, for example, derive reinforcement for their attitudes to crime control because there is usually little doubt that they have the support of the community; on the other hand, other areas of regulation may employ the devices of the criminal law without, however, removing community doubts as to the 'real' criminality of such matters. This moral ambiguity may create serious tensions in the role of officials and influence their approach to enforcement. Officials charged with regulating water pollution experience this moral ambiguity, the result being a more moderate approach to enforcement, with the emphasis on securing compliance by negotiation, at least in the early stages, rather than positive sanctions.[47]

At the more particular level, an official's moral outlook is bound to be influential at each point in his decision. Examples may again be taken from studies of pollution control. In deciding whether there has been pollution of a sufficiently serious kind to warrant enforcement action, it appears that a major consideration is the field officer's moral judgment about the case. This in turn depends largely on whether the pollution was accidental or deliberate, and the company's economic capacity to take the necessary precautions.[48] Hawkins has shown how important is the component of moral disreputability, meaning wilful or negligent rule-breaking or persistent disregard for an officer's authority.[49] Conversely, a complementary study of pollution control has pointed to the disgust shown by field officers when a genuinely accidental spillage was prosecuted.[50] Examples could be multiplied, but for our present concerns it is enough to be aware of the real and complex relationship between moral attitudes and discretionary actions.

Notes

1. See the discussion in section 1.8 [see *Discretionary Powers*]. For a study of one aspect of assumed discretion, see: Kadish and Kadish, *Discretion to Disobey*. Some of the factors that contribute to the assumption by officials of assumed discretion can be seen in: Davis, *Discretionary Justice in Europe and America*, esp. ch. 2. See also section 2.1 above.

2. The literature on this theme is voluminous; in general see: E. Freund, *The Growth of American Administrative Law* (New York, 1923); J. Dickinson, *Administrative Justice and the Supremacy of Law* (Harvard UP, 1927); J. Wilson, 'The Rise of the Bureaucratic State' (1975) 41 *PubInt* 77; Unger, *Law in Modern Society*, ch. 3; G. Poggi, *The Development of the Modern State* (London, 1980); Atiyah, *The Rise and Fall of Freedom of Contract*.
3. For an analysis of the nature of modern welfare states, see: J. Habermas, *Legitimation Crisis* (London, 1976); C. Offe, *Contradictions of the Welfare State* (London, 1984); I. Gough, *Political Economy of the Welfare State* (London, 1979); M. Brice, *Coming of the Welfare State* (London, 1968, 4th edn.); for more specifically legal aspects, see Cranston, *Legal Foundations of the Welfare State*.
4. For discussion of some of the problems associated with the exercise of discretion by professionals, see: Adler and Asquith, *Discretion and Welfare*, especially the essays by Adler and Asquith, G. Smith, H. Giller and A. Morris, T. McGlew and A. Robertson. See also the essays in N. Timms and D. Watson (eds.), *Philosophy in Social Work* (London, 1978).
5. For a general discussion of these ideas in the penal system, see P. Bean, *Rehabilitation and Deviance* (London, 1976). For a related issue, see C. Slobogin, 'Dangerousness and Expertise' (1984) 133 *UPaLR* 97.
6. For some of the background to the way views began to change, see: American Friends' Service Committee, *Struggle for Justice* (New York, 1971); A. von Hirsch, *Doing Justice* (New York, 1976). The response has been all too often a return to narrow formalism, see: Galligan, 'Guidelines and Just Deserts: a Critique of Recent Reform Movements in Sentencing'.
7. There is, however, a growing literature directed towards showing the interconnection between science and policy, and the consequences for discretionary powers. For example: L. Gostin, *A Human Condition* (London, 1976); Ian Kennedy, *The Unmasking of Medicine* (London, 1978). For discussion from a more specifically legal point of view, see M. Shapiro, 'Administrative Discretion: the Next Stage' (1983) 92 *YLJ* 1487; C. S. Diver, 'Policymaking Paradigms in Administrative Law' (1981) 95 *HarvLR* 393.
8. For discussion of this shift in power from the state to interest groups, see: Poggi, *The Development of the Modern State*, ch. vi; J. Winkler, 'Law, State, and Economy' (1975) 2 *BJLS* 103, and 'The Political Economy of Administrative Discretion' in Adler and Asquith *Discretion and Welfare*; Unger, *Law in Modern Society*, pp. 200–3; Leo Panitch, Asquith, *Discretion and Welfare*; Unger *Law in Modern Society*, pp. 200–3; Leo Panitch, 'Recent Theorizations of Corporatism: Reflections on a Growth Industry' (1980) 31 *BJSoc* 159.
9. See, generally: J. Winkler, 'Law, State, and Economy'; T. Daintith, 'Public Law and Economic Policy' 1974 *JBL* 9; G. Ganz, 'The Control of Industry by Administrative Process' 1967 *PL* 93.
10. For discussion of the British government's attempts to control private enterprise in the 1970s and the implications for law, see J. Winkler, 'Law, State and Economy'.

11. *Ibid.*, p. 106.
12. The reliance by the government on discretionary authority to achieve counter-inflation measures is discussed in V. Korah, 'Counter-Inflation Legislation: Whither Parliamentary Sovereignty?' (1976) 92 *LQR* 42; see, also, T. Daintith, 'The Functions of Law in the Field of Short-Term Economic Policy' (1976) 92 *LQR* 62.
13. This change in the nature of legal authority is one of Unger's principal themes in *Law in Modern Society*; the idea may also be seen in Hayek, *The Constitution of Liberty*. There is also a tradition of Marxist theory which detects, at a deeper theoretical level, a similar theme in the transition from a capitalist to a communist society: E. Pashukanis, *General Theory of Marxism and Law*, C. Arthur (ed.) (London, 1978). For discussion, see P. Hirst, *On Law and Ideology* (London, 1979), ch. 5.
14. Unger, *Law in Modern Society*, pp. 213–4.
15. This is how the change has been expressed by Atiyah in *From Principles to Pragmatism*; it is also one of the themes of his *The Rise and Fall of Freedom of Contract*.
16. These paradigms have been discussed in various writings of E. Kamenka and A. E. S. Tay; see, for example: 'Social Traditions, Legal Relations', and '"Transforming" the Law, "Steering Society"' in E. Kamenka and A. E. S. Tay (eds.), *Law and Social Control* (London, 1980) and 'Beyond Bourgeois Individualism—the Contemporary Crisis in Law and Legal Ideology' in E. Kamenka and R. S. Neale (eds.), *Feudalism, Capitalism and Beyond* (London, 1975).
17. *Law in Modern Society*, ch. 3.
18. Much of Luhmann's writing is not available in English translation but the following provide an introduction to his views: Niklas Luhmann, *Trust and Power* (London, 1969); 'Differentiation of Society' (1977) 11 *CanJSoc* 29; 'Generalized Media and the Problem of Contingency' in J. Loubster et al. (eds.), *Explorations in General Theory in the Social Sciences: Essays in Honour of Talcott Parsons* (New York, 1978). Useful discussions in English of Luhmann's work are Poggi, *Trust and Power* (introduction), and 'Two Themes from N. Luhmann's Contribution to the Sociology of Law' 1981 *BASLP*; also W. T. Murphy, 'Modern Times: Niklas Luhmann on Law, Politics and Social Theory' (1984) 47 *MLR* 603.
19. 'Differentiation of Society', pp. 35 ff.
20. See G. Teubner, 'Substantive and Reflexive Elements in Modern Law' (1983) 17 *LSR* 239 at pp. 272 ff. I have relied heavily on this analysis.
21. An example can be taken from American constitutional law; on one approach to the constitution which seems to be increasing in popularity, judicial review is not directed towards the substantive ends of law in a democratic society, but only towards the conditions which must be satisfied if an outcome is to be considered democratic. See J. H. Ely, *Democracy and Distrust* (Harvard UP, 1981).
22. Jurgen Habermas, 'Towards a Reconstruction of Historical Materialism' 1975

Theory and Society 287. See also: *Legitimation Crisis*. For discussion see: Tuebner, 'Substantive and Reflexive Elements in Modern Law'.

23. For discussion of this issue see Self, *Administrative Theories and Politics*, pp. 261–77 (of special importance); also D. Keeling, *Management in Government* (London, 1972); P. P. Craig, *Administrative Law* (London, 1983), pp. 235 ff.

24. The distinction is made in Self, *Administrative Theories and Politics*, p. 264. For a study of the notions of effectiveness and efficiency in the regulation of natural resources, see O. R. Young, *Natural Resources and the State* (University of California, 1981), esp. ch. 4.

25. Executive Order No. 12, 291, Sections 2 and 3(d), 46 Fed. Reg. 13, 193, 13, 194 (1981) quoted in Diver, 'Policymaking Paradigms in Administrative Law', p. 417. For a study of the many aspects of cost-benefit analysis in relation to environmental regulations, see D. Swartzman, R. A. Liroff, and K. G. Croke, *Cost-Benefit Analysis and Environmental Regulations: Politics, Ethics and Methods* (United States, 1982). For discussion of how economic efficiency might be accommodated with other goals and values, see B. Ackerman and W. Hassler, *Clean Coal/Dirty Air* (Yale UP, 1981) and R. Stewart, 'Regulation, Innovation and Administrative Law: a Conceptual Framework' (1981) 69 *CalLR* 1259.

26. In this section I am following the analysis of rules by I. Ehrlich and R. A. Posner in 'An Economic Analysis of Legal Rulemaking' (1974) 4 *JLegStud* 257. See also W. Hirsch, 'Reducing Law's Uncertainty and Complexity' (1974) 21 *UCLALR* 1233, and A. I. Ogus, 'Quantitative Rules and Judicial Decision-Making' in P. Burrows and C. G. Veljanovski (eds.), *The Economic Approach to Law* (London, 1981).

27. Ehrlich and Posner, 'An Economic Analysis of Legal Rulemaking', p. 268.

28. For useful discussion of examples and case studies, see Schuck, 'Organization Theory and the Teaching of Administrative Law'.

29. For discussion of parole decision-making, see: Parole Board Annual Reports (London); F. Beverley and P. Morris, *On Licence: a Study of Parole* (London, 1975); P. Cavadine, *Parole: the Case for Change* (London, 1977).

30. For a discussion of political constraints in the British welfare system, see Tony Prosser, 'The Politics of Discretion: Aspects of Discretionary Power in the Supplementary Benefits Scheme', in Adler and Asquith, *Discretion and Welfare*.

31. See Hawkins, *Environment and Enforcement*, pp. 70–1.

32. For a detailed study of organizational influences on the exercise of discretion, see Richardson, *Policing Pollution*. There are other factors which might be included under this broad heading, such as highly personal considerations which have affected the strategies adopted in exercising discretion. For a study which deals with such personal considerations, see Austin Lovegrove, 'The Listing of Criminal Cases in the Crown Court as an Administrative Discretion' 1984 *CrimLR* 738. For studies of the variables that operate within police forces, see J. H. Skolnick, *Justice Without Trial: Law Enforcement in Democratic Society* (London, 1966), and P. K. Manning, *Police Work: the Social Organization of Policing* (Massachusetts, 1977).

33. For discussion of how this compromise has worked out in practice in the British welfare system, see Adler and Bradley, *Justice, Discretion and Poverty*.
34. For a case study of decision-making in such an environment, see Lars Busck, 'The Family Guidance Centre in Copenhagen', in Davis, *Discretionary Justice in Europe and America*.
35. For an excellent study which considers the economic constraints on the exercise of discretionary powers, see M. C. Harper, 'The Exercise of Executive Discretion: a Study of a Regional Office of the Department of Labour' 1982 *AdmLR* 559.
36. *Ibid.*, p. 566.
37. In a study of the exercise of discretion by regional officials implementing the comprehensive Employment and Training Act, it was found that the scarcity of resources was a most significant variable affecting decisions. See Harper, 'The Exercise of Executive Discretion'. A study of pollution regulation in Britain has revealed the way in which discretion in allocating resources for enforcement are used to encourage goals which may not be required by law: Hawkins, *Environment and Enforcement*. See also K. Carson, 'White Collar Crime and the Enforcement of Factory Legislation' (1970) 10 *BritJCrim* 383.
38. See Richardson, *Policing Pollution*.
39. See Hawkins, *Environment and Enforcement*, pp. 42 ff; 99; Richardson, *Policing Pollution*, pp. 97, 124.
40. Hawkins, *Environment and Enforcement*, pp. 73–5.
41. For a study of the effect of economic constraints as one of several influences in the change of decision strategy from being rule-governed to discretionary, see David Noble, 'From Rules to Discretion? The Housing Corporation', in Adler and Asquith, *Discretion and Welfare*.
42. For general discussion of discretionary powers concerned with the regulation of economic activity, see J. Winkler, 'The Political Economy of Administrative Discretion', in Adler and Asquith, *Discretion and Welfare*.
43. See Lars Busck, 'The Family Guidance Centre in Copenhagen'.
44. Notice, however, that as administrative units increase in size, decision-making tends to be subjected to greater control by rules; see: A. A. M. F. Staatsn, 'General Assistance in the Netherlands' in Davis, *Discretionary Justice in Europe and America*. Compare J. F. Handler and E. J. Hollingworth, *The Deserving Poor* (United States, 1971).
45. The classic study is J. Hogarth, *Sentencing as a Human Process* (University of Toronto Press, 1971).
46. For useful discussion of the relationship with community morality, see Hawkins, *Environment and Enforcement*, pp. 7–15; K. Carson, 'Some Sociological Aspects of Strict Liability and the Enforcement of Factory Legislation' (1970) 33 *MLR* 396.
47. Hawkins, *Environment and Enforcement*.
48. *Ibid.*, pp. 28–30.

49. Ibid., pp. 35–6.
50. Richardson, *Policing Pollution*, p. 177.

References

ADLER, M. and ASQUITH, S. (eds.) (1981) *Discretion and Welfare*. London.
ADLER, M. and Bradley, A. (eds.) (1975) *Justice, Discretion and Poverty*. London.
ATIYAH, P. S. (1978) *From Principles to Pragmatism: Changes in the Function of the Judicial Process and the Law*. Oxford U.P.
——— (1979) *The Rise and Fall of Freedom of Contract*. Oxford U.P.
BUSCK, L. (1976) 'The Family Guidance Centre in Copenhagen' in K. C. Davis (ed.) *Discretionary Justice in Europe and America*. U of Illinois.
CRANSTON, R. (1985) *Legal Foundations of the Welfare State*. London.
DAVIS, K. C. (ed.) (1976) *Discretionary Justice in Europe and America*. U of Illinois.
DIVER, C. S. (1981) 'Policymaking Paradigms in Administrative Law' in 95 *HarvLR* 393.
EHRLICH, I. and POSNER, R. A. (1974) 'An Economic Analysis of Legal Rulemaking' in *JLegStud* 257.
GALLIGAN, D. J. (1981) 'Guidelines and Just Deserts: A Critique of Recent Reform Movements in Sentencing' in *CrimLR* 297.
HABERMAS, J. (1976) *Legitimation Crisis*. London.
HAWKINS, K. (1984) *Environment and Enforcement: Regulation and The Social Definition of Pollution*. Clarendon, Oxford.
HAYEK, F. A. (1960) *The Constitution of Liberty*. London.
KADISH, M. R. and KADISH, S. H. (1973) *Discretion to Disobey: a Study of Lawful Departures from Legal Rules*. Stanford U.P.
LUHMANN, N. (1977) 'Differentiation of Society' in 11 *CanJSoc* 29.
POGGI, G. (1980) *The Development of the Modern State*. London.
RICHARDSON, G. with OGUS, A. and BURROWS, P. (1982) *Policing Pollution: A Study of Regulation and Enforcement*. Oxford U.P.
SCHUCK, P. H. (1983) 'Organization Theory and the Teaching of Administrative Law' in *JLegEd* 13.
SELF, P. (1973) *Administrative Theories and Politics*. Toronto U.P.
TUEBNER, G. (1983) 'Substantive and Reflexive Elements in Modern Law' in 17 *LSR* 239.
UNGER, R. M. (1976) *Law in Modern Society*. New York.
WINKLER, J. (1975) 'Law, State and Economy' in 2 *BJLS* 103.

Compliance Strategy

K. HAWKINS

I. Compliance

Enforcement behaviour in pollution control is determined by the play of two interconnected features: the nature of the deviance confronted and a judgment of its wilfulness or avoidability. Field staff draw a distinction in practice between deviance which is continuous or episodic (*'persistent failures to comply'*) and that which consists of isolated, discrete incidents (*'one-offs'*). Because the persistent failure to comply is open-ended deviance, it is more amenable to detection. The one-off, on the other hand, is an unexpected discharge of relatively short duration, and hence less open to detection. It may be accidentally caused, but its unpredictability carries with it hints of attempts to evade detection.

Detectability is linked with avoidability. The field officer's common sense suggests that where a pollution is persistent, its perpetrator is easily detectable. Yet this ready detectability only rarely suggests wilful persistence. Instead the persistent failure to comply prima facie implies that the rule-breaking is unavoidable, the result of inadequate resources or knowledge. Where deviance is unavoidable, a strategy of compliance is called for to repair problems.

A one-off is a more sinister matter. The intimation of possible wilfulness, the suggestion that a pollution is preventable, prompts a penal response: there is no problem to repair, simply rule-breaking to punish. The one-off pollution, in other words, has the potential for resembling a traditional criminal act, making a conciliatory style of enforcement inappropriate, for there is a suspicion of 'trying to get away with it'. Its limited duration means that, compared with the persistent failure to comply, it is more difficult to discover the deviance, more difficult to detect an offender, and more difficult to establish that offender's motivation. Discerning cause here acquires considerable significance, since it is only a judgment about blameworthiness which distinguishes a preventable one-off from an 'accident'. Whether a pollution is regarded as 'accidental' or as something more ominous depends upon an image of the polluter. Such imagery 'explains' pollution.

Yet this is not to suggest that a penal response is never made to persistent failures to comply. Where an agency prosecutes in these cases, however, it is for a failure of the compliance process, a sanction for the irretrievable breakdown of negotiations.[1] Here, the failure of the compliance process suggests a persistence in wrong-doing whose wilfulness aligns the case with a deliberate one-off.

A notion of efficiency is the field man's major concern. The expeditious attainment of his given objectives means securing compliance at least cost to his future relationships with the polluter. With a persistent failure to comply, he must clean up a regularly dirty effluent. And though the one-off can prompt a punitive response, a concern for efficiency is expressed in the preventive work done to forestall recurrence—whether or not the incident is judged 'accidental'. In both cases the active co-operation of the polluter is required for the success of an enforcement strategy aimed at conformity.

What an enforcement agent understands by 'compliance' depends on the nature of the problem he is regulating. Persistent failures to comply are the episodic or continuing pollutions which are frequently or consistently above consent limits, either owing to inefficient treatment plant or inadequate attention to pollution control. The nature of this deviance is open-ended and persisting. It often comes to agency notice through proactive enforcement by routine monitoring and may have long been going on and only recently discovered or recently defined as pollution. Such pollutions are typically not regarded as 'serious' (most 'serious' persistent failures to comply have now been cleaned up), though field men may be anxious about their cumulative impact; indeed they are a classic embodiment of deviance as a state of affairs rather than as an act. The normal control response is to prescribe remedial measures to be applied in the course of the continuing relationship between officer and polluter.

Field staff sometimes describe persistent failures to comply as 'technical' as opposed to 'drastic' pollutions, a term reserved for serious 'one-off' cases.[2] The use of the word 'technical' aptly indicates a discharge above consent to which blameworthiness does not attach. Most such deviance is treated as routine, uneventful: 'I think people see discharges as outside consent conditions, and have been for a long time, and people don't take that much notice'. The enforcement response is modest; 'We would', said an area supervisor, 'just tend to niggle away at the bloke'. A persistent failure to comply will be treated as a more serious matter only where the discharge is regarded as substantially beyond consent limits and the pollution is noticeable.

The responses of the persistent polluter are, however, carefully monitored by field men. Here enforcement is a continuing, adaptive process. Persistence allows the construction of a career assembling past problems and past efforts at remedy made by the agency, interlinked with the polluter's responses to them.[3] To think in terms of an enforcement career is a useful device for the officer, structuring his expectation and moulding his responses. Present conduct has meaning and significance conferred upon it by past history; career creates context.

The one-off pollution is potentially of much greater significance. Almost by definition it will be serious simply because it is less open than a continuing pollution to detection through routine monitoring. The one-off pollution is a discrete event; it may be momentary or longer-lasting, but it is bounded in time with a finite beginning and end (its end sometimes the result of enforcement work). It is critical rather than chronic deviance; an 'incident', where a persistent failure to comply is a 'problem'.

The one-off normally comes to agency attention as a result of complaint from a third party; sometimes, though, it is difficult to detect: a discharge can be turned off or an effluent diluted before a field man arrives on the scene. And where a persistent pollution has overtones of the (for now) 'unavoidable', or possibly the 'careless', the one-off prompts a different moral categorization. It may be 'accidental'—a spillage, a leakage, a breakdown of treatment plant—or, more ominously, 'deliberate'—acid baths being emptied or settlement tanks flushed out.

Enforcement activity with one-offs is directed towards correcting damage done, preventing its recurrence, and deterring others. Because the one-off pollution by definition implies the existence of a discharge which is not normally made, or the presence of a pollutant in a normally clean discharge, compliance is relatively simple to secure where the effluent is still running. Field staff will demand a stop to it, and the more serious the pollution appears to be, the sooner they will expect effective remedial action. Compliance here can be instant: taps can be turned off, an effluent pipe blocked off, polluting liquid diverted or disposed of in waste tankers.[4]

Where persistent failures to comply or preventive work with one-offs are concerned, compliance is a continuing process, an organized sequence of requests and demands placed upon the regulated. Compliance here is elastic, and depends on the apparent 'progress' made by the polluter and other exigencies affecting his relationship with the field officer. The issue is not usually the discrete one of whether or not the discharger will take action, so much as whether he will act as quickly as the officer wishes and to the extent required. The stance adopted by the field man during negotiations and the sorts of demands made of the polluter will be influenced as much by interpretations placed upon the polluter's response as by the nature of the 'problem' to be remedied and the resources available for tackling it. The technical implications of the behaviour addressed by law require this fluid conception of compliance in practice. The nature of the problem has to be diagnosed, a remedy prescribed, installed, and made to work efficiently. This obtains even where the remedy may be a simple matter of digging a hole to serve as a settlement pit, for effective action is the central concern. In some cases compliance may involve installation of complex treatment plant requiring heavy expenditure of money and time. Elaborate apparatus often demands regular attention and maintenance to produce an acceptable effluent. Because compliance has an unbounded quality and is subject to constant negotiation, field men do not work to a notion of instant conformity when dealing with persistent problems. Instead compliance in one sense is measurable in months or years, and in the sense that discharges subject to continued monitoring may deteriorate or standards change, compliance is 'for now'. Where states of affairs are concerned compliance strategy requires constant vigilance.

Since water use is a continuing activity it is difficult for a field officer (unlike a policeman: Bittner, 1976b:281) to think in terms of cases being 'closed', which has implications for what he may regard as 'compliance', and for the ways in which he is able for organizational purposes to demonstrate a successful outcome to his enforcement activities. One of the important features of 'compliance' for the pragmatic field man is the existence of some visible evidence of the consequences of his enforcement activity: a cleaner effluent, a new treatment plant, or a discharge diverted from watercourse to foul sewer. Such evidence provides a considerable sense of professional accomplishment:

We drove to a colliery, whose discharges had in the past caused a great deal of trouble, to see two settlement lagoons designed to reduce the level of suspended solids in the effluent. Both had been installed at the field officer's insistence, one at a cost of £80,000, the other at £60,000. The officer thought the lagoons had greatly improved the quality of the effluent and there was no longer a 'problem'. 'I'm bloody proud of that,' he said, pointing to the larger of the lagoons. 'I look at those settlement lagoons and I think "That's me what's done that".'

The field officer's evaluation of satisfactory compliance is not geared to legal output measures[5]. So far as the one-off pollution is concerned, he will be satisfied if the discharge is stopped and precautions to prevent repetition are taken. If a one-off is categorized as accidental, the preventive work is often nothing more than some good advice about the need for greater care. A less conciliatory stance is only called for in the rare cases of major spot discharges, which are qualitatively important for their symbolic significance. The pollution which may have occurred as the result of negligence, and especially the persistent failure to comply, require a different approach. Negotiations over the introduction of new or improved pollution control facilities may take months or even years before the officer is satisfied that the discharger has 'complied'. And then compliance must be maintained.

Compliance, then, is much more than conformity, immediate or protracted, to the demands of an enforcement agent. The continuing relationship between officer and polluter, the open-endedness of problems encountered, and the pragmatism of field staff encourage a focus upon the deviant's efforts at compliance, an opportunity denied the deviant in breach of a rule in the traditional criminal code where an act committed is over and done with and beyond repair. A polluter who displays an immediate willingness to take whatever action is necessary may well discover that the gravity of the pollution itself is accorded less importance by the officer: 'it can become a secondary feature,' said one field man, 'if co-operation from the firm is complete'.

Compliance, in short, has a symbolic significance. Enforcement agents need, as much as a concrete accomplishment, some *sign* of compliance. Planning is as important as building; intention as important as action. Assessments of conformity thus tend to be fluid and abstract, rather than concrete and unproblematic. 'Attitudes' are judged as much as activities:

KH How important is the attitude of the other person?
FO Oh, I think that's the most important thing, is his attitude. Because the pollutions themselves can be so variable . . . If he's trying to solve it, I go along with him. If he's not interested in it and thinks 'Well, it will go away in time anyway,' then obviously I'm going to press him harder then. Yeah, it is *the* most single important parameter I think, his attitude. [His emphasis]

The discharger who does what the field man asks—even though he may still be polluting—will be thought of as compliant. Compliance in practice is a continuing effort towards attainment of a goal as much as attaining the goal itself. The extent to which pollution is controlled is no more significant in a compliance strategy than the extent of the polluter's good faith (Goldstein and Ford, 1972:38; Holden, 1966). How 'good' the faith is, however, depends on the kind of polluter encountered.

II. Images of polluters

It is possible to discern four working categories employed by pollution control staff, of dischargers. Their images are broadly drawn and depict both their individual contacts and the larger organization employing them. These working categories embody impressions of 'typical' kinds of polluter, as well as 'typical' kinds of problem. Polluters are identified by characteristics such as occupation, size, demeanour, and responsiveness. The characterization settled upon is important in contributing to a judgment whether a discharger can be held 'responsible' for causing a pollution (see ch. 8, s. ii) [see *Environment and Enforcement*, Ch. 8].

Most are regarded as *'socially responsible'*.[6] Polluters here comply as a matter of principle, manifesting a 'personal disinclination to act in violation of the law's commands' (Kadish, 1963:437) which spills over into the corporate identity of the firm. Profitable companies and most large undertakings, including the nationalized concerns, form part of this group (see also Katona, 1945:241; Lane, 1954:95; Staw and Szwajkowski, 1975). They are generally viewed as helpful and responsive to the enforcement activities of the agency. The epitome of the 'essentially law-abiding' or 'public-spirited' discharger was said to be one well known company which gives its employees a course on pollution work and its significance, and assigns them precise jobs if a spillage occurs. Though it may impede economic interests, compliance with the law is for some firms a fundamental tenet of company policy. 'Large industries', said an area supervisor, 'have a policy of "We conform with the law—however much it costs".'[7] Similarly, nationalized concerns are expected to be 'socially responsible' because they are particularly concerned about their reputations and also possess the resources to afford whatever remedial action is thought necessary.[8] In a large organization the presence of personalities committed to a policy of compliance in senior positions is regarded as significant. It is not that the socially responsible do not cause pollution, rather than when pollution occurs it is likely to be defined as an accident, for it will almost certainly be an isolated instance of deviance. The socially responsible discharger in these circumstances will alert the agency, co-operate fully in clearing up, and take steps to prevent a repetition.

Socially responsible dischargers, however, rarely possess unblemished virtue. Most are regarded as reluctantly law-abiding and have to be talked into cleaning up their effluents or taking preventive action. One of the arts of pollution control is to persuade polluters to bear costs in ways which are rarely justifiable commercially, hence the use of the term 'dead money' by field staff to describe expenditure on treatment plant. The image of water users as essentially law-abiding comes with field men's claims of success in achieving compliance in the majority of cases.

Most polluters, however, are half-hearted in their resistance to the demands of the water authority. Those dischargers who cause serious difficulties are in a small minority. This was not always so. The more experienced field staff have discerned a greater pollution consciousness among dischargers and public alike which has been growing since the 1950s. Staff regard this shift as helpful to the task of control. This is not to suggest that dischargers are incapable of deviance, but that serious pollution as a result of negligence or deliberate misconduct is believed now not to be widespread or to occur regularly.

The result at field level is a dual image of dischargers. Staff still expect a measure of resistance to their enforcement efforts.[9] It is portrayed as almost a ritual response to an enforcement agent. For dischargers to 'try it on' or 'try to pull a fast one' is entirely normal behaviour. Disclaimers, evasiveness, and delay are common moves in the enforcement game. Dischargers are expected to 'drag their heels': 'usually people are pretty slow to spend money,' said an area man, 'no matter which sector of the public they come from'. This is true even of those who comply in principle, because delay admits the possibility that one might not have to do what one ought (Silbey and Bittner, n.d.:9). Their hesitancy is not questioned, but is taken for granted to be a desire to avoid commercially unjustifiable costs recoverable only by higher prices. Polluters have nothing to gain from treatment measures: 'It's no skin off his nose if it carries on going downstream. It's one less problem off his mind. He's got less slurry or less whatever to deal with. Or he hasn't got another soakaway to dig'. Many dischargers regard compliance, in the words of an industrialist, as 'doing as little in the way of treatment as they [can] get away with' (Barrett, 1977).

The three remaining categories embrace cases where compliance is less than ideal. First is a group held to be '*unfortunate*'. These dischargers find it difficult to comply completely with the agency's demands, owing to technical inability or to physical or economic incapacity. Agencies recognize that many purification processes are only imperfectly understood and effective treatment is sometimes beyond the competence of available expertise. They also acknowledge that some dischargers do not possess the physical resources (in the form of available land and appropriate topography) to permit effective pollution control—quite apart from the economic costs in terms of capital outlay or potential unemployment which could be the consequence of over-zealous enforcement. Economic incapacity is recognized by field men as a 'genuine' reason for non-compliance, perhaps because for that very reason many of the agencies' own sewage works perform badly. Enforcement in these circumstances produces a feeling of resignation, even of impotence: 'I know before I get there the effluent's going to be bad, and all I'm doing is producing figures so that maybe we can present them at the end of the year. It does seem rather a waste of time in some cases'. Moral inferences as to the willingness of the polluter to comply are not drawn here, in contrast with the two remaining categories.

The third group is comprised of the *'careless'*. Many dischargers find it difficult to adopt to new ways. Many old industries, in existence long before the water authorities or their predecessors, continue to behave as they have always behaved, only to find that the invention of new regulation recasts them as deviants. Then there are other dischargers who, through sloppy management, incompetence, inadequate internal sanctions, or a negligent labour force, regularly fail to maintain their effluents to an acceptable standard, or from time to time cause pollution incidents (cf. Kagan and Scholz, 1979). For field staff the worst form of carelessness amounts to outright negligence, and is particularly exemplified by the discharger whose failure to take preventive measures recommended by them results in pollution. Such negligence is both a symbolic disdain for conformity to the law, and a predictive construct: 'It implies an attitude that will determine whether they will meet our standards and whether I need to be strict with them,' as one officer put it. It is sometimes difficult in practice to distinguish an 'accidental' pollution from a 'careless' one. Sometimes the images of the polluter which 'explain' the deviance are sharpened in the course of negotiations. Sometimes the image will be redrawn, leading to a shift in 'theory', hence in enforcement practice. Indeed it is possible for a polluter to repair by the extent of his remedial work a previously unfavourable impression, leading to redesignation of the original incident as 'accidental'.

Finally, the *'malicious'* are those who quite deliberately pollute watercourses either to avoid the costs of treatment or disposal, or in symbolic rejection of the agency's authority. Where the careless are ignorant or irresponsible, the malicious are purposive and calculating. They are capable of both isolated and persistent instances of misconduct, but are not now regarded as numerous.

While these characterizations address the reasons for any pollution, another set treat the likely response of a polluter to the process of enforcement. A field man implicitly categorizes polluters into those more able and less able to comply, the effect of which is to shape his stance in the enforcement relationship. Those thought more able to comply are treated with greater stringency and less tolerance of delay or evasion. This categorization is based on a common-sense assessment of the financial well-being of a polluter and a technical judgment about the possibility of compliance.

A more important categorization, however, one cutting across evaluations of a polluter's ability to pay, and one continually open to redefinition, is a judgment of a polluter's co-operativeness. To regard a discharger as 'co-operative' or having a 'good attitude', or, in contrast, as 'unhelpful' or 'bolshie' informs an officer's expectations about the nature of his relationship with that polluter. Co-operativeness is welcomed for facilitating the job of enforcement and for encouraging principled compliance: 'If you get on well with them, they're more likely to look at the moral issue [of complying] than the economics'. The suggestion of willing compliance from the 'co-operative' polluter announces a respect for the officer's authority and reassures him that his demands are not only reasonably put, but *legitimate*. Besides, a show of compliance is a means of coping with uncertainty, as 'something is being done'.

Characterizations emerge in everyday work and are shared (and thus transmitted to new recruits) in the area office. Informal contact with colleagues creates opportunities to learn what kinds of pollutions and what kinds of polluters 'cause trouble'. The tendency for a label applied to one polluter to be generalized to others defined as falling into the same class is recognized by some: 'You don't trust one of them and it rubs off on the others'.

Past experience is also a source of characterization, favourable or otherwise. The polluter who generates an unfavourable reputation in the agency or who has regularly been in trouble is accorded greater scrutiny and less tolerance when assessments are made: 'You know, if the guy's got a previous history you tend to look on him a bit more harshly than somebody else'.

Officers' characterizations also embody assumptions about a discharger's incentives for complying with the law. It is for this reason that larger companies, with reputations to protect, are expected to be 'co-operative' and 'socially responsible'. Some of the signs emerge from a field man's comments about a large oil company. Responsiveness and willingness to act are important: '[They are] very co-operative because they implement any suggestions we care to make, and they've got a turbo-aerator plant costing £25,000 to improve their effluent. Any suggestions you might make—they'll take note. They'll get things done which they think are necessary'.

But in smaller companies, where margins are narrow and resources meagre, it is assumed that the dictates of commercial rather than legal norms are more likely to be obeyed. Here are found the 'fly-boys' or the 'fly-by-night' polluters who are 'here today, gone tomorrow'. They fall into the 'careless' or 'malicious' categories, whose response will be expected to be 'unco-operative'. These terms are most likely to be employed of small industries in urban areas, especially where 'self-made men' are involved. Such people (it is believed) are motivated solely by profit: in contrast with those who work in large companies who are 'very interested in their image, . . . with a little firm they'd just say, "Oh, sod it" '. Some rural field men say the same about farmers: 'I find the people with money, that have not been born into it—that are jumped up to it—they're the people that hang on to it and are the most difficult'. The expectation of an 'unco-operative' attitude will be strengthened if the polluter happens to be engaged in a business which produces large quantities of effluent, or effluent which is treatable only at considerable cost, such as the wastes from metal plating.

Small companies tend to be a problem, the ones that employ ten people, operate on very tight budgets, have no money for effluent treatment, have no technical people to operate what little treatment they may afford. They can be the troublesome ones, I think . . . they're the ones that you're constantly going back to and badgering . . . It often is enhanced by a very small site. Effluent treatment and land availability often go hand in hand, so you find that small firms in . . . conurbations where they're restricted as to what they can do are the worst polluters I think, without doubt.

Farmers are an occupational group regarded by officers with rural patches as particularly troublesome, partly because of lack of resources, partly because of a characteristic stubbornness born of decades of water-use unencumbered by the attentions of any regulatory agency. They have a culture of their own which sometimes impedes the officer in doing his job: 'they're a tight-knit community,' said an officer of long standing from a large rural area, 'and they won't inform against one another. There's a lot of cover-up'. Indeed, of all the occupations regularly encountered, the farmer is the most consistently described as difficult to handle, as an area supervisor's question, posed with deliberate ingenuousness, acknowledges: 'Farmers, in particular where you get an unco-operative one, are

[some] of the worst customers that we come across. I don't know whether you've heard that before?'

The officer's understanding of the reasons why particular kinds of polluters comply helps shape his choice of enforcement tactics, especially in those cases where the field man expects or is already experiencing 'trouble'. One assumption, with profound implications for enforcement behaviour, is that dischargers are sensitive creatures whose feelings may be easily bruised if urged to do too much, too soon. To 'use the big stick' or 'crack the whip' too zealously may be counter-productive (similarly Kagan, 1980; Stjernquist, 1973). To be too eager or abrasive in enforcement work is to risk encouraging in polluters an unco-operative attitude or even downright hostility. This is a major foundation of the commitment to a conciliatory style of enforcement relying on negotiation as a means of securing compliance; 'co-operation can not be established in the atmosphere of suspicion and distrust that rigid application of the law generates'. (Nicholson, 1973:4).[10] In practical terms this assumption supports two related imperatives for field staff aimed at preserving relationships: 'be reasonable' and 'be patient'. Rather than explicitly seeking to secure compliance at the outset by coercion, officers must demonstrate an understanding of the polluter's problems by discussion and negotiation. Enforcement takes time.

Another assumption is premised on a belief in the efficacy of individual and general deterrence for organizations concerned with their business prospects and public image. Since businesses and other organizations are regarded as rational institutions which act purposively, they are held to be ultimately amenable to law as a form of control. In practice, the working concept of deterrence is perhaps broader than that contemplated by the law, which is presumably founded on a belief that compliance occurs because those tempted to cause a pollution will be deterred by the threat of the legal sanction. During the research period the maximum fine available to the courts for each pollution charge was £100 on summary conviction in a magistrates' court, £200 on conviction on indictment in the Crown Court.[11] These sanctions were universally regarded by staff of all ranks as inadequate, indeed derisory, deterrents for all polluters except the most impecunious of farmers.

This did not, however, lead to an abandonment of general deterrence as a principal stated rationale for prosecution. Deterrence, rather, resides in the threats which precede use of the formal law, or in the informal sanctions which accompany it (ch. 7) [see *Environment and Enforcement*, Ch. 7]. Because field staff intuitively perceive one-off pollutions in cases which are not satisfactorily attributable to 'accidental' causes as the outcome of rational, purposive, or negligent behaviour, they assume that dischargers are susceptible to other kinds of deterrence than the criminal sanction. Where a polluter is being less than co-operative and compliance has clear, easily-quantifiable economic costs attached to it, field staff negotiate by generating contexts in which deterrence will take on significant meaning. They draw attention, for example, to some of the difficulties which the discharger will bring upon himself if he is prosecuted. An individual's or an organization's public reputation is displayed as at risk if non-compliance continues. That aspect of the use of formal law believed to be most important for industrialists ('farmers are different kettle of fish') is the publicity associated with court appearance. Few officers suggest that there is a stigma with economic implications reflected in damage to sales and profitability which attaches to a manufacturer found guilty of polluting a watercourse. Instead it is assumed that a company will seek to protect its reputation as a good-in-itself. Some, however, believe a company's motives to comply are linked with a generalized desire to avoid the 'trouble' stemming from public embarrassment. Publicity implies the construction of reputation, and companies do not want to be seen as 'troublemakers', since the label will encourage others—especially public authorities which may have benefits to confer—to lay blame for other problems at their door. Other industrialists, it is believed, will not do business with those who engage in sharp practices. The publicity of prosecution can cause all sorts of 'trouble', as an experienced area man suggested:

> It's not so much public relations and the consumer, it's the . . . people who live in that particular area who kick up trouble about [certain kinds of industry], for example . . . In general, the public attention is focused on them. They start ringing their MP and all of the rest . . . Every firm likes to give an outward impression of responsibility to the public . . . I don't think it affects their sales, y'know, unless it gets to the point of national press campaigns. But . . . locally they're very aware of it because they know jolly well if they want to expand, if they want to make alterations and all the rest of it, they're going to get public enemies all the time . . . [The authorities] have got to investigate all public complaints . . . They know jolly well that if they don't satisfy the public, he's going to bring in his local councillor and his MP. [With some firms] you hint about publicity and, boom, that'd be done straight away, that. They're very conscious of it, it's a very useful weapon.[12]

Another kind of 'trouble' may also encourage management to comply. Internal mechanisms of control may well be an effective constraint against carelessness or misconduct in large organizations, both for senior staff and those at junior levels. 'There's a tendency for any adverse publicity to reflect on the management personally,' said one officer, who had worked in industry before joining the water authority. In other words, individuals likely to be held 'responsible' for a company's prosecution have a substantial incentive to protect their personal reputations and positions (Dickens, 1974). The favourable characterization of the large organization is again supported, as a young officer explained: 'Big companies will jump on their employees. We went round to a big American firm and the Manager said "If we don't get that interceptor within the week, I'll get the sack". So he jumped'. Although talk of the sack in this case may be an exaggeration designed to impress the officer of the urgency with which the firm was complying with his request, the sense of an internal sanctioning system to be taken seriously is clearly conveyed. Indeed, in many firms shopfloor workers held responsible for a pollution which leads to prosecution risk the sack.[13]

The important feature in all of this is the threat of public stigma associated with prosecution for pollution. It is believed to be a more powerful incentive to compliance in more suburban and rural areas where greater value attaches to reputation, and where adverse publicity is more readily transmitted, when, in the familiar phrase, 'the local paper will go to town'. Companies will go to considerable lengths, it is thought, to keep their names out of the papers. The concern about reputation can be exploited by ensuring that maximum publicity is generated about prosecutions, a supervisor explained, to assist enforcement and control in other troublesome cases:

What you would do is you'd make certain the newspapers were there—being a bit naughty. I've done this time and time again, y'know, a little tip-off to the local newspapers to go along. That used to help a lot. There used to be a splash there. You know, it helps your cause.

The real value of the stigma, however, is more symbolic than concrete. Prosecution in a compliance system is valuable not as a sanction, but as a threat, because (as the next chapter shows [see *Environment and Enforcement*]) the threatened sanctions can be made to appear more serious than they are.

III. The enforcement game

A major assumption shaping an officer's enforcement strategy and moulding his relationships with dischargers is that most polluters will ultimately do what he wants them to do.[14] Yet most polluters are expected to display reluctance and disingenuousness at the outset. Neutralization (Sykes and

Matza, 1957) by way of disclaimers and evasiveness is regarded as a normal response to pollution control work, especially early in an enforcement relationship. The enforcement of regulation, in short, is conceived of in ritual terms. The metaphor of the game is frequently invoked for descriptive purposes (cf. Edelman, 1964:44 ff; Ross, 1970:22). 'Everyone tries it on, at least to begin with. Even the big firms who will do what we want eventually,' said a field man from a mixed catchment. 'They will all say "It wasn't us", or "It was an accident".' Asked how often dischargers tried to pull the wool over his eyes, he replied:

Very regularly, very regularly. Nearly everyone to more or less a degree will try to kid you about something. Either the nature of the cause of the pollution, or how long it's been going on, or they 'weren't aware that there was a pipe there', or 'Is it really? I've never been down and looked at that watercourse for the last 20 years, I didn't realize we were causing a problem.' But nearly everyone tries some minor deception . . . They will all have a go . . . even the biggest companies where you're going to get the perfect response, but they will still try to kid you that they 'weren't aware that this was happening', or 'it was while they were on leave' or, y'know, they 'weren't doing it at all'.

Unless, however, the discharger is defined as belonging to a class considered more prone to pollute, or has otherwise displayed himself as unco-operative, his ritual reluctance will be expected to yield ultimately to compliance.

Compliance is not usually achieved without a struggle. Most enforcement problems are caused by a small number of malicious polluters and another larger group of unco-operative dischargers who continue to conduct their affairs as they wish, despite the attentions of pollution control staff. When efforts are made to have them stop or clean up their effluents they adopt protective strategies akin to those which delinquents employ to affect the outcome of their cases (Emerson, 1969:101 ff.). These 'bolshie types' are few in number but occupy disproportionate amounts of field officers' time.

Bolshie polluters employ a variety of strategies to resist, delay, or avoid enforcement. Most officers encounter dischargers who seek to evade detection, or if discovered, attempt to deflect their attentions by resort to various forms of deceit. It is all part of the game. Some polluters go to considerable lengths to pull a fast one: 'They have a man on the look out,' said one officer from a highly industrialized area, 'and as soon as you appear on the horizon, he's off. Or they dump the stuff on a Sunday evening because they know we don't work on a Sunday evening'. A colleague from a mixed catchment gave another kind of example:

I've had a certain amount . . . of y'know making discharges when they think you're not about. I had a farmer who kept me talking and feeding me with tea and biscuits while the farm labourers were emptying the silage tank that had been overflowing into the brook. And then [he] took me along to show me this tank and [said] 'Well, you see it couldn't be me, it's empty'—and it was still dripping wet!

Perhaps the tactic most familiar to field staff is for their admission to premises to be delayed so that incriminating evidence can be disposed of. 'The standard thing', said a supervisor, '. . . was . . . when the [officer] arrived, to keep him hanging about while they turned on the fire mains and all sorts of things so they diluted the effluent. So when he took a sample it was perfectly alright.'

A commoner source of difficulty for the enforcement officer, however, is to persuade dischargers to overcome their reluctance to take effective remedial action, once their pollutions have come to light. Delay is the commonest ploy adopted here. Delay is routinely expected, but difficult to establish especially where technical questions are concerned, or the polluter has to rely on others for the manufacture or supply of equipment to enable him to comply. Yet an officer cannot afford to overplay his hand; efforts to speed things up may prove to be counter-productive:

There are always delaying tactics . . . We may be talking here about a method of treating a difficult waste, so the question is how on earth do you treat it? . . . Now this can go on for ages and ages [but] they just keep on saying 'Oh well, we've written to these manufacturers and the experiments have been going on, but it broke down. . . . But you can't prove they're not getting on very quick with it . . . People are very wily and very clever and if they think they're over pushed there is a chance of . . . well, slightly delaying and procrastinating on a thing. They always come up with a story of 'Well, we haven't got planning permission yet.'

A supervising officer described another kind of delay:

You come across the technique of never really being able to . . . get a decision because you're always chasing the elusive man who can make the decision. . . . Some of [the local authorities] are much more adept at playing this game than an industry is. You get a planning committee going. They always *do* something; they're always doing things—working parties, consultants, new committees. And you go round and round and round, and in the end you just have to turn round and say 'Look you've got to do something now—or else'. [His emphasis.]

And so far as industry is concerned:

You've always got works and things going on and you're always digging the place up . . . they always employ consultants and they're always looking into problems, they have lots and lots of meetings and things and you can never really screw them down to a day when something is actually going to be finished . . . Some of the big firms do use [the] technique of hiding the person responsible . . . Once you've got as far as you possibly can go with one person and it's obvious that you're getting a bit shirty . . . then someone else will appear in the chain of command and you go and see them. And it is a job to find who was really responsible. You know, they . . . play this . . . managerial game.

In some cases the use of delaying tactics is bolstered by the polluter's attempt to display himself as the blameless victim of *force majeure*: 'They have all sorts of excuses: delay in arrival of spares and breakdowns and this and that; the factor being on holiday and not being able to produce the goods. There's stacks of that; I think that's all part of the game really'.

The use of excuses to deflect accusations of blame suggests some familiarity on the part of dischargers with field staff's conception of the rules of the game. Extensive dealings with pollution control officers make available to polluters a sense of the features of any problem which officers define as significant. The more frequent the involvement with pollution control staff, the more the polluter acquires the awareness of the 'repeat player': he learns the ropes and the procedures, and develops counter-strategies (Galanter, 1974). For instance, the officer who receives a report of a pollution from the polluter himself will be much more favourably disposed than if the pollution came to light as a result of complaints from third parties or even following routine monitoring. The polluter accordingly learns that his own early warning of the pollution is highly valued, from the requests field staff put routinely to dischargers to report any trouble to the agency immediately (Brittan, forthcoming). By alerting the agency when the pollution (however caused) occurs he has a good chance of avoiding the possibility that blame will be attributed to him: favourable moral evaluations lead to more tolerance on the part of field men and a greater reluctance to sanction misconduct (ch. 8, s. ii [see *Environment and Enforcement*, Ch. 8].[15] Someone who does not alert the agency of a noticeable pollution will be treated prima facie as 'trying to pull a fast one'. On the other hand, a polluter who presents himself as well intentioned may well be able to slow the pace at which he is brought into compliance (but if he subsequently finds himself in trouble, he may discover that his earlier reluctance is now characterized as evidence of a basic unco-operativeness, prompting a less sympathetic approach).

There are, of course, other games, of greater or lesser sophistication:

You get the attempt to use muscle on a personal level, and you get attempts to use muscle on a political level. You get both. And it works. One sees it working where there is clearly pressure being brought to bear, up in the dizzy heights, and as a result you're restricted in the things that you can do with . . . either individuals or companies. Similarly one gets intimidation on a personal level, more so from our agricultural friends than from industrialists.

In the case which is partly described in Chapter 3 [see *Environment and Enforcement*, Ch. 3]:

the officer had received a report of a pollution of a small stream in a public park in the city. The water was cloudy and foul-smelling. The pollution was traced to a small ice-cream manufacturing company, where the manager claimed that a vat had accidentally overturned, spilling the contents down a drain. The officer found the man extremely hostile and aggressive and for this reason decided not to take a formal sample. though in his view there was every reason to do so. The officer thought the manager was lying: the vat had been deliberately overturned, because the ice cream was substandard and tipping it down the drain was the most convenient way of getting rid of it.

Discussing this case later, the officer suggested the firm had probably escaped prosecution because of the manager's belligerence, which made him think first of his personal safety.

FO It wasn't that we didn't want to take a stat; we were anxious to do so, but because this gentleman had been throwing [his product] at us and threatening other forms of physical violence, it was just impossible to get onto his premises to take a statutory sample. In cases like that, one is advised by the authority not to get a broken nose but to retire gracefully and seek a magistrates' warrant, and return. Which is fine if it's a long-term problem, but in the case of a one-off then the intimidation works, and by the time you return then the problem has passed.
KH . . . If you are sufficiently belligerent following a one-off pollution then it's very difficult for the officer to enforce the law in effect?
FO Yeah, yeah, . . . it can be made virtually impossible if you have to enter land to get the sample.

The bribe is doubtless the best known means of evading law enforcement, though as a taboo subject it is difficult to research (cf. Blau, 1963:187–93). Subtly employed, it can be a useful delaying tactic: 'We used to think that a good lunch would put it off for another three months,' said an officer who had worked, before joining his authority, in an industry notorious for its pollution problems. Sometimes, it seems, polluters are less discreet and attempt to persuade field officers to accept a gift in return for favourable treatment:

I was once taken to a showroom of a very reputable company when dealing with a pollution. I'd just taken legal samples, and my biggest problem was to get out of the showroom and get on with dealing with the pollution, when he said to me, 'Is there anything you want?' Y'know he didn't say 'Is there anything you want to drop the prosecution or throw away the samples?', or—he just said, 'Is there—if you want anything, just say'. And he put it as simple as that. And the cheapest thing . . . they made was about £400 . . . He made it perfectly clear that, y'know, I could have had what I wanted.

IV. Bargaining

The voluntary compliance of the regulated is regarded by the agencies as the most desirable means of meeting water quality standards. For the agencies it is not only viewed as the most effective strategy, it is a relatively cheap method of achieving conformity.[16] For agency staff it is a means of promoting goodwill, a matter of profound importance in open-ended enforcement relationships which must be maintained in the future. Compliance takes on the appearance of voluntariness by the use of *bargaining*.[17] Bargaining processes have 'a graduated and accommodative character' (Eisenberg, 1976:654, italics omitted) which draw their efficacy from the ostensibly voluntary commitment of the parties. The more legalistic style of penal enforcement with decision-making by adjudication and the imposition of a sanction risks, according to agency staff, continued intransigence from the guilty polluter. Bargaining is central to enforcement in compliance systems; control is buttressed for it is derived from some sort of consensus (Schuck, 1979:31). Bargaining implies the acquiescence of the regulated, however grudging. And it inevitably suggests some compromise from the rigours of penal enforcement.

The essence of a compliance strategy is the exchange relationship (Blau, 1963:137 ff.), a subtle reminder of the mutual dependence which Edelman (1964:47) regards as central to the conception of the game. The polluter has goodwill, co-operation and, most important, conformity to the law to offer. The enforcement agent may offer in return two important commodities: forbearance and advice.

The offer of forbearance is the opportunity for another display of the officer's craft. He will not ask for costly remedies unless the problem is a major one or the polluter is undoubtedly wealthy. He will recognize inherent constraints facing the polluter, such as lack of space. He will respect a previously co-operative relationship. Most important, he will offer a less authoritarian response than that legally mandated. He offers the polluter time to attain compliance, for bargaining strategies 'are based on the principle that success in pollution control is "bought" by giving up some of the demands that are fixed in the legal norms to be implemented' (Hucke, 1978:18).

Bargaining is possible, then, only *because the law need not be formally enforced*. Rules are a valuable resource for enforcement agents since, as Gouldner has observed (1954:174), they represent something which may be given up, as well as given use. The display of forbearance is valuable in obliging the polluter to take action in response to the show of leniency:

instead of leaving the impression that you're some jumped-up little upstart from an office using the law to tell him what he must do, if you talk to him right, you finish up leaving him with the view that 'Well, he's a damn good chap . . . I could've been prosecuted for this. I'm breaking the law, but he's obviously going to shoot it under the carpet and let me get away with it.' So . . . he does what he has to do, with goodwill, and everybody's happy.

Or, again:

I've said 'Alright, well, look, I've got this stat sample here and this could be used against you in legal proceedings, but provided that you play ball with me and get done what I want to get done, then I'll get rid of it.' And [I've] very often poured it out in front of his eyes. And you'd be amazed. This has done the trick on every occasion that I've ever dealt with. They're *so grateful* . . . [His emphasis.]

Discreetly coercive bargaining is equally useful in preventive work:

You can say 'If you build this bund wall or you take these steps, we'll drop the stat.' You can achieve a lot with this—*they* feel they've got you off their back, and *we've* got them to do something. [His emphasis.]

A sense of mutual trust is important in sustaining the bargaining relationship: trust that the polluter will not 'pull a fast one', trust that the officer will not penalize theoretically illegal conduct. Field staff generate a sense of trust by showing how 'reasonable'—that is forbearing—they are, polluters by displaying a willingness to conform and a readiness to report 'problems'. Forbearance aids the detection of pollution by encouraging self-reporting whenever there is an escape of effluent. The polluter himself is, after all, in the best position to discover and control the pollution and prevent it from becoming a public matter. 'You do tend to learn an awful lot more,' said a supervisor based in a major conurbation,

particularly if they know what your actions are going to be. And also if they realize that when things do go wrong, if they tell you . . . you're going to react in a sensible sort of way. And when something goes wrong that they could be prosecuted on, they would still tell you because they know that . . . you're going to be more reasonable with them, because you know that they haven't hidden things in the past.

The field man will do all of this in recognition of the belief that without forbearance, compliance is the most difficult to attain. What he will demand in return for his forbearance is some show on the polluter's part of compliance with his requests:

FO I always explain to them . . . that they can do it in active co-operation with us, with us showing forbearance, and we give them as much advice and help as possible. Or they can be awkward and they can do it having been prosecuted . . . and with us watching them like a hawk forever afterwards. Which option would they choose? And when I sell it to them that way which way would anyone choose? They are obviously going to opt for the easier option. Now they might try backsliding, but they can procrastinate once, but then we put dates on them. And legal samples are taken.

KH When you say you put dates on them—?

FO Dates for them to meet the standards. If they say 'Yes, we'll gladly spend the money required' and time goes by and they keep coming up with excuses and it is *clear that they are not in earnest*, then in that case you put a date by which they are required to meet the consent. And if they do not then I feel no compulsion in taking a legal sample. After all, I have tried my best to get them to do it the easy way. And with good feeling all round. [My emphasis.]

The symbolic properties of compliance are clear in these remarks: what is important is the display of good faith.

Since compliance enforcement is practised in a continuing relationship, 'fixing dates' by which a certain degree of compliance must be displayed provides a method of assessing the polluter's degree of conformity. If a period of time is fixed by the field officer (or, better, agreed upon with the polluter) and articulated in the form of a calendar date, the polluter is presented with an unambiguous target by which he should have stopped the discharge, dug a hole, installed equipment or replaced existing plant. Time is a good, exact index of compliance, more useful for enforcement purposes than the prescription of work to be done which is inherently open to negotiation. Deadlines can nevertheless be made flexible to provide the appearance of a softening of demands. Since compliance is often a lengthy and costly process, the field man may spin out the time allowed and present his demands in stages (cf. Kaufman, 1960:225; Thompson, 1950).

Deadlines aid an enforcement officer's appreciation of the polluter's career as deviant. A deadline met is an index of progress, of headway, while a polluter who fails to meet a deadline can be mutually recognized as a rule-breaker, even as unco-operative. In presenting such unambiguous evidence of non-compliance the field man inevitably casts most polluters into a defensive posture, forcing them to account for their failure to live up to their side of the bargain. If the field man can convincingly portray himself as having been generous in the amount of time he has allowed, he can increase the polluter's sense of obligation to comply with future demands. At the same time, failure to meet a date is a breach of the implied bargain and gives the officer grounds to be less forbearing towards the polluter, if this is tactically desirable, unless the latter can offer some plausible account for his failure. What constitutes a plausible account usually involves establishing financial or technical impediments to compliance; thus, to quote an example frequently treated as plausible by field staff, if the polluter can claim the late delivery of equipment he is usually freed from imputations of fault, and has an 'excuse' demanding forbearance from the officer.

The realization that in pollution matters conformity with the law involves more than simply refraining from proscribed conduct is reflected in the second commodity a field man can offer in bargaining. An officer's willingness to act as expert consultant recognizes the often costly and complicated business of compliance. Technical advice will be given whether the field man is dealing with persistent polluters, where the advice will be about the suitability of particular remedial measures, or with one-off cases, where he will want to prevent a recurrence. Advice is always given tentatively, since the officer must protect himself and his agency against any repercussions which may follow heavy expenditure to little effect. Field agents' preoccupation with ends ensures that advice is realistic and negotiable. The remedial action suggested will not simply depend on technical issues centred around the nature of the 'problem', but will also be geared in particular to the officer's perception of the abilities of the polluter to pay. Industrialists are considered to be economically capable of supporting more elaborate measures than other dischargers: 'I think the expensive jobs are loaded on industry and I think industry can bear the cost'. By 'industry' is normally meant companies with large numbers of employees; minor concerns are regarded as 'small men' and *ipso facto* assumed less capable of bearing the costs of control. This categorization between rich and poor is an important distinction, for it provides the officer with a means of establishing whether the troublesome discharger has reasonable grounds for dragging his heels. As most dischargers are not large private companies or nationalized industries, field staff are accustomed to offering advice about remedial action which is as cheap to effect as possible. The more modest the remedy, the more reasonable the demands seem, and the easier the task of enforcement: 'With those that can't afford it I'll say, "Look, just dig a bloody hole in the ground—that'll serve the purpose." I don't like people's money being used needlessly.' And the cheaper the action proposed, the less embarrassing it will be for the officer if his suggestion proves less than successful. The following conversation nicely illustrates the officer's stance, focused as it is on getting the job done 'for now':

FO If there is a problem, a problematical discharge at the site and there are a variety of solutions, one invariably tends to lean towards the least costly, providing it does the job . . . simply because you want the industrialist to go along with you and if you go for the most expensive one, you're less likely to get a satisfactory response. The pollution's going on longer while he's umming and ahing and perhaps at the end of the day you've got to take a stat and go through all that machinery, whereas perhaps at the first or second meeting you can agree on a much cheaper, but perhaps equally satisfactory solution. So . . . the primary consideration is stopping a pollution as quickly as possible, and if by demonstrating that there is a cheap and effective way of doing it as well as putting an expensive plant in, then in the long-term perhaps it's not as desirable, but by getting that in and getting it cleaned up, you've got breathing space to then keep wearing at them and trying to get them to improve that facility. And in the long-term maybe getting the better plant.

KH So you are saying then that to do your job it's important to know the financial state of a firm?

FO Well you need to know two things really. It's very necessary to be, within a few pounds, aware of the cost of treatment facilities, so that you're not asking for the ridiculous, as well as being able to assess whether the company can afford it, so that you know where to pitch. If there's a plant costing £50 and a plant costing £500 and a plant costing £5000, it helps to know those three costs and to know which one the company can afford, so that you get the right response when you say, 'This is necessary to improve your effluent, you really ought to put this in or put that in.' [You tell them to] get the stuff organized and get it in. But again you're thinking of financial constraints and consultants' fees and, without committing the authority, if you can in general terms give an indication of the plant that they need, give them an on-the-spot consultation if you like, for free, and get them moving . . . they're more disposed to doing it than if you say, 'Well, if you ring a consultant he'll come and see you in a fortnight and he'll charge you £200 for driving in through your gate,' and this sort of thing. It's all adding to cost. Obviously one doesn't—can't—give clearly defined guidelines as to treatment facilities because ultimately if the treatment facility fails, then you can't go back to them and say 'It's not good enough,' because they'll say, 'Well it was put in subject to your recommendations'.

V. Postscript

Compliance is often treated as if it were an objectively-defined unproblematic state (e.g. Nagel, 1974), rather than a fluid, negotiable matter. Compliance, however, is an elaborate concept, one better seen as a process, rather than a condition. What will be understood as compliance depends upon the nature of the rule-breaking encountered, and upon the resources and responses of the regulated. The capacity to comply is ultimately evaluated in moral terms, and is of utmost importance in shaping enforcement behaviour. A greater degree of control is likely where a discharger is regarded as able to bear the expenditure for compliance; this issue is still a moral one, fundamentally, not one of economics.

Compliance is negotiable and embraces action, time, and symbol. It addresses both standard and process. It may in some cases consist of present conformity. In others, present rule-breaking will be tolerated on an understanding that there will be conformity in future: compliance represents, in other words, some ideal state towards which an enforcement agent works. Since the enforcement of regulation is a continuing process, compliance is often attained by increments. Conformity to this process itself is another facet of compliance. And when a standard is attained, it must be maintained: compliance here is an unbounded, continuing state. It is not simply a matter of the installation of treatment plant, but how well that plant is made to work, and kept working. And an ideal, once reached, may be replaced or transformed by other changes—in consent, in water resource or land use, for example—which demand the achievement of a different ideal (ch. 2, s. iii) [see *Environment and Enforcement*, Ch. 2]. Central to all of this is the symbolic aspect of compliance. A recognition of the legitimacy of the demands of an enforcement agent expressed in a willingness to conform in future will be taken as a display of compliance in itself. Here it is possible for a polluter to be thought of as 'compliant' even though he may continue to break the rules about the discharge of polluting effluent.

A strategy of compliance is a means of sustaining the consent of the regulated where there is ambivalence about the enforcement agency's mandate. Enforcement in a compliance system is founded on reciprocity, for conformity is not simply a matter of the threat or the rare application of legal punishment, but rather a matter of bargaining. The familiar discrepancy between full enforcement and actual practice is 'more of a resource than an embarrassment' (Silbey and Bittner, n.d.:5). Compliance strategy is a means of sustaining the consent of the regulated when there is ambivalence about an enforcement agency's legal mandate. The gap between legal word and legal deed is ironically employed as a way of attaining legislative objectives. Put another way, bargaining is not only adjudged a more efficient means to attain the ends of regulation than the formal enforcement of the rules, bargaining is, ultimately, morally compelled.

Notes

1. About half the prosecutions brought in the southern authority are for persistent failures to comply (agency communication).
2. Similarly Ross and Thomas (1981:12) report that housing inspectors regard violations as 'chickenshit' or 'important'.
3. The concept of career is treated in Rock, 1973b and Roth, 1963.
4. Stopping (rather than cleaning) the discharge is sometimes the ultimate (but not the immediate) aim with persistent failures to comply: some polluters are encouraged to recycle waste water or dispose of it to foul sewer.
5. So far as the agency is concerned, the nature of the behaviour to be regulated and the realities of the enforcement strategies adopted mean, in effect, that it is difficult to display 'compliance' or other measures of success or impact in any meaningful fashion (cf. Wilson, 1968:57). River surveys will indicate broad changes in water quality over the years which agency staff may attribute to more extensive and efficient enforcement. But it is extremely difficult to sustain this claim with any precision owing to problems in disentangling enforcement practices from other shifts which have had an impact on water quality, such as changing patterns in land use or changes in public expenditure on sewage treatment.
6. Similarly, businesses are regarded in the consumer field as essentially law-abiding (Cranston, 1979:29). Principled compliance may be commoner in Europe than the USA, as Kelman's data seem to suggest (1981). See generally Anderson, 1966. Britain's study of discharges (forthcoming) supports the perception of a majority given to principled compliance.
7. Field staff assume that larger businesses have a reputation to protect and are thus 'better' or 'more responsive', a notion suggested by Lane (1953).
8. Brittan (forthcoming) suggests that officers' assumptions about large companies and nationalized industries are correct.
9. Compare the unquestioning compliance which police can expect in regulating traffic behaviour (Bittner, 1967a:702)
10. Kagan and Scholz (1979:15) observe that legalistic enforcement 'seems to have been a primary factor in stimulating political organization by regulated firms and attempts to attack the agency at the legislative level', while a more legalistic approach by the California Occupational Safety and Health authorities has led to an increased number of appeals (*ibid.* 14). See also Barrett, 1979; Kagan, 1980; Kelman, 1981.

11. See further ch. 1 [see *Environment and Enforcement*, Ch. 1]. These penalties have since been increased by the Control of Pollution Act 1974 and the Criminal Law Act 1977—but the sanctions are still generally regarded as modest by field men despite the provision for terms of imprisonment. Most officers, however, only had a very hazy notion of the availability of this sanction.
12. In the USA, the Environmental Protection Agency has consciously used adverse publicity as an enforcement tool: Gellhorn, 1973:1401 ff. See also Rourke, 1957.
13. See further Brittan, forthcoming. Cases have been known to occur in which shop-floor workers have manipulated the system of internal organizational control to encourage an unwilling management to comply. They have acted in a deliberately unco-operative manner, ultimately forcing the field officer to threaten the company with prosecution. This is normally enough to ensure that management make the resources available to improve their pollution control arrangements.
14. The proportion of dischargers presumed to be utterly unco-operative varies according to the nature of the field officer's patch, with those working in urban industrialized areas expecting more deviance and greater reluctance to comply than those whose areas are predominantly rural. This is despite the view of the farmer as 'troublesome'. The apparent contradiction may possibly reflect the fact that farm effluents, in contrast with many industrial discharges, are all relatively familiar and easily traced. One supervising officer from a largely urban area suggested that about 70 per cent of dischargers were 'co-operative, willing to comply'; about 20 per cent were 'slow to comply'; the remaining 10 per cent or so he classified as 'having to be forced to comply'.
15. This raises the question of the techniques adopted by field staff for categorizing dischargers into those 'genuinely' reporting a pollution, and those taking unfair advantage of pollution control staff, an issue addressed in ch. 8 [see *Environment and Enforcement*, Ch. 8].
16. A former Chairman of the Illinois Pollution Control Board has observed that 'if no one complied until prosecuted, enforcement costs would surely strangle the program' (Currie, 1975:390).
17. Bargaining is one of the central characteristics of legal processes: see Hagevik, 1968; Holden, 1966; Jowell, 1977a; 1977b; Ross, 1970; Silbey, 1978; Stjernquist, 1973:1-9; and generally Gouldner, 1960; Strauss, 1978.

References

BARRETT, J. W. (1977) 'We're Good Boys Now—But Can We Stay in Business?' in *Product Finishing* March.

BITTNER, E. (1967) 'Police Discretion in Emergency Apprehension of Mentally Ill Persons' in *Social Problems* 14.

BLAU, P. M. (1963) *The Dynamics of Bureaucracy. A Study of Interpersonal Relations in Two Government Agencies*. University of Chicago Press, Chicago.

BRITTAN, Y. (n.d.) *The Impact of Water Pollution Control on Industry: a Case Study of Fifty Discharges*. SSRC Centre for Socio-Legal Studies, Oxford.

DICKENS, B. M. (1974) 'Law Making and Enforcement—A Case Study' in *Modern Law Review* 37.

EDELMAN, M. (1964) *The Symbolic Uses of Politics*. University of Illinois Press, Urbana.

EISENBERG, M. A. (1976) 'Private Ordering Through Negotiation: Dispute-Settlement and Rulemaking' in *Harvard Law Review* 89.

EMERSON, R. M. (1969) *Judging Delinquents: Context and Process in the Juvenile Court*. Aldine, Chicago.

GALANTER, M. (1974) 'Why the "Haves" Come Out Ahead: Speculations on the Limits of Legal Change' in *Law and Society Review* 9.

GOULDNER, A. W. (1954) *Patterns of Industrial Bureaucracy*. Free Press, New York.

HUCKE, J. (1978) 'Bargaining in Regulative Policy Implementation: the Case of Environmental Policy in the Federal Republic of Germany' paper prepared for the Workshop on Inter-organizational Networks in Public Policy Implementation, European Consortium for Political Research, Grenoble, April.

KADISH, S. H. (1963) 'Some Observations on the Use of Criminal Sanctions in Enforcing Economic Regulations' in *University of Chicago Law Review* 30.

KAGAN, R. A. (1980) 'The Positive Uses of Discretion: the Good Inspector' paper presented to the 1980 Annual Meeting of the Law and Society Association, Madison, Wisconsin, June.

KAGAN, R. A. and SCHOLZ, J. T. (1979) 'The Criminology of the Corporation and Regulatory Enforcement Strategies' paper presented to the Symposium on Organizational Factors in the Implementation of Law, University of Oldenburg, May.

KATONA, G. (1945) Price Control and Business. Field Studies Among Producers and Distributors of Consumer Goods in the Chicago Area, 1942–44. Principia Press, Bloomington, Ind.

KAUFMANN, H. (1960) *The Forest Ranger. A Study in Administrative Behavior*. John Hopkins University Press, Baltimore.

LANE, R. E. (1954) *The Regulation of Businessmen*. Yale University Press, New Haven.

NAGEL, S. (1974) 'Incentives for Compliance with Environmental Law' in *American Behavioral Scientist* 17.

NICHOLSON, N. J. (1973) 'Water Pollution' unpublished paper.

ROSS, H. L. (1970) *Settled Out of Court: The Social Process of Insurance Claims Adjustment*. Aldine, Chicago.

SCHUCK, P. (1979) 'Litigation, Bargaining and Regulation' in *Regulation* July/August.

SILBEY, S. S. and BITTNER, E. (n.d.) 'The Availability of Legal Devices' unpublished paper.

STAW, B. M. and SZWAJKOWSKI, E. (1975) 'The Scarcity-Munificence Component of

Organizational Environments and the Commission of Illegal Acts' in *Administrative Science Quarterly* 20.

STJERNQUIST, P. (1973) *Laws in the Forest. A Study of Public Direction of Swedish Private Forestry.* C. W. K. Gleerup, Lund.

SYKES, G. M. and MATZA, D. (1957) 'Techniques of Neutralization: a Theory of Delinquency' in *American Sociological Review* 22.

THOMPSON, V. A. (1950) *The Regulatory Process in OPA Rationing.* King's Crown Press, New York.

Adjudication in Local Offices

J. BALDWIN, N. WIKELEY, AND R. YOUNG

The legal regulation of the adjudication officer's work

On the face of it, the legal rules governing the disposal of benefit claims appear to place a heavy burden on both the Department of Social Security as a whole and its officers as individuals. The law requires that a claim for benefit be 'submitted forthwith' to an adjudication officer for determination, who shall then 'take it into consideration and, so far as practicable, dispose of it... within 14 days of its submission to him'.[1] In fact the interpretation accorded to these provisions by the Court of Appeal in *R. v. Secretary of State for Social Services ex p. CPAG*,[2] decided in 1988, means that the duty is somewhat less onerous than it might seem.

In that case, the applicants[3] were concerned about the long delays being experienced by many claimants in having their claims decided, especially those for single payments. These delays were caused by a number of factors, not least the shortage and high turnover of staff and the large volume of claims. The applicants' argument was that, properly interpreted, the statute obliged the Department to transmit a claim to an adjudication officer as soon as it had been received, and that the officer then had fourteen days in which to make a decision. This limit could, it was conceded, be exceeded for reasons peculiar to the particular claim, if, say, further enquiries needed to be made of a claimant's former employer. The Court of Appeal, however, held that the relevant parts of the statute could not be construed in so rigid a fashion. The duty of the Secretary of State to refer a claim forthwith only arose when the claim itself was in a fit state for determination, so a claim lacking some essential information did not require immediate referral. Furthermore, the reasons for exceeding the fourteen-day time limit were not confined to matters relating specifically to the claim. They also included external factors, such as the number of staff available and the overall volume of claims. The Court also ruled that the Secretary of State was not under any open-ended obligation to appoint as many adjudication officers as were necessary to comply with the statutory timetable because the Act made it clear that appointments were subject to the consent of the Treasury.[4]

Decision making by adjudication officers is based entirely on the papers, that is, a person's claim form and any other relevant correspondence and documentary evidence. Claimants are not entitled to an oral hearing in front of the officer, a facility which must be offered by social fund officers if they are minded to turn down an application for assistance. Traditionally, adjudication officers (especially on the contributory benefits side) have avoided any face-to-face contact with individual claimants, although this may sometimes occur, as for example when they double up as section supervisors and see claimants at the public counter in that capacity.

In deciding a claim, adjudication officers have three options. Claims can be allowed, disallowed, or referred to an appeal tribunal for it to decide the case, the last option being rarely used. In the case of claims for unemployment benefit and the other contributory benefits, claimants must be notified of the decision in writing with the reasons for that decision, whereas in income support cases, in the absence of an express request by the claimant, there is no legal requirement to give reasons. This reflects the continuing dichotomy between the two sides of the Department, two differing traditions that continue to have a marked bearing upon the way that claims are determined.

The 'construction' of claims for benefit

Adjudication officers are usually drawn from the executive officer level of the civil service hierarchy, but they rely heavily on more junior staff within the office to produce the materials which form the basis of the claim on which they must adjudicate. This preliminary activity by clerical staff has been the subject of two observational studies which highlight the importance of this stage of the decision making process.

In a detailed examination of a social security office in Northern Ireland, Howe (1985) observed that there was a large gap between departmental policy on administering welfare benefits and local office practice. Guidelines required officials to provide every claimant with an explanation of entitlements under the social security scheme, to give courteous and prompt attention to all claimants, to ensure that the claimant was asked about any possible needs, and to give proper consideration to exceptional circumstances. The basis of local practice, however, was to place the onus on claimants to ask for an explanation of their entitlements and to draw attention to their needs and circumstances. The preoccupation of staff was to discourage claims, not to facilitate them. Interviews were accordingly perfunctory and concerned with recording the information presented by claimants. Little information or explanation was offered on what other benefits or allowances might be available. To put it bluntly, if claimants didn't ask for a benefit, they didn't get it.

Howe identified two main factors which explained this phenomenon. First, the time available to officers to conduct interviews with claimants was grossly inadequate because there were simply too many cases for them to handle satisfactorily. Consequently staff had to keep interviews as short and as uncomplicated as possible, aware that, if they imparted more information to claimants, it might trigger further queries or claims, thereby prolonging the interview and adding to the already massive workload within the office. Secondly, and providing a justification for the practice of discouraging claims, there was a set of beliefs and attitudes held by officers about claimants. Primary amongst these was the view that fraud and abuse were rife and that to provide more information about benefits was to invite an avalanche of unmeritorious claims. Officers saw abuse in very wide terms, encompassing those who claimed their entitlements to the full range of benefits in a persistent or assertive manner. Claimants who were passive and did not seek to claim all their legal entitlements were seen as 'deserving'. But most were classed as 'undeserving' since, as Howe puts it, 'the average claimant is not perceived as a pawn manipulated by an oppressive and complex system but as someone who actively and cunningly exploits it' (p. 64). Howe's analysis of the consequences of the practice fuelled by this belief was that:

many claimants get less than they should under the law, whilst only some receive their full legal entitlement . . . local practice effectively penalizes most claimants, either by creating circumstances in which it is unlikely that needs will be exposed or, if they are expressed, by not examining them in sufficient detail. (pp. 67–8)

A similar conclusion was drawn by Cooper (1985) in the only other study in the literature involving observation of local office procedures. As in Howe's study, Cooper focused on the administration of the supplementary benefit scheme, and noted that 'most of the time, the emphasis was on getting the work done at all, rather than doing it well' (p. 8). He detected practices which closely resembled those that Howe described. For example, visiting officers, making calls to homes to assess claims for single payments, faced the possibility that a more lengthy and general reassessment of the claimant's position might be necessary should the claimant, during the course of the visit, draw attention to changed circumstances or further needs. But, as Cooper observed:

> In practice, the rapidity of interviews, a reluctance to make general welfare enquiries and the adoption of a minimalist approach appeared to discourage requests that added to the workload and may, subconsciously, have been intended to do so. (p. 25)

The distinction between 'genuinely needy claimants' on the one hand and 'unscrupulous exploiters of the system' on the other was again noted by Cooper. He argued that 'the treatment of claimants could vary according to a judgement made about them by officers, on the basis of very little information and a brief acquaintance' (p. 51). Sympathetic treatment would be given to needy claimants making single payment claims whilst those regarded as unscrupulous would, with varying degrees of subtlety, be discouraged from pursuing their claims. Thus, it was apparent that moral evaluations of claimants could be just as important as fine legal distinctions or technical regulations in making decisions. Cooper concluded that the 1980 reform of supplementary benefit, which began the process of replacing a discretionary scheme with a regulated one, had not removed the possibility of personal judgement playing a major part in welfare administration.

These findings demonstrate the importance of studying the administration of welfare claims in terms of the interrelationships between the various stages. Adjudication officers can only make decisions on the claims which are put before them, yet many claims are strangled at birth by lower level staff. Such staff lack the training and experience of adjudication officers and any notion of their playing an independent or impartial role is alien to them. This means in practice that an inestimable number of claimants fail to get the money to which they are legally entitled. An assessment needs to be made as to how far the beliefs and personal judgements held by those within social security offices affect the process of adjudication itself. It should be noted that the single payment visiting officers whom Cooper observed were all adjudication officers and they often determined claims with reference to moral evaluations. Since Cooper completed his study, however, single payments have been abolished, the amount of visiting undertaken by adjudication officers has been curtailed and the new income support system is more tightly regulated than the supplementary scheme it replaced. The scope for personal judgement in adjudication, although it has by no means been eliminated, appears to have been much reduced.

The relative paucity of literature on first-tier adjudication is due to part to the Department having been in the past protective to the point of secrecy about the work of its officers. It also stems from the fact that until recently researchers concentrated their attention on the more visible processes of adjudication before the appeal tribunals. Far fewer problems of access arose, and there was more of a tradition of researching the behaviour of courts and tribunals to draw upon.[5] Mashaw's influential book, *Bureaucratic Justice*, published in 1983, represents a landmark in reorientating interest towards administrative decision making.[6] The idea that criteria based on the concept of 'administrative justice' can be developed and used to evaluate the performance of bureaucracies is novel and has enormous potential, and the present study has been much influenced by this new direction in research.[7]

Not much, then, is known about the work that adjudication officers do, despite the fact that their decisions profoundly affect the quality of life of many thousands of families every day. Yet the little material that is available suggests that their performance leaves much to be desired. Although Cooper's (1985) study was not of adjudication *per se*, it provided many useful insights into first-tier decision making procedures. In particular he drew attention to the infrequency with which adjudication officers consulted the legal regulations upon which their decisions were supposedly grounded. As he put it:

getting the job well done, with people waiting in the queue, or a pile of cases to be dealt with today, did not seem to include checking carefully on the regulations applying to each decision. Officers gave a clear impression of knowing, and consciously applying, the main clauses of the most frequently used regulations; but equally, their knowledge of the finer points of detail appeared very hazy. (p. 17)

The most important source of information about how adjudication officers perform is found in the Annual Reports of the Chief Adjudication Officer.[8] Adjudication officers have been criticized year after year by the Chief Adjudication Officer whose reports, based as they are on independent monitoring of samples of decisions, serve to indicate the shoddy standards of adjudication that appear to obtain in most local offices. This unsatisfactory standard has improved little since the first Annual Report was published in 1985. In is report for 1989/1990, for instance, the Chief Adjudication Officer concluded that 'adjudication standards remain low on many benefits' (para. 1. 14) and 'appeals work in DSS local offices mostly deteriorated further this year, with many deficiencies in submissions to appeal tribunals' (para. 1. 13). The catalogue of failings identified in this report parallels those noted in earlier years: officers basing decisions on insufficient evidence, making wrong findings of fact, applying the law incorrectly or applying the incorrect law, and a general lack of knowledge of changing legal provisions.

The Chief Adjudication Officer has identified in successive Annual Reports those factors which he believes have the greatest bearing on improving these poor standards: simpler rules, less frequent change in regulations, better guidance, greater specialization, and better supervision. In the 1988/9 report, he included another factor—the creation of 'a climate in which good adjudication practices are actively pursued, measured and recognized by management at all levels' (para. 1. 14). He added that 'too often management regards adjudication as subsidiary to, or even standing in the way of, the main job of seeing that claimants get their benefit' (para. 1. 15). The following year he felt compelled to note that 'the Departments' progress on this has not been lit with great enthusiasm' (para. 1. 14).

So much for the views from above, what do adjudication officers themselves say about their situation? How do the officially prescribed concepts upon which the principles of adjudication are based get translated into the daily decision making of adjudication officers? Do they translate at all? It was with these questions in mind that we set out to talk to adjudication officers across the country.[9]

The research findings

We were largely given a free hand in interviewing adjudication officers. The interview schedule that was devised ranged widely over many aspects of the work that adjudication officers do, including the main problems they experienced, the way they coped with the pressure of work, the level of supervision they received, the adequacy of their training, the practical meaning of independence, and the sources of guidance available to them.[10] Given the confidentiality surrounding particular claims, we were not able to ask about individual cases, although during the course of the interview, respondents frequently illustrated their answers with reference to specific cases with which they had dealt. Almost all of the respondents agreed to be tape recorded, so that a complete record of each interview was available. In our discussion we shall draw freely on the qualitative materials that were collected.[11]

It will come as no surprise to those with any familiarity with the vast socio-legal literature of recent years to learn that we found a marked discrepancy between the formal legal position of adjudication officers and the way that they said they carry out their work. This body of literature shows clearly how the behaviour of officials such as police officers, lawyers, court officers, and tribunal personnel very commonly deviates from the formal rules.[12] The same was true of the adjudication officers to whom we spoke in the course of this study.

Before we describe our findings in detail, a word should be said about our selection of material from the vast amount that we collected. There is no doubt that some adjudication officers, usually those working with fewest distractions in the smaller offices, find the task of adjudication easy and uncomplicated. They said to us that they could cope with their work with ease, found the guides, manuals, and regulations simple to operate, received good and sufficient training, and had adequate guidance and support within the office. For a majority, however, things are very different and most adjudication officers reported a lengthy catalogue of problems. It is known that standards of adjudication are generally poor, and we noted earlier how the Chief Adjudication Officer has been outspoken in his criticisms of these standards in recent years. We shall therefore concentrate in this chapter on why this is so and what kind of difficulties adjudication officers face in their work. The material that we quote from interviews is intended to give a good idea of how the average adjudication officer goes about the business of adjudication.

Bureaucratic rationality or conveyor belt justice?

In the previous chapter [see *Judging Social Security*], we discussed the competing models developed by Mashaw (1983) to provide a framework within which to analyse administrative systems. Mashaw contends that in particular contexts one of these models will tend to dominate. His study of the disability programme in the United States suggested that 'bureaucratic rationality' was the dominant model—that is, that accuracy and fairness were achieved through the application of detailed rules and guidelines combined with internal systems of management and control to achieve consistency. Mashaw himself argued that this model provides a better guarantee of justice for claimants than the 'moral judgement' model which emphasizes the value of external systems of review and appeal such as courts and tribunals provide. Ogus (1987), while welcoming the development of these ideas, has cautioned against naïvety in their application by pointing to the practical difficulties which impede the achievement of bureaucratic rationality by officials operating welfare systems. He writes:

With the current high levels of unemployment and public spending cuts, the DHSS staff have been stretched mercilessly in processing social security claims. To envisage greater information flows and 'cultural engineering' in such circumstances may not be realistic. (p. 315)

The difficulties of achieving accuracy when under pressure to process claims quickly was perhaps the strongest point to emerge from our own survey of adjudication officers. In the interviews, we pressed them about how they tried to balance speed and accuracy. How did they maintain reasonable standards of accuracy whilst processing claims with expedition? Only about a half replied that they gave priority to accuracy and more than a quarter immediately conceded that their priority was speed. The interesting point here was that there were very marked variations between the different types of adjudication officer. While less than 10 per cent of contributory benefit and unemployment benefit officers indicated that speed was their priority, fully a half of income support officers admitted that accuracy took second place to speed. Table 1 shows these remarkable differences between types of adjudication officers.

TABLE 1: *Striking a balance between speed and accuracy according to type of work*

	Department of Social Security CB AOs		Department of Social Security IS AOs		Department of Employment UB AOs		Total	
	No.	%	No.	%	No.	%	No.	%
Accuracy given priority	33	80.4	13	24.1	14	66.7	60	51.7
Accuracy and speed given equal priority	4	9.8	14	25.9	6	28.6	24	20.7
Speed given priority	4	9.8	27	50.0	1	4.7	32	27.6
	41	100.0	54	100.0	21	100.0	116	100.0

Note: Column headings refer to types of adjudication officer.
(CB = Contributory Benefit, IS = Income Support, UB = Unemployment Benefit)

We also asked adjudication officers how difficult it was for them to balance these two main objectives. It was again significant that, whereas 39 per cent of contributory benefit officers said they encountered no difficulty at all in balancing speed and accuracy, only about 15 per cent of both unemployment benefit and income support adjudication officers claimed the same thing. Another notable difference was that, while a half of income support and unemployment benefit adjudication officers said they had constant problems in striking the right balance, only a quarter of contributory benefit adjudication officers said they experienced such difficulties. It is not, in our view, that contributory benefit adjudication officers find it easier to strike the right balance: rather it is that they have a different attitude to their work. While they tend single-mindedly to plod through their work and not to worry overmuch about striking a balance with speed, income support and unemployment benefit adjudication officers are more concerned with getting work done quickly but are perturbed about sacrificing accuracy to achieve this. They see their work as piling up on a relentless conveyor belt of cases. The main difference between the two latter groups is that income support officers are, it seems, far readier to make such sacrifices to keep the conveyor belt moving than are unemployment benefit officers.

We carried out various analyses to see whether these differences in approach could be explained by other factors, such as the officers' length of service or the region in which the office was located. None of the factors, however, produced any statistical differences, and we were forced to conclude that the key variable in explaining these differences in attitude was simply the type of work adjudication officers carried out. It will be important, therefore, in the ensuing discussion to focus on the context in which each category of adjudication officer works. Consideration will be given first to adjudication officers working in the department of Social Security, and we shall pay particular attention to income support officers, since they formed the largest group of respondents and appeared to experience the greatest problems in carrying out their work. We shall then examine the differing experience of contributory benefit adjudication officers, before analysing the extent to which both these types of adjudication officer adhere to the notion that they are 'independent' adjudicating authorities. Finally, the distinctive arrangements for adjudication in the Department of Employment will be discussed.

(i) *Income support adjudication officers: running to stand still*

Within the Department of Social Security, adjudication officers are divided into two main groups—those who adjudicate on contributory benefits and those whose staple work is income support. In addition to acting as adjudication officers, where they are applying the law, they take various decisions relating to claims for benefit on behalf of the Secretary of State, where they must follow departmental guidance. Adjudication officers are integrated into the local social security office so that their independent status is not at all obvious to an observer. Contributory benefit adjudication officers have traditionally had a private room away from the distractions of the open-plan office and have been able to concentrate on adjudication work. On the income support side, on the other hand, an adjudication officer (an LOI in departmental parlance) supervises a team of clerical officers (LOIIs), all grouped around adjoining desks. In practice, income support adjudication officers must field a constant stream of enquiries (and complaints) from claimants and staff alike throughout the day. This inevitably means that their ability to give considered attention to adjudication is seriously impeded.

It quickly became apparent as we interviewed staff that many income support adjudication officers regarded the process of adjudication as no more than checking or authorizing the work of the clerical staff working under them. This practice is encouraged by the existence of department targets requiring adjudication officers to carry out checks on the provisional assessments made by their clerical team. All new claims are meant to be checked, but officers are only required to check 10 per cent of those cases where the assessment is being reviewed for some reason, such as a change in the claimant's circumstances. Thus, the emphasis in an income support adjudication officer's job is on checking, not adjudication, and respondents tended to see accuracy in these terms. This means that accuracy is not an issue in 90 per cent of review cases, as is illustrated by the following quotes drawn from our interviews with officers working on income support:

All you're supposed to do is check all the brand new cases and 10 per cent of the others that involve changes. But of course I find that I can't cast more than a cursory eye over what is there. I suppose you can do a passable job, what's expected of you, but you can't do as good a job as you'd want. You can't give the quality in the time that you've got. Sometimes, you just sort of close your eyes and sign it, and that's all. (Interview 43, North East)

You can't aim for accuracy because you are only checking one in ten of the review cases. You really can't guarantee accuracy. The bottom line for us is that we have to pay benefit and occasionally you have to let accuracy slip to get the benefit paid. (Interview 60, North London)

All I do on claims is authorize them. I have clerical staff who look at the forms, check that they've been completed fully and do the assessment. As far as accuracy is concerned, I'm only required to do a 10 per cent check on existing cases. So there you are. (Interview 67, South London)

We have to do a 10 per cent check on all change of circumstance type things and a 100 per cent check on all new claims. The rest of the time, if you want to, you can just sign the stuff without really looking at it. A lot of the time you are just signing things and you don't even have time to check the mathematics. You are just a rubber stamp. That isn't the way it's supposed to be. It's not the way you are taught to do it at adjudication courses. But you just don't have any choice. You have to remember that this is not a busy office. The situation will be ten times worse in a busy inner city office. Some offices are literally months behind and the truth of it is that they are forced to chuck checking out of the window. You can't even do the 100 per cent check on new claims. You just have to sign the stuff and even then that's not enough. If we can't do it, what chance do they stand? (Interview 78, South West)

The truth of the matter is that most income support adjudication officers do very little adjudication: they simply authorize with their signature the decisions taken *de facto* by junior staff. The responsibility for taking decisions is in practice delegated to a level lower than is officially intended. The likelihood that mistakes will be made is great, especially since these low level clerical staff receive no special training in adjudication, in how to use the Adjudication Officers' guide, or in how to interpret the complex regulations to be applied. It is not a case of adjudication officers seeking to shirk their responsibilities: they simply do not have the time to carry out more than a superficial check, and then it involves only a minority of all the claims passing through their sections. In this situation, to expect an adjudication officer to consider the claimant as a 'whole person'[13] and investigate all possible needs, circumstances and entitlements is quite unrealistic. Officers do not even have time properly to investigate and consider those specific claims that are put before them. As one respondent conceded:

Speed unfortunately comes first, the accuracy tends to come later. Sometimes you know that a mistake is going to happen and that you've got to try and pick it up at a later date because you just don't have time to go thoroughly into the case. Most adjudication matters need a lot more evidence than we've got time to find out about with the pressure of the work here. But if we are going to meet our targets, and if people are actually going to get their money, then we've got to adjudicate wrongly on the basis of incomplete information. It's normally because people don't fill the forms in correctly. If we spot something, then perhaps we can clear it up. But you get so many boxes unticked, and we are supposed to clear claims within five working days. (Interview 73, South London)

Adjudication officers often raised in interview this question of targets set by superiors, by office managers, and by regional or head office. But revealingly none made reference to the statutory requirement of dealing with a claim within fourteen days 'so far as practicable', nor did they show any awareness of the stipulation that claims, once received by the local office, should be submitted to them 'forthwith' for determination. This demonstrates that the statutory yardsticks have minimal impact on the behaviour of adjudication officers when compared to the practical importance of departmental goals. The increasing importance attached to 'performance indicators' serves only to heighten pressure upon staff and to make the tension that exists between speed and accuracy the more acute. As one officer observed:

> In these days, where there is a tremendous amount of pressure and work, certainly there is very much of an impetus on to achieve productivity to reach performance indicators and it's got to be to the detriment of decision making. With the pressures and importance of performance indicators now, where it's related to office performance and even linked, or will be in the long term, to merit pay, it's obviously not in the interests of quality decision making. (Interview 77, South West)

In his 1988/9 report, the Chief Adjudication Officer recommended the institution of a new performance indicator to measure the achievement of accuracy. The introduction of yet another target to be met by hard-pressed adjudication officers seems as likely to promote nervous breakdowns as accuracy in local offices. The stress of working in social security offices under such pressure should not be underestimated. The responsibility of interpreting legal regulations weighed heavily on many of the adjudication officers to whom we spoke, and they frequently pointed out that they held no legal qualification to do such work. One income support adjudication officer, when asked what qualities were needed for the job, replied:

> A lot more than I've got! I think we need some basic legal training. I don't really think that people like me, educated up to O level standard, are really capable of interpreting the law because of the way it's written. When I first joined the Department, I didn't expect a clerical officer with seven O levels to be required to perform this kind of work. To be quite honest about it, I often have sleepless nights because of the responsibility. It can spoil your whole weekend if a difficult case drops on your desk on a Friday afternoon. It's too great a responsibility for someone like me. (Interview 57, Midlands)

It will only be by slowing down, taking more care, and investigating claims more thoroughly that decisions taken by adjudication officers involved in income support will become more accurate. If other factors remain constant, accuracy will inevitably continue to take second place to speed, and new performance indicators will not change that situation.

The need for speed is deeply entrenched within local office culture, however, and in an important sense it is right and proper that it should be so. Speed is vital to claimants and this is recognized by adjudication officers. As one officer from Wales remarked, 'With income support we are the last straw, the last place people can go. The necessity for speed is always there, so we accept it as second nature without saying that something is urgent. Everything is urgent.' People who are obliged to claim income support are in dire financial straits, frequently lacking the means even to buy food or keep a roof over their heads. A ponderous system providing 100 per cent accuracy has little merit in such circumstances. Naturally, claimants also want to receive their full entitlement so have an interest in accuracy too. Given the fact that the resources available to the system are limited, some compromise between speed and accuracy has to be struck, and a central question confronting policy makers, managers and adjudication officers themselves is how a satisfactory compromise can best be achieved. What level of inaccuracy in decisions should be tolerated in order to maintain acceptable expedition?[14] With current levels of staffing on income support, the balance in our view is too heavily tilted in favour of haste. 'We still pay lip service to adjudication,' commented one experienced officer acidly, 'we are here basically to pay people benefit: if we are lucky we can get the assessment right at the same time.'

Two further factors exacerbate the problems that income support officers face in coping with their workloads: the demands of acting as supervisors and the limited supervision that they themselves receive from superiors. Adjudication officers dealing with income support were often scathing about having to combine the roles of adjudication officer and supervisor. They told us that in practice the duties of acting on behalf of the Secretary of State generally took precedence over the requirements of careful adjudication. The following quotes give a flavour of the stressful environment in which these officers work.

The problem with an adjudication officer on IS is that the majority of your time is spent purely as a checking officer and organising the staff that you've got. You need to take up work and ensure that it's done quickly and got out. You've got the responsibilities of supervisor and everything that entails for the five staff you've got: influx of work, checking on boxes, doing reports, and all the major and minor crises that come up as when somebody goes off sick or whatever. So I find that there's very little time to adjudicate properly. You hope that you are adjudicating correctly but, because of the time constraints on you, you are in some cases just sitting there, crossing your fingers and signing, and hoping it's right. We should have adjudication and supervision totally separate. They don't marry at all. (Interview 48, North East)

It is extremely difficult. You find that there are constant interruptions all day and you don't have the time to study the really difficult case, plus you lack somewhere quiet to study a case since we don't have an office of our own. Your staff have to ask you things because they have people in the reception area who want an immediate decision about whether they can get payment or not. In those circumstances you are almost obliged to make an instant decision, especially if it is late in the afternoon. The same applies to telephone calls. If people think they are entitled to money and they haven't got any, they are obviously in a desperate situation so you are almost forced again to give a hasty decision. (Interview 57, Midlands)

In making decisions, income support adjudication officers received little support from their superiors and considerably less than officers adjudicating on other benefits. While a quarter of contributory benefit and unemployment benefit adjudication officers said that their superiors had little or no involvement in their work, twice as many of officers working on income support made this claim. As a further illustration of their differing experiences, nearly four times as many contributory benefit officers as income support officers indicated that their superior was someone to whom they could turn on a day-to-day basis for advice and guidance. Income support officers much more often saw themselves as forced to be self-reliant.

Many of the superior officers (who hold the rank of higher executive officer, commonly referred to as an HEO) in income support have little or no experience of assessing claims under the new system, and they are in any event preoccupied with managerial responsibilities. In consequence, adjudication officers tend simply to call across the desk to a colleague for advice. To seek advice from an HEO, let alone from adjudication specialists based externally in the Regional Offices and within the Office of the Chief Adjudication Officer (OCAO), is too time consuming and the rewards are rarely seen as being worth the effort. No more than 7 per cent of respondents working on income support (compared to 14 per cent of unemployment benefit officers and 39 per cent of contributory benefit officers) had sought specialist advice or guidance from either the Regional Office or OCAO on more than a handful of occasions in the previous twelve months, and nearly 20 per cent had never approached these sources. Working under intense pressure, income support officers want speedy and definite answers to their queries, and they tend to be put off from seeking specialist assistance either because they have to submit their request in writing or through a superior or else because the advice they receive is often equivocal, no doubt out of respect for the independent status of adjudication officers. These points were commented on in forthright terms by the income support officers we interviewed, as can be seen from the following quotes:

It's become more difficult to go to the OCAO for further guidance. Before, if you wanted to do that, you sounded out your HEO first, and if they couldn't throw any light on it, it was a question of picking up the phone and asking for the appropriate person. Now we have to go through line management, and we have to do a submission on why we want the information, and they're not as accessible as they once were. I think that's a shame because it is a complicated subject and you can never know it all and time is a great factor. People want their money and they need it now. So to make things more difficult to get is not a good thing. You need the expert advice to be as accessible as possible. It's quite rare that I go to the Regional Office or OCAO nowadays because it's become such a rigmarole. (Interview 62, North London)

I don't think we always get the full support [from HEOs] that I would like, mainly because they are harassed and they find it time consuming. And often it comes back, 'Well, what do you think? You are the adjudication officer. Get on with it.' Beyond that, I've contacted Region more often than OCAO but often you don't get the degree of co-operation you would like their either. They don't come back to you quickly enough; they don't seem to realize the urgency of our job. I don't know whether they have experience of working in a local office, I think it would help if they did. Their general attitude is that we should know and why are we bothering them? (Interview 70, South London)

As you are probably aware, the higher up you get, the less familiar you become; you lose touch with adjudication. so they [HEOs] would probably admit that they're not as good at it as we are. They do a monthly check on a selection of cases, but there's not an awful lot of involvement really. (Interview 82, North West)

We have all these sources of advice or guidance but, I have to be perfectly honest, as LOIs we don't use these things all that often. Unless you have a really complicated case, the LOIs will just discuss things amongst themselves. (Interview 94, Scotland)

So it is rare for decisions to involve sources outside the immediate section of the office in which the adjudication officer works. Officers try to get by on their own or else enlist the assistance of those working alongside them. It seems that the pressure of work is such, at least for officers working on income support, that they do not even have time (or indeed, the inclination) to consult the law, and only exceptionally turn to their guide to this law—the Adjudication Officers' Guide (AOG). This guide, produced by OCAO, was often referred to as 'the Bible' during our visits to local offices, yet just 20 per cent of officers working on income support were unreservedly positive about the guide. The same proportion regarded the guide as of only very limited value.[15] In practice, a high premium is placed on experience, the instant help available from colleagues, and on getting the work done.

There are serious dangers in this situation. Relying on each other's experience may simply perpetuate or exacerbate bad patterns of decision making and cause erroneous interpretations of the law to spread through the office. In addition, the failure to refer to the standard reference works or manuals containing law and guidance in relatively straightforward cases means that adjudication officers remain unfamiliar with these sources and are incapable of making effective use of them when they encounter more complex matters.[16] Those officers who make frequent use of the Adjudication Officers' Guide tend not to read the law and guidance in tandem but rather to treat the guidance as the final word on the subject. These points are vividly illustrated by the following passages drawn from our interview material:

We have the Blue Book which contains the law relating to IS but you will find that the IS AO on a section very rarely has regard to that. We mainly have regard to our AOG and our IS manuals which are our lifeline in the running of our IS sections. (Interview 46, North East)

Basically you rely a lot on experience. You're only going to refer to Codes and legislation if you come across an area you're unsure of, or if there are facts about a particular case that you can't deal with straight off. Although you are an adjudication officer when you authorize a claim, you're really dealing with it in terms of the Secretary of State in the sense that it's all straightforward and done automatically, more or less, unless there is something you need to check up on. (Interview 51, Midlands)

Because you don't do adjudication very often, it's difficult. I suppose we are adjudicating all the time in theory, but we don't think of it in those terms, we just sign and authorize the claims. Although that strictly speaking involves a number of decisions, you do it automatically. It's when you get a difficult case that you have to go to the books and start looking things up, and because you don't do that very often, you tend not to be very good at it. We don't look things up very often because we don't have the time. (Interview 59, North London)

The evidence we collected indicates that, while the quality of guidance provided by the Office of the Chief Adjudication Officer is not a major factor affecting the standards of accuracy achieved by adjudication officers, the lack of time to make effective use of that guidance is. Nor did it seem to us that a lack of good training in adjudication techniques lay behind the infrequency with which the law and the attendant guides were consulted. Rather, methods taught in training could not realistically be followed given the limited time that was available to officers to deal with each claim. One respondent summarized the dilemma well:

I'll try and look at the mechanics of everything, but, if you are too busy, you just can't do it. The actual adjudication work has to be done mainly out of your head and the only time you end up going to the law or the AOG is if you get something you don't know the answer to. When we go on our adjudication courses, we are told that with any decision we should start with the law, to go through all the Acts, which you just can't do in practice. The course itself is very good. But when you actually get back to the office, in practice you can't do what they are telling you to do which is to sit down and dissect every single aspect of the case and make a separate adjudication decision on each applicable amount or on each fact. (Interview 78, South West)

Income support adjudication officers develop ways of coping with their situation as best they can, and one of these mechanisms deserves mention—the belief that the vast majority of claims are straightforward and can be dealt with on the basis of experience alone. That this belief is mistaken was explicitly recognized by one respondent from the north-east who told us that in training 'away from the office, you could see that it's a very complicated subject, which tends not to come across in the office'. Or, as an officer based in north London put it, 'training opened your eyes to how much you don't know rather than how much you do: you realize that you're just scratching the surface, and you find that it's actually complicated and so involved'. However, the pressures within the system are so great that, if adjudication officers were to treat every claim put before them as a complex matter requiring careful attention, the system of adjudication would soon break down under the weight of all the accumulated cases on the sections. Adjudication officers are well aware of this, as was made clear in an internal review of appeals handling in the Department of Social Security: 'A widespread view pervading [local offices] was that it would be neither possible, nor cost effective, for AOs to treat every case at the first tier adjudication level as if it was a potential appeal' (DHSS, 1987, para. 3. 3).

It might be thought that the position of the income support officer has become much easier since the system of income support was introduced in 1988. This was intended to be simpler to understand and to operate than the supplementary benefit regime that it replaced. Two qualifications, however, need to be made about this. First, the removal of a broad band of discretion in decision making has meant that decisions will now tend to be unequivocally wrong or right. Whereas under supplementary benefit, a rough and ready view of a case could be taken by an adjudication officer with little risk that the decision might be impugned as clearly wrong, this is no longer the case. Once the facts have been established, the rules either require payment of benefit or forbid payment of benefit, with no margin for superficial consideration of claims to be disguised as a legitimate exercise of discretion. In this sense, income support poses more, not less, of a problem for adjudication officers. Secondly the new system may have been designed to be straightforward, but the execution and development of that blueprint have undermined much of the intended simplicity.

As discussed in Chapter 1 [see *Judging Social Security*, Ch. 1], welfare policy making is a highly political activity, and is subject to a number of conflicting pressures. The frequent and often rushed amendments to the income support regulations are testimony to this and to the difficulties that are faced by those responsible for designing a scheme which will not provoke a political outcry. There is inevitably an acute tension between having a system which is easy to operate and one that is fair in the sense of taking full account of a claimant's circumstances. Any simple scheme will be rough and ready, and a just scheme will require an elaborate and complex body of regulations to operate. Governments striving to strike an acceptable political balance face a difficult task. The implementation of a relatively straightforward system like income support is bound frequently to throw up gross and manifest injustices, and particularly so when the rates of benefit are less than generous. Adjudication officers find themselves at the sharp end and must daily bear the brunt of these underlying political tensions. They are the ones who are expected to cope with the tangle of new regulations, and, as the following quotes show, it is not easy for them:

> One of the things which is difficult for AOs on the ground is the wealth of information which you have to absorb, and the amount of changes that occur. It's often very difficult to keep pace with it because you can have various circulars advising you of changes, and just trying to keep tabs on them all is extremely difficult. There's the Blue Volumes, there's the AOG, there's IS circulars, you get internal circulars and various other things. You tend to find that you see a case and you think, 'Oh, I know something about that somehow. Now where the hell did I see that?' (Interview 51, Midlands)

The Government tends to spring changes on us without much warning, for good political reasons in their eyes. Like the pensioners' enhanced premiums coming in at the end of October. Many pensioners were worse off when income support came in so the Government has bowed to political pressure and is paying extra premiums for those over seventy five from October. That creates a lot of extra work for us, and we haven't got any extra staff to do it. This Department knows it's a political football and we are prey to such decisions. (Interview 67, South London)

I sometimes feel that the AOG (especially the most recent one) hasn't been written as well as the previous guides. The English is so poor in the Guide, it has double meanings and you can interpret it in different ways. With the income support legislation being rushed through, the Guide was written as fast as they could, and I don't think the drafting is as good as it should be. (Interview 86, North West)

The number of changes we get is one of the major problems. The staff are just used to working one way and then we get something else coming through. That doesn't help. With income support, we were told that there were going to be less sort of changes now and that doesn't seem to be the case. (Interview 95, Scotland)

To summarize, income support adjudication officers tend not to regard adjudication as playing more than a minor or even a marginal part in their daily work. They conceive of themselves as performing largely administrative tasks, chiefly organizing, checking, and authorizing the work of their sections. They view the processing of most claims for benefit as 'straightforward', 'automatic', and 'routine'. Their priority is to 'get people paid', 'shift the work', 'meet targets'. Accuracy 'comes later' or is achieved 'through luck' or 'by chance'. They work largely 'out of their heads' with scant reference to the official channels and sources of advice and guidance, let alone the law which they are obliged to apply. These factors, in our view, go a long way to explaining the official figures on the poor standards that pertain in income support adjudication.

(ii) Contributory benefit adjudication officers

Officers adjudicating on contributory benefits are fully integrated into local Department of Social Security offices but their outlook is the product of a very different tradition from that which underpins the work of their colleagues working on income support. As we have noted, they say that their first priority is accuracy and that they are much less concerned with striking a balance with speed than their counterparts working on income support or unemployment benefit. Similarly they consider themselves to be better supported by their superiors within the office, and are certainly more prepared to seek guidance from the Regional Office and the Office of the Chief Adjudication Officer. The preoccupation with achieving accuracy can in part be attributed to the more highly developed legalistic tradition of these officers. Furthermore, the frenetic atmosphere which pervades income support work is largely absent here, because people claiming contributory benefits tend not to be in such dire financial straits as those seeking income support.

One might expect, then, to find that standards of accuracy in the adjudication of contributory benefits would be considerably higher than those achieved in income support. Yet this does not seem to be the case. The Chief Adjudication Officer concluded in his 1989/90 report that the 'overall standard of [contributory benefits] adjudication was variable, with an improvement in disablement benefit offset by a continuing unsatisfactory standard on short-term benefits' (para. 1. 9). Moreover, the statistics in that report indicate that there is no real difference between the two types of adjudication officer working in the Department of Social Security in this respect.[17] This suggests that those features which income support and contributory benefit adjudication officers have in common may be more significant than those which distinguish them.

Chief amongst these common characteristics seemed to be the difficulties of combining adjudication with supervisory duties, described as a major problem by over half the officers working on contributory benefits to whom we spoke. In addition, like their income support colleagues, they are constantly struggling to keep up with the frequent changes to the regulations they must apply. Contributory benefit officers evidently have an even more complex job than their counterparts in income support inasmuch as they must master and apply the law relating to a whole range of benefits.[18]

In some offices, we found that contributory benefit adjudication officers had been released from supervisory duties, and this seemed to have made the task of adjudication much simpler and relieved the pressure so as to make the attainment of accuracy more likely. Yet in other offices the trend was towards making adjudication officers responsible for supervising a contributory benefit section. The latter policy jeopardizes good adjudication practices, not least because of the threat that such integration presents to the independence of adjudication officers.

(iii) The 'independence' of adjudication officers
Both types of adjudication officer within the Department of Social Security regarded the notion of their independence as extremely problematic. Independence in this context involves the weighing-up of evidence

and the application of the law in an objective and impartial manner. It means also that adjudication officers should resist the temptation to make moral judgements and should ignore any prejudicial comments that are made about particular claimants within the office. Finally, it means that officers should resist any departmental policies or pressures which conflict with their duty to apply the law as fairly and as accurately as possible.

When we asked staff to explain what independence meant to them in practice, we found a wide spectrum of views among adjudication officers, varying from those who appeared to have a good understanding (even if tempered with some scepticism) to others to whom the idea seemed entirely foreign. In the light of what has already been said about differences between adjudication officers dealing with the different types of benefit, it is not perhaps surprising that the highest levels of confusion and misinterpretation were found among officers working on income support. An appreciation of the idea of independence and its importance in decision making seems to be most highly developed amongst those working in Department of Employment officers, as Table 2 shows.

It is worth exploring the main sources of confusion and misinterpretation, and this discussion will focus on adjudication officers working within the Department of Social Security, as their experience of the problems of acting independently is quite distinct from those of unemployment benefit adjudication officers working for the Department of Employment.

The most common way that officers misinterpret the notion of independence is to assume that it just means that no one within the Department can change any of their decisions, though the precise nature of this confusion varies a great deal from officer to officer. The concept of independence at best rests uneasily in the minds of adjudication officers as they are engaged in their daily office routine. Officers inevitably find it difficult to preserve a lofty detachment from the rest of decision making in a local social security office, not least because they are physically part of that office and are very commonly working immediately alongside those from whom they are expected to be detached. Indeed, the very people from whom adjudication officers should be independent are those most ready and accessible when it comes to seeking advice or help with particular decisions. In most cases adjudication is just one part of the job, and is not seen as needing in any way to be differentiated from other duties or to be removed from the general hurly-burly of office life. This was particularly true of those income support adjudication officers who, working on sections, had especially onerous supervisory duties to perform. As Table 2 shows, about one in five of these officers had no idea what independence

TABLE 2: *Understanding of the meaning of 'independence' among adjudication officers in different departments*

	Department of Social Security				Department of Employment		Total	
	CB AOs		IS AOs		UB AOs			
	No.	%	No.	%	No.	%	No.	%
Full understanding of meaning of independence	14	34.1	10	18.5	17	81.0	41	35.3
Full understanding but sees it as compromised	11	26.9	9	16.7	0	0.0	20	17.2
Misinterpretation of meaning of independence	14	34.1	25	46.3	4	19.0	43	37.1
No idea that independence plays a part in the job	2	4.9	10	18.5	0	0.0	12	10.4
	41	100.0	54	100.0	21	100.0	116	100.0

Note: Column headings refer to types of adjudication officer.
(CB = Contributory Benefit, IS = Income Support, UB = Unemployment Benefit)

meant in relation to their job. Only just over one-third had a proper understanding of the concept. Some officers, indeed, appeared dumfounded that we should be asking them questions about it at all.

As with similar organizations,[19] an office culture develops, and it seems to us quite unrealistic and artificial to expect adjudication officers to remain aloof from it. Yet the opinions, beliefs, and prejudices that are traded throughout the office contaminate the process of adjudication. 'You're indoctrinated in the Department's way of thinking' is how one officer in Scotland expressed it to us, and another in the Midlands described himself as 'part and parcel of the group'. A third officer expressed her reservations more graphically:

I think it's possible to be independent to a degree, but basically I think it's a bit of a joke and I'm not surprised that claimants don't accept it. I am paid by the Department, work for the Department, my supervisor works for the Department. I am, to all intents and purposes, a DSS employee. (Interview 12, Contributory Benefit Adjudication Officer, Midlands)

Other officers, particularly those working on income, support, conflated the concept of independence with that of discretion. This group argued that decision making is nowadays so hide-bound by the legislation that any scope for independent decision making is in effect stifled. These officers understood independence to mean that they had the freedom to take whatever decision they thought proper, and that, if they so wished, their personal assessment of the claimant and the claim might be a legitimate consideration.

It must be said that adjudication officers rarely express the view that they are under pressure to toe a particular Departmental line in decision making:[20] rather it is that officers, insofar as they consider the question at all, think that the potential for independent decision making is heavily circumscribed by a combination of factors, prominent amongst which are constraints imposed by ever-tightening legal regulations, daily contacts with junior staff with whom they share an office, and by the insidious pressure of simply being an employee of the Department (feeling the Department, as one adjudication officer put it, 'sort of sitting on your shoulder'). Some impression of the reservations that adjudication officers themselves held is given by the following quotes from their interviews:

It's difficult to be independent because you are in the local office and people come to us and say they have got a case involving someone who causes a lot of trouble by always ringing up and complaining. I used to work out there anyway so a lot of the names I know. You'd have to ask claimants about how independent we can really be. Really you haven't got very much leeway in what you do. The law is laid down and you've got to apply it. There's not much discretion to be exercised. I'd be rather sceptical myself about all the talk about 'independence', to be honest. When a claimant rings up and is told an adjudication officer is someone who works in the office but is independent, it must be very hard to believe. (Interview 17, Contributory Benefits Adjudication Officer, North London)

Independence often doesn't mean that much, though I do try not to be influenced by other people on the section. I just make sure that I only consider the evidence and apply the law. If you don't do it like that, you're not worth having there. But sometimes I feel that I am out of step. I get criticized a lot in the office for decisions I make. There's no way that sitting behind a desk here I am independent of other staff in the office. I come into contact with them every day of the week yet, supposedly to preserve my independence, I'm cut off from the claimants. That must in a lot of cases make it impossible for an AO to be unbiased. I was on the dole myself for nine months so I know what it's like not to get your giro or to have got troubles with the Department. You can do without someone who's supposedly independent being biased against you. (Interview 31, Contributory Benefits Adjudication Officer, North West)

I don't think independence means a great deal really. Obviously we've got the freedom to make decisions, but on the other hand we're bound by the regulations. So really we have no discretion. At one time there was a much greater amount of discretion. We weren't bound to the same extent by the rules and regulations that we are now. (Interview 47, Income Support Adjudication Officer, North East)

I don't see how you could be independent. There's the law, that's the law and you just apply it. There's no question of saying that you'll change the law on a particular day. The law is there and you apply it as it's written down. That's especially true nowadays where the law is very much black and white. There used to be more leeway for personal interpretation prior to income support. Nowadays in 99 per cent of cases, the law is cut and dried. People either qualify or they don't qualify, regardless of how I might feel about it. (Interview 86, Income Support Adjudication Officer, North West)

We have noted that adjudication officers are not under much pressure from senior levels in the Department to make decisions in a particular way. Indeed, one might wonder whether it might not be desirable for there to be greater involvement of more senior staff in routine decision making. It is much more common for there to be no real involvement of superiors in officers' decision making (reported by 37 per cent of those we interviewed) or else for there only to be involvement when an adjudication officer takes the initiative. Consultation with either the Regional Office or the Office of the Chief Adjudication Officer is likewise a very infrequent occurrence for most adjudication officers, with 79 per cent saying that these sources of guidance were rarely, if ever, used in their experience. Routine monitoring of a proportion of adjudication officers' decisions is, it is true, carried out by the local office. But again, if the perceptions of the adjudication officers themselves are to be believed, this rarely has any bearing on the way they approach their task. No more than 14 per cent told us that such monitoring was an important influence upon their decisions, whereas two-thirds said that it had no effect whatever.

In this sense, the independence of adjudication officers might be more appropriately viewed as an isolation from superiors. The natural consequence is that, in the average busy office, it is junior colleagues who fill the vacuum. One of the most revealing points to emerge from the interviews we conducted concerned the procedures that adjudication officers follow when they get stuck on cases. Most officers indicated that they would first of all consult others working in the same room. This seemed an almost universal response, almost too obvious to be worth discussing. Yet adjudication officers are not for the most part highly trained individuals.[21] The essence of the job for the majority is to pick it up as they go along, sitting with a more experienced colleague to learn the ropes, and then struggling on their own as best they can with the numerous guides, volumes of legislation and Commissioners' decisions. Being 'thrown in at the deep end' is how adjudication officers commonly characterized this process. In these circumstances, it is not surprising to find that two-thirds of the sample saw the experience they acquired working on the job as more important than the training they received. It is no surprise either to find that, in what is for many a disagreeable and stressful position, adjudication officers are thrown upon each other or upon the junior staff for mutual support and assistance rather than upon their superiors. It is in this way that a potent office culture develops, and the notion of 'independent decision making' comes to look decidedly precarious and unrealistic. For most, indeed, independence is a luxury that they cannot seriously entertain.

(iv) *Department of Employment adjudication officers*

The situation is very different amongst adjudication officers working in the Department of Employment. These officers had a much more highly developed sense of what independence means, and not one of them viewed the concept as compromised by influences within the Department.[22] Official recognition of the independent status of adjudication officers is also much more marked in the Department of Employment. Since 1987 adjudication has been carried out in sector adjudication offices, physically and organizationally separate from the unemployment benefit offices at which claimants are interviewed, lodge their claims for unemployment benefit, and sign on each fortnight. Unemployment benefit office staff are responsible for the initial collection of information on a claim and for transmitting it to the sector office for an adjudication officer to consider. Adjudication officers may then make further enquiries in writing or ask the local benefit office staff to interview the claimant in question to elicit more details before taking a decision.

Adjudication officers in the Department of Employment specialize in adjudication: they do not have any supervisory or other Secretary of State functions to perform. This means that they are not subject to the kinds of conflicting interests and pressures experienced by their counterparts in the Department of Social Security. It is true that many adjudication officers began their careers in local unemployment benefit offices, but they seem swiftly to adopt the different perspective needed for adjudication on leaving those offices, as the following quotes well illustrate:

You obviously need to be fair because, initially, all the AOs come from the UB office. When you work there, even if you are not actually anti-claimant, there is a feeling, when you send a case to adjudication, that you want to see the claimant disallowed or disqualified. When you come on to adjudication, you have to separate that away totally and you've got to give a decision based solely on the evidence before you. If that means allowing someone who, from the tone of their replies, you don't particularly like, you've got to put that aside and simply apply the law. To my mind it's imperative that adjudication officers be independent. Although we are not truly independent in the sense that we are still employed by the Department and housed in offices of the Department of Employment, and we're all people who worked in UB offices, I think it's as much as anything something that you've got to set in your own mind. If you believe in your own mind that you are independent, then you give your decisions according to the Acts and regulations, based only on the facts that are before you. (Interview 109, Wales)

In practice our independence means that we've got to be objective in looking at what's presented to us. And we've got to remember as well that, although we're employed by the Department, we don't take the Department's side in giving decisions. We have learnt to view evidence provided by the local Unemployment Benefit Office quite critically really. All of us here have worked in UBOs and we know the sort of things that go on there, and we know the sorts of misleading information which is put out by them to claimants. We're not on the side of the UBO, but we should be fair to the claimants who claim benefit. (Interview 114, North West)

The physical separation of adjudication officers from local office administrative staff clearly encourages officers to think and act independently. Furthermore, unlike their counterparts in the Department of Social Security, adjudication officers working on unemployment benefit commonly perceive themselves as having to judge which of two opposing stories, the employer's or the employee's, should be believed. In determining entitlement to unemployment benefit, it is often crucial to decide whether the claimant left a previous job voluntarily. In doing so, the claimant's evidence must be weighed against any provided by the former employer. As one adjudication officer put it, 'you have to gain all the available information from both sides, claimants and employers, and make a fair decision on the information in front of you'. The distinctive nature of unemployment benefit adjudication appears, then, to reinforce a sense of playing an independent quasi-judicial role.

We have already noted that adjudication officers working in the Department of Employment attach great importance to accurate decision making. The great majority told us that accuracy must take precedence over speed. It is arguable, however, that this emphasis on accuracy serves merely to disguise the reality of everyday decision making by hard-pressed adjudication officers. Departmental targets concerning the amount of work that each officer is meant to clear each day are at odds with the expectation that a high standard of accuracy be achieved. One adjudication officer described these conflicting pressures:[23]

> On the regional staffing scheme, which determines our staff complement, we are allowed sixteen minutes per case, which isn't realistic at all . . . Sometimes it can take sixteen minutes to read through it, just to devise your first batch of questions. We are also charged with gathering all the evidence we possibly can, not to leave out any evidence that is necessary. That goes against speed as well . . . So yes, we do need more time. I don't like to give duff decisions. But it's a difficult balance and you have to trade off accuracy against speed at the time. (Interview 104, South London)

In addition to the sheer pressure of work, other factors militate against accurate decision making. Much of what has already been said about Department of Social Security adjudication officers applies with equal force to those working on unemployment benefit: they must cope with frequent changes to the relevant legislation; they receive little support from superiors; they rarely seek specialist guidance from Regional Office or the Office of the Chief Adjudication Officer, and they tend to rely heavily on their own or their colleagues' experience. We have already argued that this situation is not conducive to good adjudication. Indeed, despite the greater recognition of the independent role played by adjudication officers and their emphasis on accuracy, standards of adjudication within the Department of Employment are currently lower than those achieved by adjudication officers working in the Department of Social Security.[24]

Conclusion

The adjudication of claims for benefit is not easy work, and it is unrealistic to expect that adjudication officers, who are after all relatively low grade civil servants, will be able readily to master the art. It is, in essence, a quasi-judicial function, which at the very least involves having a good understanding of how to weigh evidence, how to remain impartial in determining the merits of a claim, and how to apply the appropriate standard of proof. These are difficult concepts and to assume that relatively junior officials will be able to apply them is in our view quite fanciful. As we went around the country, visiting a range of different types of office and talking to adjudication officers of all levels of experience, it became evident to us that such a judicial approach was signally lacking. Many adjudication officers, it is true, are able to recite the standard textbook prescriptions taught to them on a course. But in many cases this reflects only a superficial understanding and is very different from a proper absorption of all that is meant by playing a judicial role.

While it may be unrealistic to expect adjudication officers to fulfil the lofty expectations set for them, it is surely not unreasonable to expect higher standards than those that seem presently to apply. Adjudication officers make decisions which have a great impact on the livelihood of millions of individuals who find themselves in serious financial difficulty. For them, decision making is not routine or administrative. But while it seems that for most adjudication officers good training programmes and reliable sources of advice and guidance are available, what is lacking is the time to make effective use of them. This, together with the complexity of the system they have to operate and the frequent legal changes that have to be mastered, make it difficult for staff to reach a competent standard of adjudication, particularly when they are having to combine this work with supervisory duties. All in all, it is a sorry picture. And it emerges not just from our own observations and impressions: it is also well supported by the detailed monitoring conducted by the Office of the Chief Adjudication Officer to which we have referred in this chapter.

Our findings suggest that an awareness of the need to think independently and a concern with accuracy are necessary but not sufficient preconditions of good adjudication. In practice the efforts of adjudication officers to adhere to the independent role that they are meant to play are systematically undermined by a variety of factors, not least of which is the failure of successive governments to fund the number of officers needed to adjudicate fairly and accurately on much obscurely drafted legislation rushed through Parliament at frequent intervals. Poor decision making is not the fault of the individuals concerned, most of whom doubtless strive to do their work competently. It is rather a natural consequence of a system which is under-resourced and in which hurried decisions are not merely tolerated but expected. High quality decision making will remain an elusive goal as long as the political will is lacking to make the system work properly. Yet those groups in society affected by this system tend to be those whose voices, if they are raised at all, can most easily be ignored.

Notes

1. Social Security Act 1975, ss. 98(1) and 99(1).
2. [1989] 1 All E.R. 1047, C.A. See Wikeley (1988).

3. The Child Poverty Action Group, the National Association of Citizens Advice Bureaux, and two inner London boroughs.
4. Although the application for judicial review was unsuccessful, the applicants none the less gained a considerable amount of publicity to highlight the problem of delays in local offices, and the Court of Appeal was itself at pains to stress the importance of expedition in dealing with benefit claims.
5. It was not until the 1960s, with the development of a separate discipline of socio-legal studies, that the work of courts and tribunals attracted the critical attention of large numbers of legal and sociological researchers in this country.
6. See Ch. 1 [see *Judging Social Security*, Ch. 1] for discussion of Mashaw's ideas. In 1986 the Economic and Social Research Council launched a major review in this area: see further the valuable review of socio-legal research on administrative justice conducted by Rawlings (1986).
7. For a rigorous and illuminating application of Mashaw's thesis in a British welfare context, see Sainsbury (1988).
8. A useful analysis of these reports, together with a review of the first four years of the Office of the Chief Adjudication Officer, is contained in Sainsbury (1989).
9. An outline of the research sample is given in Ch. 1 [see *Judging Social Security*, Ch. 1].
10. The interviews also dealt with review and appeals work: see Ch. 3 [see *Judging Social Security*, Ch. 3].
11. Since we were able to tape record most of the interviews we conducted, the extracts from interviews that we reproduce in this and subsequent chapters are cited *verbatim*, modified slightly only where we seek to preserve the anonymity of individuals or cases.
12. See the provocative discussion of this tradition in McBarnet (1978).
13. The term derives from one of the objectives set out in the Department of Health and Social Security's operational strategy published in 1982 (DHSS, 1982). The idea is that claimants should not be treated in a compartmentalized manner but as people who could be helped in a range of ways through the social security scheme.
14. As Sainsbury (1989) notes, the Chief Adjudication Officer, has shied away from specifying what he would consider to be an acceptable standard for adjudication, preferring to talk in rather vague descriptive terms such as 'unsatisfactory' or 'disappointing'.
15. This compares with around 45 per cent of contributory benefit and unemployment benefit adjudication officers who expressed great satisfaction with the Adjudication Officers' Guide, and 5 per cent of these two groups who saw the guide as of little or no value to them.
16. The Guide was also seen as too cumbersome to use on a routine basis. It comes in several volumes, and officers commonly complained that it was difficult to find the information that they required and that the index was poor with subjects listed under obscure or confusing headings.

17. OCAO 'raises an adjudication comment sheet' if the process of adjudication on a decision is deemed to have been deficient in some way. The ratio of comment sheets to decisions provides a guide to adjudication standards in each area of work, albeit a rough and ready one: see Sainsbury (1989). The comment ratio for income support and contributory benefit work currently stands at 37 and 39 per cent respectively: Chief Adjudication Officer (1991) App. 1 and 6.
18. These include disablement, invalidity, maternity, and sickness benefits as well as retirement and widows' pensions.
19. There has been a long tradition in sociological research which has examined the complex interrelationships between formal and informal structures in influencing the behaviour of individuals and groups. This work covers a variety of different groups from police officers to factory workers. See, for example, Skolnick (1966), Cole (1979), and McConville and Baldwin (1981).
20. Only three of the adjudication officers to whom we spoke said that they felt they were put under overt pressure to act in particular ways.
21. It is apparent that the training that adjudication officers receive varies enormously from office to office and that the amount of training they are given depends upon a number of factors, such as the staffing levels in a particular office, pressure of work at a given time and the length of time officers have spent in the job. A third of respondents viewed the training they had received as inadequate, and it was disturbing to find that five officers claimed to have received no training whatsoever and that a further 25 (16 per cent) said they had only received tuition within their own office, perhaps supplemented by a self-instruction package.
22. Adjudication officers in the Department of Employment appeared to have received better training than their counterparts in the Department of Social Security and to be considerably more positive about the content of the training programme.
23. The 15 minute time allowance referred to by this officer has now been abandoned for one which varies according to the type of question involved.
24. The latest figures available show that adjudication decisions on unemployment benefit attract a comment ratio of 53 per cent (see n. 17 above). It is worth noting that the central offices of the Department of Social Security, set up to specialize in particular benefits, generally achieve much higher standards of adjudication than the local offices (with the exception of family credit); see Chief Adjudication Officer (1991), para. 1. 11 and statistical tables set out in the appendices.

References

CHIEF ADJUDICATION OFFICER (1991) *Annual Report of the Chief Adjudication Officer for 1989/90 on Adjudication Standards*. HMSO, London.

COLE, G. F (1979) *The American System of Criminal Justice*. Duxbury, Mass.

COOPER, S. (1985) *Observations in Supplementary Benefit Offices*. Policy Studies Institute, London.

DHSS (1982) *Social Security Operational Strategy: A Framework for the Future*, London.

—— (1987) *Appeals Handling in Local and Central Offices*. London.

HOWE, L. E. A. (1985) 'The "Deserving" and the "Undeserving": Practice in an Urban Local Social Security Office' in *Journal of Social Policy* 49–72.

MASHAW, J. L. (1983) *Bureaucratic Justice: Managing Social Security Disability Claims*. Yale University Press, New Haven.

MCBARNET, D. J. (1978) 'False Dichotomies in Criminal Justice Research' in Baldwin, J. and Bottomley, A. K. (eds.) *Criminal Justice: Selected Readings*. Martin Robertson, London.

MCCONVILLE, M. and BALDWIN, J. (1981) *Courts, Prosecution, and Conviction*. Clarendon, Oxford.

RAWLINGS, R. (1986) *The Complaints Industry: A Review of Sociolegal Research on Aspects of Administrative Justice*. ESRC.

SAINSBURY, R. (1988) 'Deciding Social Security Claims: A Study in the Administrative Theory and Practice of Social Security' unpublished Ph. D. thesis. University of Edinburgh.

—— (1989) 'The Social Security Chief Adjudication Officer: The First Four Years' in *Public Law* 323–41.

SKOLNICK, J. H. (1966) *Justice Without Trial: Law Enforcement in Democratic Society*. John Wiley and Sons, New York.

WIKELEY, N. J. (1988) 'R v. Secretary of State for Social Services, ex parte Child Poverty Action Group and Others' in *Journal of Social Welfare Law* 269–74.

Release From Prison Decision-Making within Special Hospitals

G. RICHARDSON

Application decisions: fixed-term sentences

Up until 1992, and even under the new system, the process of parole decision making in England and Wales was, and remains, largely a paper exercise, Under the old system the main actors were: the local review committees (LRCs), the Parole Board, the parole unit, and the Secretary of State. Every prison establishment containing eligible inmates had to have an LRC.[1] The members were nominated by the governor and appointed by the Secretary of State. The Parole Board, by contrast, has always been a national body. Its members are appointed by the Secretary of State and must include a judge, a psychiatrist, a probation officer and a criminologist or penologist.[2] Originally the Board had 16 members but, by 1990, had grown to approximately 70. The parole unit within the Prison Department of the Home Office manages the parole system and, under the old system authorised each individual grant of parole on behalf of the Secretary of State. Only a small proportion of the most sensitive cases were referred to either a junior minister or to the Secretary of State.

For fixed-term prisoners the pre-1992 process began when the governor identified them as eligible for parole and prepared the papers prior to reference to the LRC. In theory all cases were referred in sufficient time to ensure that all prisoners knew the outcome at least three weeks before the due date. In practice the delays were such that 'quite a number [of prisoners] hear nothing until they have gone past their date'.[3] Each case was considered by a panel of LRC members. The prisoner could make written representations, and would be interviewed by one of the panel members. The interview was not meant to be an assessment, but rather to provide an opportunity for the prisoner to put forward his case. The LRC member's account of the interview was then attached to the dossier together with any written submissions from the prisoner. The dossier would also include: a statement of prison history; an assessment by prison staff, including a medical report if relevant; a record of any prison offences; a report by the Probation Service; a home circumstances report; details of any previous convictions, and a police report of the present offence.

According to the Carlisle Committee, LRC panel meetings lasted from about one to three hours, at which anything from six to ten cases were considered. Officially, the same criteria were used as those applied by the Parole Board, but the Carlisle Committee observed 'considerable diversity of approach and a good deal of confusion over the test to be applied'.[4] The LRC's recommendation was then recorded on a standard form and, although the LRCs were encouraged to explain their reasoning, very little space was provided for them to do so. It is clear, however, that the vast majority of recommendations were unanimous: an exercise completed for the Carlisle Committee on the 1986 figures showed that all but 3.5 per cent of LRC decisions were unanimous.

Following consideration by the LRC, all papers were sent to the parole unit. Release could be ordered directly without reference to the Parole Board in the case of some positive recommendations, by virtue of section 35 of the Criminal Justice Act 1972. All other positive recommendations had to go to the Board. Negative recommendations in section 33 cases rarely went to the Board, but longer-term cases did go, either if the negative decision was by bare majority of the LRC, or if the prisoner's reconviction prediction score (RPS) was good. The RPS is a measure of statistical probability of reconviction within two years, and is calculated on the basis of 18 variables relating to the prisoner's criminal and social background. The use of such statistical techniques in relation to the assessment of risk in individual cases is discussed further in chapter 12 [see *Law, Process and Custody*, Ch. 12].

About 32 per cent of cases reviewed by the LRCs came before the parole Board and were considered by panels, typically consisting of four members. The Board considered the same papers as were available to the LRC, with some occasional additions: an RPS might be added, for example, or a recent police report. Each meeting lasted between three and four hours, the papers having been circulated in advance, and considered about 28 cases. Some cases were disposed of in a matter of minutes while others took up to half an hour. Special panels with a reduced case load were convened to consider life-sentence cases. Like the LRCs, the Board panels usually reached unanimous decisions. During two weeks in October 1987 all but 7 per cent of the 325 cases considered were decided unanimously.[5] At the meetings, panel members had a list of the six criteria and would record which criteria formed the basis of any decision not to recommend release.

After consideration by the Parole Board the papers were returned to the Parole Unit. All positive recommendations were then screened to ascertain whether there was any danger of public criticism, or any conflict with normal policy. A handful of sensitive cases would then be identified for reference to ministers. All other positive recommendations were authorised without further scrutiny. In a tiny number of cases annually, usually within single figures, the Secretary of State turned down the Board's positive recommendation. According to Wasik and Pease, there is some evidence in the case of fixed-term prisoners that the ministerial veto was used more frequently in electorally sensitive years, and more frequently by Tory Home Secretaries than by their Labour counterparts.[6] Within anything up to three months after the start of proceedings, or even longer if the case had been referred to the minister, the prisoner was informed of the outcome. A negative outcome was conveyed on a standard form. The prisoner was merely told that his or her case had been given full and sympathetic consideration, but that parole had not been authorised. No reasons were given.

A prisoner granted parole was released on licence, and was therefore subject to recall. Under the 1967 Act recall could be ordered by the Secretary of State acting on the advice of the Parole Board, or in urgent cases by the Secretary of State acting on his own (section 62). In all urgent cases, and in any other where the prisoner so requested, the case had to be referred to the Board, and if it ordered immediate release, the prisoner had to be so released, section 62(5). The prisoner was also entitled to be given reasons for his recall, section 62(3). Similar provisions are included in the 1991 Act, section 39 of which provides for the recall of long-term and life-sentence prisoners.

There is little empirical data available in this country concerning the way in which parole decisions are reached in practice. Keith Hawkins, drawing on research conducted in the United States, emphasises the importance attached to the idea of 'time from crime': had the prisoner been in custody long enough to reflect the gravity of the offence?[7] In responding to that question, the individual Board members would assess the prisoner's level of 'wickedness', but, as Hawkins remarks, individual members could easily interpret factual evidence in different ways. The importance of 'time for crime', or in this country the demands of the tariff, was also stressed by Maguire, Pinter and Collis in their study of the review of lifers prior to the 1983 changes.[8] Analysing the decisions reached on lifers between 1977/8, Maguire et al. discovered a marked consensus between the professionals involved in assessing individual cases, and emphasised the significance of the role played by the secretariat responsible for compiling the dossiers. The preference for consensus matches the 'strong momentum towards consensus' within both the LRCs and the Parole Board, discovered more recently by the Carlisle Committee.

This brief account of the pre-1992 procedures applying to the application of parole policy to individual fixed-term prisoners, reveals the old arrangements as falling far short of those required by full process in a number of respects. In the first place, as has already been explained, there existed no formal statutory criteria to guide the decisions of any of the actors involved, and there is evidence that those 'informal' guidelines which did exist were not always fully understood. The Parole Board's plea for clear statutory guidance is particularly revealing here since it does no more than reflect the minimum demands of full process. It is clear that the Board regarded both the pre-1992 criteria and Carlisle's own proposals as insufficiently specific. It was the Board's view that, if it were to be given executive powers to release prisoners, the criteria to be applied to those decisions would have to be articulated with far greater precision, particularly with regard to the nature of the initial presumption.[9]

Secondly, under the original procedures, the prisoner's participation in the process was limited to the right to make written representations and to meet one member of the LRC. The prisoner had no right to the disclosure of the written material on which the parole Board reached its decision, no right of direct access to the Board, and achieved no reasons for the final decision. Admittedly, in formal terms the Board's role was advisory only, but it was nevertheless required to assess each case, and any notion of full process would have demanded the prisoner's full participation, particularly since the prisoner was, and still is, entirely excluded from participation in the process of ministerial review.

Finally, the role of the Secretary of State served simply to compound the problems. It is hard to find any convincing justification for the involvement of a senior political figure in individual decisions of this nature, where political self interest will almost invariably point against release. The public interest must of course be represented, but private consideration by a politically vulnerable member of the government does not provide an appropriate means of ensuring the full reflection of the public interest. A decision making structure which required the open application of clearly articulated criteria, and before which the Secretary of State was entitled to be represented in individual cases, would be more appropriate

These decision making procedures were severely criticised in the Carlisle Report, which recommended: full disclosure of reports to the prisoner; the development of 'parole counsellors' to help prisoners with their submissions to the board; an obligation on the Board to give their reasons for the refusal of parole, and a right in the prisoner to complain to the chairman of the Board.[10] On the question of an oral hearing, the Carlisle Committee was divided. A minority were of the view that in cases where, after consideration of the papers, the board was inclined to refuse parole, the prisoner should be entitled to an oral hearing. The majority, however, for largely financial and logistical reasons, felt that a paper hearing following disclosure would be sufficient. The Parole Board's initial response to these proposals was mixed and cautious.[11] They expressed the familiar fears that disclosure would result in less candid reporting and that the giving of reasons would be difficult in practice and would lead to challenge from disappointed prisoners. It was therefore encouraging that the White Paper, recognising the potential of greater openness to improve the quality of decision making, committed the government to moving towards disclosing reports made to the board and the board giving reasons for its decision.'.[12] On the other hand, the White Paper joined the majority of the Carlisle Committee in rejecting an oral hearing.

As was described above, the 1991 Act leaves the specification of procedures to the Secretary of State. Section 32(5) empowers him to issue rules 'with respect to the proceedings of the board'. These procedures have now been supplied by way of Circular Instruction, CI 26/1992, and will apply to the discretionary conditional release of all prisoners sentenced from October 1992 to fixed terms of four years or over.[13] The review process will remain a largely paper exercise but some significant improvements have been introduced. The parole dossier will be disclosed to the prisoner, although the governor may recommend to the chairman of the Parole Board that certain items be withheld in the interests of national security, for the prevention of disorder or crime, for the protection of information received in confidence, or if it is felt necessary to withhold information on medical grounds. A panel member will visit the holding establishment and interview the prisoner, the panels will be required to give reasons for their decisions, whether they be negative or positive, and to assist them in this a list of factors which should be taken into account are given in the 1992 directions, reproduced in C126/1992. Thus, while there are still no oral hearings, prisoners will be better informed and thus better able to make representations to the panels. Nevertheless some reservations remain concerning the operation of the criteria for parole contained in the 1992 directions.

As described above, these directions place considerable emphasis on the risk of re-offending, and require the Parole Board to balance the risk of re-offending during the period when the offender would otherwise have been in custody against the benefit to be derived from supervision. There is, however, no guidance on the level of acceptable risk, beyond the stipulation that the risk of violence is to be regarded as more serious than the risk of non-violent offending. In practice, no doubt, a variety of factors will be taken into account in reaching individual decisions and some panels will be more risk-averse than others, but the absence of clear guidance in the directions reflects the difficulties inherent in using risk assessment as a guide to early release. Indeed the Carlisle committee itself attracted some criticism for giving insufficient consideration to the assessment of risk. Adverse comparisons were drawn between the efforts being made in Canada to develop sophisticated statistical analyses of risk and the relative lack of rigour with which the issue is tackled in this country.[14]

The directions in CI 26/1992 also require the Parole Board, before recommending parole, to be satisfied that 'the offender has shown by his attitude and behaviour in custody that he is willing to address his offending and has made positive efforts and progress in doing so'. Strictly, the stipulation would seem to require evidence of positive improvement. It will no longer be sufficient for the panel to be assured that the prisoner has kept out of trouble. The prison reports in the dossier will therefore take on a far greater significance, and it is to be hoped that the panels will be prepared to challenge inadequate reports in order to uncover any positive evidence that might exist, rather than merely to accept silence as indicating in absence of 'positive efforts'. The stipulation will also question the true 'voluntariness' of any participation in treatment programmes which purport to address offending. The problem of 'coerced' consent is considered further in the context of special hospitals [see *Law, Process and Custody*].

* * * * *

The validity of consent—participation in practice

First, the question of capacity and participation. In all cases the responsible medical officer (rmo), and in the case of the possibly non-volitional patient the second-opinion doctor, must assess whether the patient is capable of understanding the nature, purpose and likely effects of the treatment, and if so, the relevant doctor must then consider whether the patient has in fact consented. The phrase 'nature, purpose and likely effects' also appears in section 57, in relation to those treatments for which both consent and certification are required, and in that context it was considered by the High Court in *Ex parte W*.[15] While the decision in that case did not turn on the interpretation of the phrase, the court appeared to suggest that, for the purposes of Part IV, consent rests not on actual understanding but merely on the patient's abstract intellectual ability to understand. Such an interpretation would have been at odds with established second-opinion practice, according to which doctors were advised to consider both the patients' intellectual capacity and their actual understanding. It would also dilute the notion of consent. In the light of further legal advice, however, the Mental Health Act Commission has not altered its interpretation of consent, and continues to advise second-opinion doctors to consider both capacity and actual understanding.[16]

The 1983 Act requires the second-opinion doctor to consult a nurse and one other professional who has been concerned with the patient's 'medical treatment', section 58(4). This obligation to consult is imposed in cases where the patient is regarded either as incapable of consenting or as withholding consent. It is not expressly required when the second doctor forms the opinion that the patient is capable and is consenting, nor is it expressly imposed on rmos. Nevertheless, the Code of Practice states that the consultees should comment on 'the proposed treatment and the patient's ability to consent to it'.[17] There is thus formal recognition of the view, expressed by Bean, that the assessment of capacity should not be the exclusive preserve of psychiatry.

In practice, however, second-opinion doctors can experience difficulties in achieving consultation with the 'third professional': the non-nurse. Sometimes, owing to the absence of a multi-disciplinary approach at the detaining hospital, such a person simply does not exist.[18] In such circumstances the second-opinion doctor should not certify the treatment. Certainly, where such gaps are found, both the doctor concerned and the Mental Health Act Commission comment adversely to the hospital, and to a considerable extent such gaps have been filled. The 1989–91 Biennial Report of the Commission suggests that the difficulties encountered by second-opinion doctors in this respect now arise primarily from poor organisation by the hospitals, not from the complete absence of the relevant personnel.[19] Thus, to some extent it can be said that a formal procedural requirement—the obligation to consult—has contributed to an improvement in the standard of decision making by forcing psychiatrists to include other disciplines, such as social work and psychology, in their decisions concerning patient care.

The level of information available to a patient was the second aspect of consent emphasised by Bean. The patient's ability fully to consent, and thus to participate, will be significantly affected by the nature and quality of the information provided by all those involved. If consent is to include not merely the intellectual capacity but also actual understanding, then the provision of adequate information is of crucial importance. At present there is some debate as to whether the requirement that the patient understand the nature, purpose and likely effect of the treatment is merely declaratory of the common law on the question of the information that must be given, or whether it requires more. The common law merely insists that the patient be given that amount of information that a reasonable body of professional opinion would give.[20] Whatever the precise relationship between the 1983 Act and the common law, however, there appears to be some agreement that the nature of the required information will vary from 'broad terms' to great detail, depending on the patient's ability to understand and the complexity and risks of the proposed treatment.[21] It is up to the doctor to ensure that adequate information has been given to enable the patient to consent.

In addition to the provision of adequate information, Bean also emphasises the importance of specificity. In order for consent to be real it must relate to a specific course of treatment: the patient must know exactly what he or she is consenting to. When the consent provisions were first introduced it was common, particularly in some special hospitals, for the treatment allegedly consented to to be described in very broad terms by the rmo. The rmo might, for example, certify that the patient was consenting to 'a course of treatment'. Such certifications are now the exception, but the problem of specificity still exists. The 1983 Act itself does not deal directly with the issue, and the forms provided by statutory instrument require merely a 'description of treatment or plan of treatment'.[22] The issue is addressed with greater precision by the Code of Practice, which states that the r.m.o. should indicate on the certificate the drugs proposed, by the classes described in the British National Formulary (indicating the dosages if they are above the B.N.F. advisory maximum limits) and the method of their administration.[23]

However, such a prescription does not necessarily impose sufficient specificity to enable true consent to occur. In the first place, under the requirements of the Code of Practice, the rmo can certify that the patient is consenting to the administration of any number of an extensive class of drugs, each one up to the advised limit. Thus, in combination, the patient is regarded as 'consenting' to a very high dose indeed. It would seem unlikely that most allegedly consenting patients are aware of this. Further problems occur where the certified doses are above the BNF limits. Here the Code of Practice, which although indicative of good practice is not directly binding in law, is slightly ambiguous in its requirements, enabling those rmos who are hostile to any interference with their clinical judgment to evade the spirit, if not the letter, of the regulations. Whatever the nature of the formal requirements, however, from the point of view of the validity of the patient's consent, the greatest practical specificity should be encouraged. A patient cannot be said to consent unless, within the limits of his or her capacity, he or she knows the precise nature of the treatment proposed.

Finally, the issue of coercion is an enduring problem in relation to consent. In practice its impact is felt early in the decision process. A patient who is 'coerced' into consenting will never be seen by a second-opinion doctor: he or she will be certified as consenting by the rmo. During routine visits to special hospitals, members of the Mental Health Act Commission discuss treatment with patients, and in the course of such discussions may conclude that the consent of a particular patient is of doubtful validity. A patient may, for example, have 'consented' to treatment because a Mental Health Review Tribunal was due and he or she did not wish to alienate the rmo at such a sensitive time by appearing uncooperative. In many such cases the commissioners will raise the issue with the patient's rmo, who will be advised to seek a second opinion, but quite frequently the patient will ask that the matter be not mentioned to the rmo. Such requests appear typically to reflect the patient's fear that any failure to co-operate with treatment will be regarded as evidence of continuing disorder, and militate against favourable reports from the rmo. While these fears may be exaggerated, they emphasise the difficulty of assessing 'consent' against a background of indeterminate detention and within an environment such as a special hospital, where coercion is subtle and pervasive.

The statutory structure for seeking and certifying the patient's consent to treatment has led to some welcome adjustments to practice, most notably in the involvement of other professionals and in the greater specificity with which treatment plans are now presented to patients. But it is evident that the effectiveness of the system is largely dependent on the co-operation of the rmo. If the rmo certifies that the patient is capable and is consenting, the second-opinion procedure is never initiated unless the case is picked up on, for example, a Commission visit to the hospital.

This feature of the system is all too clearly illustrated by the attitude of medical staff at Broadmoor to the consent provisions when first introduced. In the first two years of the operation of Part IV, 89 second opinions were requested at Broadmoor, as opposed to 413 at Rampton, and 262 at Moss Side and Park Lane (now Ashworth).[24] The patient numbers at the three hospitals over the relevant period were 494, 590, and 562, respectively. In their second Biennial Report the Mental Health Act Commission refer to the fact that the clinical teams at Broadmoor would not consider a patient ready for discharge or transfer to a less secure hospital if the patient was not co-operating with medication. Patients were, no doubt, only too aware of this attitude, and consequently reluctant to withhold consent. Further, among certain clinical teams there was a tendency to certify that patients were consenting to 'medication', for example, or to 'a course of treatment'. As suggested above, such certifications were of no value as evidence of true consent; but, more significantly, they served in practice successfully to evade the need for a second opinion.

More recently the position at Broadmoor has improved,[24] although there are still problems with achieving universal compliance with the Code of Practice. Nevertheless, the experience of the first few years is arguably just one example of the exploitation by medical staff of a feature inherent in the system. Where detention is indeterminate and the rmo immensely influential, many patients are likely to incline towards acceptance of the rmo's proposals.[26] They will therefore be certified as consenting unless very strenuous efforts are made to neutralise the inevitable 'coercion'. In other words, from the perspective of full process, it is hard to ensure true and free participation where the patient occupies such a dependent role.

Overriding the absence of consent

The second-opinion doctor, as explained above, also has to assess the validity of imposing treatment in the absence of consent. If a patient is either not consenting or is regarded as incapable of consent, the second-opinion doctor must determine whether the treatment should be given, 'having regard to the likelihood of its alleviating or preventing a deterioration of [the patient's] condition', section 58(3)(b). It is here, particularly in the case of a patient withholding consent, that the second doctor is being asked to balance the principles of autonomy and beneficence—in other words, to test the justification for imposing treatment in the absence of consent. In theory, at least, the second-opinion doctor should be aware of the significance of the decision that he or she is required to make. Both the guidance from the Mental Health Act Commission and the Code of Practice implicitly emphasise the relevance of autonomy, and the importance of questioning the need for the proposed treatment.[27]

In practice, however, it seems that second-opinion doctors seldom refuse to certify a course of treatment. In the period up to June 1985 certification was refused in only 4.6 per cent of cases throughout England and Wales.[28] The Commission freely admits that this represents a high level of agreement, and that further research is required to examine the reasons for it, but certain possible explanations present themselves. In the first place, as the Commission points out, it is not clear from the figures how many certifications have emerged from a process of negotiation between the rmo and the second-opinion doctor. It is possible that rmos have been prepared to agree alterations to their proposals in order to achieve approval. A revised report form enabling second-opinion doctors to record the presence of such negotiations has now been introduced by the Commission.[29] Secondly, there is the question of professional loyalty and the nature of the test itself. The second-opinion doctor is asked merely to assess whether 'the proposed treatment is reasonable in the light of the general consensus of appropriate treatment for such a condition'.[30] It is thus a review of reasonableness, rather than an appeal. Thirdly, although the second doctor is encouraged to search for less intrusive forms of treatment and to discuss the possibilities with the 'statutory consultees', it is possible that, however receptive the second doctor might be to non-drug-based options, the availability of viable alternatives is limited. Finally, it is possible that the existence of the second-opinion system has deterred rmos from proposing anything but the most orthodox treatments in the case of patients who do not consent—treatments which, given the test applied, are almost bound to receive approval.

Whatever the reasons, however, it would appear that the system of independent peer review introduced by section 58 has done little overtly to challenge the authority of the rmo. The patient's genuine participation, short of control, is not guaranteed, and the second-opinion system seems rarely to question the rmo's assessment of the need to impose treatment in the absence of consent. Beneficence, it seems, will tend to outweigh autonomy. Perhaps this is not surprising since the crucial weighting is being performed by another psychiatrist. Certainly, many patients regard approval of the proposed treatment by the second doctor as almost inevitable and, as a consequence, their confidence in the system is minimal. It is relevant, therefore, to ask what alterations to the process might encourage a more searching system, a system better designed to ensure both the fullest participation of the patient and the most stringent examination of the justification for the imposition of treatment.

The question, to some extent, presupposes an acceptance of the demands of autonomy but, as explained above, such a position flows naturally from the whole approach to legitimate authority adopted here, and from the characteristics of full process that flow from it. Any changes to the current system would need to address the difficulty of ensuring true consent, and the need rigorously to question the justification for overriding the patient's wishes.

The second point is perhaps more readily dealt with than the first. The most obvious change to the system would be the introduction of multi-disciplinary review before a patient's lack of consent, whether volitional or non-volitional, could be overridden. Social workers, psychologists and lawyers, while appreciating the benefits to be gained from treatment, might bring with them a deeper sympathy for autonomy. The suggestion is not new, but the experience of the first few years of the compromise introduced by Part IV,[31] at least, indicates that peer review has tended merely to legitimate the professional preference for beneficence. However, such multi-disciplinary review once the absence of consent is recognised does not meet the difficulties posed by the initial assessment of consent. Here it is essential to ensure that no patient can be certified as consenting unless the treatment for which consent is sought is specific with some precision, unless the patient is given, in an accessible form, sufficient information about its nature, purpose and likely effects, and unless adequate steps are taken to guard against inevitable 'coercion'. In this respect, therefore, the Code of Practice must become significantly more stringent in its requirements, and compliance with those requirements must be assiduously monitored, as a matter of routine, by an independent body.

Restrictions on a patient's freedom

The decisions discussed in this section concern the regulation of a patient's freedom in areas which are not directly related to the administration of treatment in the narrow sense. They are, nevertheless, decisions which impinge directly on the residual freedoms of the patient, and as such require some specific justification.

As was explained in the previous chapter, a patient is detained in hospital under the Mental Health Act 1983 because he or she is deemed to be in need of treatment for mental disorder, compulsorily administered if necessary. Additional restrictions relating to discharge etc. can be imposed in the case of patients from whom society is thought to require protection from serious harm. The infringements of the patients' freedoms which are expressly authorised by the Act must therefore be assumed to be justified by reference to the demands of treatment and security. The infringements flowing from the decisions considered in this section, however, are not specifically authorised by the 1983 Act, and no such assumption can therefore be made. Such infringements can only be justified if clearly required by the need to treat or to maintain security, and the internal decision making processes should be designed to ensure that they are so justified. Individuals are not compulsorily detained in hospital for any other purpose. As in the context of imprisonment, however, the need to maintain security—that is, successfully to segregate individuals from the rest of society—carries with it the need to maintain order or control within the secure environment.

Notes

1. Section 59, Criminal Justice Act 1967 and SI 1967/1462.
2. Originally the composition of the Parole Board was governed by the Criminal Justice Act 1967, section 59 and schedule 2; it is now covered by section 31 and schedule 5 of the Criminal Justice Act 1991.
3. *The Parole System in England and Wales* (The Carlisle Report) (1988, Cm 532, London: HMSO), para. 121.
4. *Ibid.*, para. 126.
5. *Ibid.*, para. 134.
6. M. Wasik and K. Pease, 'The Parole Veto and Party Politics' (1986) Crim. LR 379. In 1991 the Home Secretary rejected 13 recommendations for release in relation to fixed-term prisoners (the figure had been 36 in 1990) and 12 in relation to life sentences—*Report of the Parole Board 1991* (1992, London: Home Office).
7. K. Hawkins, 'Assessing Evil' (1983) 23 Brit. Jo. of Crimin. 101.
8. M. Maguire, F. Pinter and C. Collis, 'Dangerousness and the Tariff' (1984) 24 Brit. Jo. of Crimin. 250.
9. See *Report of the Parole Board 1988* (1989, London: Home Office), for the Parole

Board's response to Carlisle's proposals. That response was reproduced in the 1989 Annual Report following the 1990 White Paper, *Report of the Parole Board 1989* (1990, London: Home Office), para. 11.
10. The Carlisle Committee's proposals with regard to procedures are contained in The Carlisle Report *supra*, n. 3, ch. 8.
11. *Supra*, n. 9, paras 9–11.
12. *Crime, Justice and Protecting the Public* (Cm 965, London: HMSO), para. 6.26.
13. CI 26/1992, 'Criminal Justice Act: The Discretionary Conditional Release Scheme'.
14. See, particularly, N. Polvi and K. Pease, 'Parole and its Problems: a Canadian–English Comparison' (1991) 30 Howard Jo. of Crim. Jus. 218. The attitude of the courts to these problems is considered below in the context of life sentences. The issue is also discussed in chapter 12 [see *Law, Process and Custody*, Ch. 12].
15. *R v. Mental Health Act Commission ex p. W* (1988) *The Times*, 27 May.
16. MHAC Fourth Biennial Report 1989–1991 (1991, London: HMSO), p. 32.
17. *Code of Practice* (1990, London: HMSO), para. 16.35.
18. See the First, Second and Third *Biennial Reports* of the MHAC (1985, 1987 and 1989, London: HMSO), at pp. 40, 23 and 23, respectively.
19. *Supra*, n. 16, p. 32.
20. See: *Chatterton v. Gerson* [1981] W.B. 432 and *Sidaway v. Governors of Bethlem Royal Hospital* [1985] AC 871; and, for further discussion, see P. Fennell, 'Sexual Suppressants and the Mental Health Act' (1988) Crim. LR 660, and P. Fennell, 'Inscribing Paternalism in the Law: Consent to Treatment and Mental Disorder' (1990) 17 Jo. of Law and Soc. 29.
21. See *Ex p. W*, *supra*, n. 15, and Mental Health Act Commission, *Third Biennial Report 1987–1989* (1989, London: HMSO), pp. 24–5.
22. See, sch. 1 to the Mental Health (Hospital, Guardianship and Consent to Treatment) Regulations, SI 1983/893, forms 38 and 39.
23. *Supra*, n. 17, para 16.11.
24. MHAC, *First Biennial Report 1983–5* (1985, London: HMSO), pp. 43–4.
25. MHAC, *Third Biennial Report 1987–1989* (1989, London: HMSO), p. 39.
26. The importance of the rmo is discussed further in chapter 12 [see *Law, Process and Custody*, Ch. 12].
27. *Code of Practice*, supra, n. 17, para. 16.37.
28. MHAC, *Second Biennial Report 1985–1987* (1987, London: HMSO), p. 22.
29. Its introduction was promised in MHAC, *supra*, n. 16, para. 6.3.d.
30. Code of Practice, *supra*, n. 17, para. 16.37.
31. See the previous chapter [see *Law, Process and Custody*] for the history of Part IV.

Discretion in a Behavioral Perspective

R. LEMPERT

The empirical study

This chapter is about the adjudicative discretion which Hawaiian state law gives a public housing eviction board. It is concerned not only with discretion as a quality of behavior but also with the sense that adjudicators have of their discretion. The two are related, for an adjudicator's sense of discretion can shape the way discretion is exercised. Not only are adjudicators likely to respect the law where it appears to limit their discretion, but, despite legal discretion, adjudicators may establish norms that lead them to feel that they have no discretion in particular cases. In this chapter I shall look at a variety of ways in which the eviction board I observed has exercised discretion, and I shall try to identify forces that shaped the board's discretion in particular cases and changed the pattern of discretionary decisions over time.

The eviction board I studied hears the cases of almost all tenants whom the Hawaiian Housing Authority (HHA) seeks to evict from its public housing projects on the island of Oahu.[1] The board was authorized by state law in 1949 and established in 1957. However, I shall focus on the board as it existed from 1960 on, which is when a board composed of three authority officials was replaced by one composed of five citizen volunteers.

My investigation into the HHA's eviction board occurred in two stages. The first stage, which involved three months of field research during the summer of 1969, examined the eviction board from its inception until that time.[2] The second stage, which involved fieldwork during the summer of 1987, examined the eviction board from 1966 until that point. During both stages I received the full co-operation of the HHA. I was able to interview the great majority of those people, except for tenants, who had been involved in the Authority's eviction process since 1960. These interviews included eviction-board members, Authority officials, including those responsible for prosecuting the Authority's cases before the board, project managers, and private legal-aid attorneys who had defended tenants before the board. I sat in on more than thirty eviction hearings, all those held during the two summers of my field-work. I read the full transcripts of more than a hundred additional hearings, most involving cases from the early 1960s. I perused Authority records for any official documents or other materials relating to evictions. I collected and coded information from the records of more than 1,400 eviction actions. And I read all the Federal and state statutes and regulations relating to the eviction process that I could identify.

I found that, during the eighteen years between my two visits, some aspects of the eviction process had remained the same, but others had changed—sometimes dramatically. The board's status, jurisdiction, and powers were officially the same. In both 1969 and 1987 the board was composed of citizen volunteers who were paid only a nominal sum ($10 a member a meeting in 1987) for their services. Although the board members were appointed by the Authority, they were independent of it. The board's chair was a board member, and neither the board nor its chair had to answer to the Authority for its decisions. The Authority was required to bring before the board any tenant it sought to evict, and the tenant had a right to a 'full and fair hearing', which included the rights to know in advance why the Authority sought to evict, to present witnesses or documentary evidence, to cross-examine opposing witnesses, and to be represented by counsel.[3] At both points in time the board had the power to acquit tenants, in which case the tenants had to be allowed to remain; to evict tenants, in which case the Authority was granted a writ of possession without further litigation; or conditionally to evict tenants, in which case an eviction order would be issued but its execution would be held in abeyance and eventually cancelled so long as the tenants complied with the conditions specified.[4]

The types of cases the board heard and its procedures for hearing these cases also looked much the same in 1969 and 1987. In both years and every year in between actions brought for non-payment of rent dominated the docket. This was the sole charge in about three-quarters of the cases, and it was charged together with some other offense in an additional 5 per cent of the actions. Other cases the boards heard involved what I call 'trouble behavior'. This includes such things as income falsification,[5] fighting, parking more than one car or a car that does not run, keeping pets, and allowing unauthorized guests to occupy units.

Hearings were held around the same long table in the same conference room in 1969 and 1987, and in many ways they looked similar. Lawyers were seldom present, rules of evidence were relaxed; conversation was informal; tenants who did not spontaneously excuse themselves would be invited to tell their stories, and board members would not only question tenants but might advise them on how to deal with their problem or lecture them on their moral deficiencies. The hearings ordinarily lasted as long as the parties had something to say. Most took between twenty and thirty minutes, but a number took somewhat longer, and cases lasting an hour or more occurred.[6] Decisions were reached by the board in a brief discussion following the close of the case, and the tenant and manager were immediately informed about what the board had decided.

In other respects, however, there were marked differences in the situations I observed in 1969 and 1987. Many of these were not observable from the hearing but rather concerned the Authority's project management and its officials' views of the appropriate scope of board discretion. In 1969 considerable discretion was granted project managers with respect to rent collection on the projects. Managers were free to 'work with' tenants in financial difficulty, and it was largely up to the manager to decide if and when to bring a tenant before the eviction board. Thus, when non-payment tenants were brought up for eviction, they commonly had three, four, or more months rent owing. In 1987, thanks to computers the central management staff knew as soon as the project managers which tenants were behind on their rent, and project managers had to justify decisions not to seek eviction when tenants were more than six weeks in arrears. Thus many non-payment tenants who faced the board in 1987 owed two months rent or less, and a number of them owed nothing because they had cleared their debts after being subpoenaed.[7] In 1969 the latter group would have had their cases cancelled.

In 1969 the Authority's central office officials, including its Supervising Public Housing Manager (SPHM), who was in charge of presenting cases to the eviction board, saw the board's independence as a virtue,[8] did not question the board's discretion to withhold eviction despite finding a lease violation, and regarded the conditional deferral as an appropriate decision when tenants owed rent.[9]

In 1987, by contrast, top Authority officials regarded the board as an awkwardly independent cog in the Authority's efforts to maintain peaceful, smooth running projects. While the board's power conditionally to defer was recognized, its discretion to do so was not respected, and during the preceding seven years steps had been taken to minimize the occasions on which such discretion would be exercised. In 1979 a training session had been held for the board at which the Authority's rent-collection needs were emphasized. In 1982 the board chairs had been sent to 'judge's school' in Reno, Nevada, in the hope of promoting more legalistic decision-making, and in the same year another training session was held for all board members. Also beginning about 1980 fixed terms were established for board members. Several members were not reappointed because they were regarded as too pro-tenant, and new appointments were made with an eye to whether they would appreciate the Authority's point of view.[10]

Some changes between 1969 and 1987 were visible just from observing hearings. The most obvious was that the board in 1987 consisted of fourteen members rather than five. In 1970 two tenants were added to the eviction board to create a seven-member panel and in October 1979 a second seven-member panel was created, with its own chair, so that eviction actions could be heard every week rather than every other week, thus allowing the Authority to process cases more rapidly for eviction. As the panels never got together except for one or two parties a year, the situation was one of two seven-member eviction boards rather than one fourteen-member board.[11] Another difference was that the Authority's cases were presented by an attorney, whom I shall call the DAG,[12] rather than by the SPHM. However, in many respects the attorney proceeded at the hearing in much the same manner as the SPHM had in 1969. Both acted informally. They avoided legal jargon except at the outset when the cause of action was explained, and they conversed with the tenant to make sure that his or her story came out. The SPHM, however, tended to leave the presentation of the Authority's case to the project manager, while the DAG presented the details of the manager's report himself and relied on the project manager for confirmation and further information.

A more subtle difference between the hearings of 1969 and 1987 was that the board members in 1987 seemed less sympathetic to the tenants than they had in 1969. In 1987 the board members were less prone to delve into ways that the tenant might solve his or her problems and almost never questioned the adequacy of the project manager's efforts to 'work with' the tenant.

Finally, the possibility of an appeal to the Authority's Board of Commissioners was often mentioned during the 1987 hearings—both before and after the board's decision was rendered—but was seldom if ever mentioned in 1969. During his case presentation or summation the DAG emphasized the possibility of an appeal to remind the board members that, even if they voted to evict, the tenant would not necessarily be forced to leave.[13] After an eviction decision, the tenant was told how to appeal and what he or she would have to do to be successful. In 1969 such explanations were seldom necessary, for tenants were almost always allowed to stay.

From the Authority's point of view the stress placed on the appeal process was made possible by a 1980 amendment to the Act establishing the eviction board which provided that appeals had to be based on 'new facts or evidence pertinent to the case which could not have been presented and were not available for presentation' to the eviction board.[14] Before the law was amended the Commissioners had to hear appeals *de novo*, and any system that encouraged appeals would have been untenable. Indeed the burden of deciding whether an appeal presented new facts and evidence was eventually deemed excessive, and in 1984 this responsibility was delegated to the HHA's Executive Director. The result was that after 1984 appeals almost never reached the Commissioners unless it was a foregone conclusion that they would be allowed. Indeed, the Commissioners typically did not hear appeals but instead ratified 'stipulated agreements' negotiated between the housing staff and the tenant which noted as a new fact that the tenant had fully corrected the problem giving rise to the board's eviction order (usually by paying on outstanding rent debt) and stipulated that, in exchange for the withholding of the eviction order, the tenant agreed to comply fully with all lease provisions for a period of one year and to waive all rights to a hearing should any lease provision be violated within that time.

The changes that occurred between 1969 and 1987 did not, of course, occur at the same time. Yet, for purposes of investigating changes in the board's exercise of discretion, there are two watershed years. The first is 1975, which marks a dramatic change in the leadership of the Authority as well as the commencement of a lawsuit that at one time appeared to threaten the existence of the eviction board.[15] Before 1975 cases were handled as they had been in 1969 or, for that matter, in 1961, and the outcomes were the same. The second is 1979. This is when a secretary was assigned full time to handle the paperwork of evictions, and a full-time specialist was hired to process and prosecute eviction cases. It also marks the appointment of the second eviction panel, which was formed by dividing the old panel into two and adding three new appointees to one group and four to another. After the appointment of the second panel, the eviction process came to look much as it looked when I observed it in 1987. The period between 1975 and 1979 was not so much a period of gradual change as a period of upheaval and uncertainty (Lempert, 1990;

Lempert and Monsma 1988). Hence we shall not focus on these years when we discuss the transformation of discretion.

With this information as background, we are now ready to examine the discretion the board exercised. First, I shall discuss several varieties of discretion that are illustrated by board decision making. Then we shall see that discretion may not only be influenced by external forces, but may be systematically transformed.

Varieties of discretion

The case of the house that burnt

In the 1960s, before the income limits for continued occupancy in federally aided low-income housing had been abolished, a family with eight children, let us call them the Teofilos,[16] exceeded the income limits and contracted to build a home. The day before the Teofilos were supposed to move, and after the grace period which federal law gave them to find a home had expired, their new house burnt to the ground. The Authority was not anxious to press the case for eviction, but felt that federal law required that action. The eviction board refused to issue the order. On several occasions it remanded the case to see if anything could be worked out and to give the Teofilos more time to find a home. One member who was involved in real estate went so far as to search for housing for the family in his own time. The board knew what the law required, but its members wished to avoid the force of the law. Indeed, in discussing what to do with the Teofilos, one member said he would not evict, no matter what the law required. Eventually the case was resolved when the Authority transferred the family to no-income-limit housing it operated for the Navy. Doing this breached both policy and regulations, for the family had no Navy connection and was too large for the unit available, but the Authority apparently felt that it was less important to conform to these rules than to federal housing regulations.

The board was able to exercise discretion effectively in this case because its actions were not reviewable. The board was given no legal authority to do anything other than evict, but it could effectively refuse to evict because the law establishing it did not provide an avenue by which the Authority could appeal to a court or other higher tribunal, and the same law did not allow the Authority to secure a writ of possession except by prevailing before its eviction board. The board was well aware of its power and that, but for it, the family would have been without decent shelter.

What is most striking about this case is that it is apparently unique. While I could not look at every case the board heard over thirty years, my perusal of several years of case transcripts did not turn up any other case in which the board knowingly did something it was not empowered by law to do. Furthermore, I did not see such a case during the two summers I sat in on board hearings, nor did I hear of such a case in my interviews with Authority officials, board members, or project managers. The latter's silence is quite telling, for they freely complained about board decisions that in their view exceeded the board's proper authority.

The board's more usual attitude in over-income cases is expressed by the chair's statement to a couple with seven children, one of whom was a mute. This family had been unable to find a house because the private rental market provided little housing for moderate-income families with more than a few children: 'I regret very much to inform you of the decision we came to arbitrarily; it's one that we have no other recourse [sic] on account of the qualifications of the law governing a case such as yours. We have to order eviction because there is no way we can do otherwise.' The difference in the attitude expressed here and the attitude expressed in the case I first described cannot be explained by board composition since many of the same people sat on both cases. It is explained, I believe, by the extraordinary nature of the tragedy that befell the first family. Almost all over-income cases the board heard involved families who had been successful by middle-class standards and who could not find suitable housing because of the tight nature of the housing market that confronted large families. The first case involved a family that had solved the housing problem in the most culturally approved fashion—buying a home—only to have their house unexpectedly taken from them.

It appears from both the philosophical and sociological literature that discretion in the sense of unreviewability is relatively common, since it is easy to provide examples of adjudicators who have discretion to make decisions that are unreviewable and continually use that discretion (see, e.g., Rosenberg 1970–1). Consideration of the Hawaiian data suggests that there is a further important distinction to be made. Typically, when an adjudicator like a multi-judge court has unreviewable discretion, it also has discretion in the rule-oriented sense that the authority to exercise judgment is entrusted to it; that is, it is authorized to choose from a wide range of outcomes, any one of which is permissible. Indeed, a major reason for making an exercise of discretion unreviewable is that it is unlikely that a reviewing agency will be able to exercise better judgment. The eviction board's members did not have discretion in this sense. The law did not allow them to exercise judgment about whether families that were over the income limits should be evicted once the statutory grace period had expired. They were mandated to evict in these circumstances. Their only discretion was whether to comply with their mandate. This discretion was effectively allotted them only because the Authority could not appeal from their decision.

It is a mistake to think that the law authorizes this type of discretionary decision-making.[17] Rather the law establishes structural or legal conditions[18] which ensure that a particular adjudicator's decisions will be complied with while not providing a way effectively to remedy errors through appeal. Such conditions give decision-makers the power to force actions that do not comport with legal norms, although as a matter of law they lack the authority.

What the law can do for judges, other structural features do for other decision-makers. Thus the cop on the beat who has stopped two youths in 'the wrong neighborhood' may arrest the one who 'gives him lip' and let off the one who is respectful (Werthman and Piliavin 1967; D. Black 1971). The cop can do this, not because the law provides that disrespect or being in the wrong neighborhood is a crime, but because the officer's power *vis-à-vis* youths is such that they must comply with his decisions, and the officer's 'credibility advantage' coupled with the low visibility of the encounter means that, if he later concocts an account of the encounter legally sufficient to justify arrest, he rather than the youth will be believed. The eviction board that spared the Teofilos was much like the cop, except that its flouting of the law was more visible. However, the Authority shared the board's sense that the Teofilos had done nothing that was unreasonable, and it was pressing for eviction only because it could be held accountable if it did not enforce federal income limits. Had the Authority been unsympathetic to the Teofilos, it might have found a way to impel the board to conform to the law.

Discretion in the sense of unreviewability is common at the street or factory level (Lipsky 1980; Bardach and Kagan 1982), and it might seem that it is common among adjudicators also. Looking at court decisions from the outside, there appear to be many situations in which adjudicators are able to enforce their will only because their decisions are unreviewable. But consider the matter from the adjudicator's point of view. The members of the eviction board in the Case of the House that Burnt could and, in effect, did say, 'That law required me to evict, but I exercised my discretion,' meaning discretion in the sense of the power of unreviewability. Usually, however, it will appear to the adjudicator

Discretion in a Behavioral Perspective

whose decision sticks because of unreviewability that his discretionary action (which just happens to be unreviewable) is in accord with a legal mandate to exercise judgment and is not an exercise of lawless power. The Teofilo case may be unique in my data because adjudicative discretion that exists only because unreviewability confers a power is, at least phenomenologically, rare.[19]

The case of the beans that burnt

This case involved a Korean woman, whom I will call Mrs Park, who lived in one of the Authority's high-rise buildings for the elderly. On three separate occasions over two years, while boiling down beans with ginseng for an ethnic dish she enjoyed, Mrs Park had forgotten she had beans cooking and left the apartment. On each occasion the beans boiled dry and then burnt, sending smoke into the halls. After the third such incident the project manager sought to evict Mrs Park because he thought that she was likely to forget again and that an overheated pot or burning beans might in some way spark a fire. He cited the woman for violating lease covenants relating to (a) not damaging the dwelling unit or causing insurance premiums to increase, (b) keeping the unit in a safe and sanitary condition, (c) using facilities only in a reasonable manner, and (d) conducting oneself so as not to disturb the neighbors' peaceful enjoyment of their accommodations and maintaining the housing project in a decent, safe, and sanitary condition.

When I discussed this case with the Authority's prosecutor the day it was to be heard, we both expected Mrs Park to be growing feeble minded as well as old and increasingly incapable of living alone. At the hearing, however, a different picture emerged. Mrs Park came to the hearing with a lawyer, a minister who translated for her since she did not speak English, a Korean-speaking neighbor, and a petition signed by many of the building's tenants saying that they wanted her to remain and would look after her. She looked as if she were about 70 and quite capable of looking after herself. The evidence at the hearing was that she was well regarded by the neighbors and was an active volunteer in her church, who was there almost every day and could greet each of the five hundred or so church members by name. While the project manager made a convincing case that the building was constructed so that a fire would be especially dangerous, Mrs Park's advocates showed that, with the exception of one occasion when the hot pot was apparently dropped, scorching part of a rug, no damage had been done by the several incidents. Except for these incidents, the project manager agreed, Mrs Park was a good tenant and a pleasant person. The Korean-speaking neighbor who lived across the hall was a friend who said that she would look in on Mrs Park daily and boil beans for her once a month, the schedule Mrs Park had followed.

Perhaps the key to the hearing was that the Authority's prosecutor did not think that Mrs Park should be evicted, nor did he think that the board would be willing to evict her. Thus he shaped the discussion so that it focused on steps that could be taken to ensure that Mrs Park would not pose a fire hazard if she stayed. Mrs Park's lawyer had the same agenda, hence the offer by the neighbor to look in on her. There was also some discussion about whether Mrs Park would be willing to give up her stove and cook with a microwave instead.

The prosecutor had perhaps misjudged the board. At the start of its deliberations one member, a real-estate manager who had just joined the board, moved to evict. Eventually the motion was defeated, and a conditional eviction—a form of probation—was voted The conditions were that the Authority remove Mrs Park's stove within a week, that Mrs Park secure a microwave to replace it, and that there be no further incidents for three years. It is impossible to say whether this decision, as opposed to an outright eviction, would have been reached had the prosecutor not obviously favored a compromise of this sort. The member who moved for eviction was intelligent and articulate and he might have persuaded a majority of the board to go along with him had the Authority's representative been pressing for the same action.

The board in this example is exercising three kinds of discretion. First the board has discretion to determine whether there is a lease violation. The determination is discretionary in the sense that the judgment is entrusted to the board and the board must pick out those facts that bear on its decision-making task (Barak 1989:13). If burning beans does not violate any of the cited lease provisions, not only will the woman avoid eviction, but she may go on burning beans to her heart's content. The second, which is dependent on finding a lease violation, involves deciding whether to allow Mrs Park to remain in housing despite the lease violation.[20]

The third locus for discretion, which is dependent on both finding a lease violation and determining that alternatives to eviction should be explored, is in deciding the conditions under which the woman will be allowed to remain. This discretion too is not clearly confided by the statute authorizing the eviction board, but is firmly rooted in the board's 'common law' and would be regarded by both the board members and the Authority as a necessary concomitant of the board's power to withhold eviction when the Authority has presented a legally sufficient case.

The existence of analytically distinct forms of discretion does not, however, mean that all forms will be equally salient to those involved in the decision-making process. Ordinarily only two discretionary decisions are salient, but they are not the same for tenants and project managers on the one hand and the board members on the other.[21] For the parties, the first two types of discretion—determining whether there is a legal cause for eviction and if so whether eviction should follow—are lumped together and important while the third is distinct and subsidiary. Thus both the tenant and the project managers are interested in whether the tenant will be evicted immediately or allowed to stay. It does not matter whether the tenant is allowed to stay because no lease violation has been found, or whether, despite a lease violation, the tenant is not expelled. When the tenant avoids immediate eviction, the tenant regards the decision as a victory, and the manager regards it as a loss, regardless of the conditions that are set and the implications that these conditions have for the tenant's prospects of avoiding eviction in the long run. The tenant's attitude is like that of the criminals described in a number of plea-bargaining studies who focus on the sentence which might be received and are relatively indifferent to whether the sentence is a result of charge or sentence bargaining, even though the charge pleaded to will become a matter of record that may have substantial future implications.

From the board's standpoint it is the first type of discretion on the one hand and the second and third types on the other which are distinct. The board must decide whether there has been a lease violation and, if so, how to dispose of the case. The former determination seldom poses any difficulties. But in dealing with the latter issues, the decisions on whether to allow the tenant to stay and on the conditions to be imposed if the tenant does stay are inseparable. Assuming a lease violation has been proven, the more likely it is that the tenant can cure that violation and not violate again, and the more likely it is that the tenant will be given another chance conditional on the cure and subsequent good behavior. Conversely, even a sympathetic tenant may face eviction if it appears unlikely that future violations can be prevented. Thus, had there been no way to meet the Authority's concerns regarding the fire hazard Mrs Park posed, she would not have been allowed to stay. Indeed, at one point it appeared that the board's decision might become unraveled because it was unclear that federal regulations allowed the removal or disconnection of a tenant's stove even when all parties desired it.[22] Similarly the board has evicted families for damages caused by their children when it appeared that the families, despite their best efforts, could not control their children's actions.

From an analytic perspective the board's discretion to determine the conditions under which Mrs Park could stay seems stronger, in the sense of being less law bound, than its discretion to determine whether she should be allowed to stay subject to conditions (cf. Dworkin 1977*b*). The law provided no guidance to the board members as they creatively sought to determine arrangements that would prevent Mrs Park from posing a fire hazard, but, in deciding whether Mrs Park should be allowed to stay despite her actions, the board was constrained by its need to respect the goals of the lease clauses Mrs Park was shown to have violated.

This analytic distinction, however, makes no sense from a behavioral perspective. The two determinations cannot be separated, for it is the board's creativity in establishing conditions that determines whether it can allow a tenant to stay, while still respecting the goals of the lease provisions that it is called on to enforce. Discretion is often intertwined in this way, and efforts to limit or extend discretion of one analytically distinct sort may affect how discretion of another analytically distinct sort is exercised. Thus Heumann and Loftin found that the Michigan legislature's effort to prevent judges from sentencing gun-carrying criminals to less than two years in prison affected the charging discretion of prosecutors and the discretion that judges had to accept or reject plea bargains (Loftin, Heumann, and McDowall 1983). It is for similar reasons that Abel (1982*a*, 1982*b*) and others argue that institutions of informal justice may extend state control. When police or prosecutors have the discretion to refer disputes to institutions of informal justice, they may pursue matters that they would have dropped had pursing the matter necessarily placed it in formal court. Focusing on the discretion that inheres in particular rules may miss important ways that discretion constrains and frees choices. Individual rules must be examined as parts of applied rule systems.

The board's other discretionary decision, the decision about whether there has been a lease violation in the first instance, is, of the decisions entrusted to the board, the one most closely confined by law. In reaching this decision, the board members are to examine the facts and determine whether they make out a lease violation. From a legal–analytic perspective this narrow task may none the less involve substantial judgmental discretion, since the factual determination may be quite difficult, and lease provisions may require interpretation.

From a behavioral standpoint, however, the situation is different. The fact of the lease violation in Mrs Park's case appeared so unproblematic that the board in its discussion did not even address the issue. Rather the members turned immediately to the question of whether there were any conditions under which the woman could be allowed to remain without posing a threat to her neighbors. In the eviction setting this is almost always the case. Lease violations are ordinarily clear,[23] and the board has no discretion, except in the sense that, as in the Teofilo case, they may ignore the law, to find otherwise. Thus, what is conceptually a major locus for the exercise of board discretion is behaviorally almost never the occasion for discretionary decision-making. The question whether there has been a lease violation seldom merits discussion.

The case of Mrs Park reveals one other way in which the board's discretion is affected and, in effect, limited. In this case, the DAG, despite the manager's position to the contrary, did not seek eviction. Rather he participated with the board members and Mrs Park's lawyer in a discussion of arrangements that would remove the threat of a fire yet still allow Mrs Park to cook her beans, and he concluded the Authority's case by stating: 'If the board feels that there has been sufficient corrective actions . . . then I would see no problem with some kind of conditional deferment that there be no further forgetfulness of boiling beans down or whatever. Some type of condition; that is what I would recommend. Let her stay on probation'. This prosecutorial concession further limited the board's discretion. While the board might have decided to let Mrs Park stay, even if the prosecutor had sought her eviction, when the prosecutor is willing to accept a conditional deferment, the board as a behavioral matter is unlikely to offer less. The point applies generally. In an adversary system, whatever the discretion of the decision-maker, a party is unlikely to do worse than the opposing party requests.[24] Thus, the board did not seriously consider the motion of one member to evict Mrs Park. Had the prosecutor's concession not been made, the motion certainly would have divided the board and might well have carried.

* * * * *

The strength of law

These observations lead to two final questions that should be addressed in an effort to understand discretion from a behavioral perspective. The first is when is a legal mandate strong enough to foreclose adjudicative choice? The second is how do extra-legal factors come to constrain the decisions of adjudicators vested with legal discretion? The questions are obviously related, for the influence of the law will vary inversely with the influence of extra-legal factors, and vice versa. For this reason a law that will strongly influence the decisions of some legal actors may have little or no influence on the behavior of other such actors. Thus the prohibition against illegal searches and seizures in the US Constitution may lead most judges to discard certain types of evidence, although it might not prevent most police officers from acquiring it.

This case study of an eviction board was not aimed at answering the question what makes a law influential, but it does illustrate some factors that are likely to have this effect. First, the mandate and clarity of the law as understood by the decision-maker seem important. Before 1980 the statute establishing the eviction board did not require the board to evict simply because it found a violation, and the board developed a pattern of not evicting. One of several amendments passed in 1980 could arguably have been interpreted as mandating eviction whenever a lease violation was found, but it does not clearly require this and the Authority has not so interpreted it. Indeed, in training sessions both before and after the passage of the 1980 amendments, board members were told that their discretion extended to withholding eviction even if they found lease violations, and for five years after the passage of the amendments the board sometimes did this.

The board has, however, almost always complied where a legal mandate appeared clear. Thus, the Case of the House that Burnt was a unique act of rebellion against the law. In other cases, the board has regretfully evicted tenants who were over the income ceiling, citing federal law that required such tenants to move within six months of the over-income determination. In similar fashion the board, beginning in the mid-1970s, limited the period over which tenants were allowed to repay their rent to six months, because they were told that that was the limit which federal law provided for the repayment of back charges. Tenants who had no prospects of repaying their rent debts in six months were evicted.

A related factor which affects the binding power of law on an adjudicator is his role conception. An attitude toward the binding nature of statutory language and precedent is usually an important aspect of judicial role conception. The eviction-board members are somewhat similar to a jury in the way they define their role. The members feel that they are to exercise common-sense judgment but that in doing so they are bound by the law. Thus, if board members believe a particular action is legally required, they comply. It follows that one way to affect the board's exercise of discretion is to convince its members that certain actions are or are not legally permissible.

A major difference between the board and a jury is that board members serve sufficiently lengthy terms that some come to feel that they are experts on what the law requires. For example, in one case I observed in 1987 the prosecutor erred slightly in making his customary speech. Rather than telling the tenant that the board's usual procedure in cases like hers was to evict, he suggested that the law gave the board no choice but to evict. The board chair, who had served for about a decade, interrupted the prosecutor to emphasize that the board had discretion to refuse eviction regardless of its usual practice. In a later case this chair's panel granted a tenant a conditional deferment despite an outstanding rent debt, the first time in almost a year that it had been lenient in this fashion. One member commented jokingly that it must have been my influence. He may have been right. In questioning board members about their usual practice and changes in it over time, I reminded them of their discretion to defer evictions despite rent that was outstanding and of the fact that they once exercised it.

The board is like a jury, however, in that the salience of other values affects the law's actual binding authority. The Case of the House that Burnt illustrates this. The Teofilo family's situation induced so much sympathy and respect that the board refused to evict, despite its understanding that this was what the law required. I observed a similar conflict in 1987, except that strongly held sentiments clashed not with the demands of external law but with the requirements of the board's by then well-established precedent. The case in which this occurred was a nonpayment action involving a divorced woman, let us call her Mrs Sua, with ten children. Mrs Sua had not cleared her debt by the time of the hearing but said that she expected soon to receive a special welfare grant to pay it. In 1967 or 1977 the board would have deferred eviction on the condition that the debt be paid by a certain date; in 1987 the board regularly evicted on such facts, relegating the tenant to her right to appeal. In Mrs Sua's case the board did neither. Rather it continued the case for two weeks to allow her to secure her grant without the stress of an outstanding eviction order and the need to proceed through an appeal that would have left her vulnerable to an eviction without a hearing for the slightest defalcation over the ensuing twelve months. In the DAG's judgment the woman's large family was the factor that led to this special treatment. Judges are supposed to be better able than lay decision-makers to ignore personal values when these clash with legal interpretations. Perhaps they are, but judges too balance the importance of the values affected by their decisions with their understanding of what law or consistent practice requires.

Canalized discretion

Since the law as understood by the eviction-board members gave the board considerable leeway in deciding how to dispose of cases, our investigation can address the second question, which asks what shapes the exercise of authorized or rule-given discretion. One important factor is that, when a decision-maker is repeatedly confronted with cases of a particular type, there is a tendency toward what Professor Sanders and I call 'shallow' decision-making (Lempert and Sanders 1986). That is, there is a tendency to eschew a deep probing of circumstances and to rely instead on a few key facts that can be used to fit cases to stereotypes.[26] There are no doubt many reasons for this, including psychological mechanisms[27] and the efficiency that routine-processing allows. This tendency is complemented by a common element of judicial role conceptions, the sense that, regardless of the range of outcomes that discretion allows, cases that are similar in relevant particulars should be decided in the same way. Thus we can expect adjudicators to see cases as similar on the basis of a few particulars and to dispose of cases that are seen as similar in a similar fashion.

As a consequence, adjudicators who have discretion to decide a series of similar cases will generate a pattern of decisions which is sufficiently regular to call into question the actuality of their discretion.[28] Indeed, it may be, as was apparently the case with the eviction-board, that adjudicators with broad discretion to decide will feel in most cases that their decisions are tightly constrained despite their knowledge of the leeway law gives them. Thus to understand how extra-legal factors come to constrain the decisions of those vested with discretion, we must explore those conditions that lead an adjudicator to feel that cases of a certain type should systematically be decided in one way or another.

In the case of the eviction board, the predominant factor leading to the early precedent of never ordering an immediate eviction may have been the values that the original board members brought to their work. The original five-person board included a social worker and a minister among its members and was dominated by a chair who did considerable volunteer work on behalf of the poor. Moreover, in establishing the independent board, the Authority conveyed the impression that special sensitivity to the interests of the poor was appropriate. The choice of members reflected the notion that the impoverished tenants were a constituency with interests that deserved representation.

In addition, the Authority's original prosecutor was untroubled by leniency in cases where it appeared that tenants would be able to repay their rent. The members' natural sympathies coupled with their difficulty in deciding which of the tenants who promised to pay back their rent could be believed fostered the development of a precedent that allowed all tenants who said they would pay back their rent a second chance.[29]

Other factors also contributed to this outcome. One was probably the desire of board members to avoid the responsibility for evicting tenants with innocent young children, even when the parent's failure to pay rent was blameworthy. The strategy of deferring eviction placed the responsibility back on the tenant. For tenants had their eviction deferred only if they promised to repay their rent. If they then failed, they were not only shirking their responsibility, but were also breaking their word, and the subsequent eviction could easily be seen as their own doing rather than as the result of the board members' refusal to accept the sympathy-inducing story they were likely to have heard at the initial hearing. Moreover, the board members would not have to confront the tenant again, but would take the Authority's word that conditions were not being met and would vote to execute the deferred order.[30]

The forces that establish a precedent are not necessarily those that keep it in motion.[31] Members, like a retired project manager, who joined the board with neither an inclination to sympathize with financially troubled tenants nor an optimistic view of their prospects for repaying their debts none the less respected board precedent and voted to defer eviction despite their doubts. Other factors also served to keep the precedent alive. One was the appointment of a 'bleeding heart' (the managers' term) board chair in the mid-1960s who served for sixteen years. As chair he dominated the discussion. Moreover, as the years passed and new board members were appointed, this chair's experience gave him a special claim to expertise about how different types of cases should be decided.

Another factor that may have helped maintain the pattern of lenient decision-making is the feedback that the board members received during the 1960s and much of the 1970s. When the board gave tenants a second chance, they were often warmly thanked by the tenants. Managers did not thank the board in the cases in which the board evicted, and they usually hid the depth of their displeasure when the board failed to evict. The board members also learnt about what happened after they deferred eviction, since they voted to cancel eviction orders when debts were cleared or voted to evict or set new conditions if tenants failed to live up to the conditions of their initial deferral. More often than not, tenants cleared their debts, and even those that did not often repaid a portion of their debt before again falling behind. Thus the board members felt that, when they were lenient, they were usually right.[32]

Finally, the attitude of the Authority officials who prosecuted cases was important, for these were the officials who regularly met with the board and presented the Authority's positions. The prosecutors during the 1960s and early 1970s not only respected the board's authority but were also relatively passive in presenting the Authority's case. While in some cases involving behavioral violations, like fighting or harboring unauthorized guests, the SPHM or other prosecutor might press hard for eviction, in non-payment cases they conveyed the impression that it was for the board to decide what was to become of the tenant. Indeed, this was the attitude that the Authority's central office staff conveyed to the project managers when the managers tried to get their superiors to press the board to evict more often. The staff's attitude was that the board was given the power to decide cases as it saw fit, and that the board's pattern of leniency was tolerable. One reason for this attitude was that the Authority's prosecutors during the 1960s, at first the HHA's Assistant Executive Director and later in most cases the SPHM, devoted relatively little of their attention to evictions. Handling evictions was one duty among many, and, given the nature of these officials' other responsibilities, their role in managing evictions could not have seemed particularly consequential.

The transformation is this 'second chance' pattern to a pattern of always evicting is interesting because the transformation required a 180-degree change in precedent. The attempt to turn the board around began in 1979, and it took about seven years before the transformation was complete. It was spurred by the Authority's serious financial troubles,[33] by a sense that the Authority was losing a substantial amount of money in unpaid rent[34] and by the feeling, confirmed by HUD auditors, that the Authority's lenient eviction system was largely responsible for this.[35] It also reflected a different, less welfare-oriented attitude at the highest levels of the HHA toward the task of housing poor tenants.

In 1979 the HHA, in part responding to pressure from HUD and in part as a result of its own increasingly businesslike (as opposed to welfare) orientation, decided to get its 'eviction house' in order. There were two basic elements to its strategy. One was to rationalize the eviction process and make it more efficient. The second was to transform the board so that it was more appreciative of the Authority's concerns and stricter in dealing with non-payment tenants.[36]

The effort to rationalize the eviction process began in 1979 with the appointment of an administrator whose primary responsibility was to handle eviction actions and the appointment of a secretary to work full time on the paperwork of the eviction process. Prior to these appointments responsibility for evictions at both the secretarial and the administrative levels were part-time duties of staff members who had other tasks that both they and the Authority deemed more important.

One of the first tasks of the new administrator was to study the eviction process in order to respond to HUD's position, which was that the Authority should abolish its eviction board and use the ordinary judicial process when it wished to force tenants out. The administrator found, as I had found a decade before, that the eviction board seemed to save the Authority money by securing time payments from the majority of those tenants whom the managers (and a court) would have immediately evicted. Thus the decision was made to retain the eviction board but to increase the efficiency of the eviction process. The major changes are mentioned earlier in this chapter [see *the Uses of Discretion*], where I note the differences between the eviction process I studied in 1969 and that which I observed in 1987. I will recapitulate briefly.

First the then existing board was split into two seven-member groups which allowed weekly eviction hearings. Secondly, the HHA drafted and the Hawaiian legislature passed amendments to the Act establishing the eviction board that removed a requirement for in-person service of process in eviction cases and limited appeals to the HHA's Commission to cases which alleged that relevant 'new facts and evidence' had become available only following the board hearing. Later the Commission delegated the task of determining whether such new facts existed to the Authority's executive director, who in turn delegated it to the Director of Housing Management (DHM), and it was the DHM's policy never to find new facts and evidence in non-payment cases when rent was outstanding. Thirdly, the HHA reformed its system for recording rent payments and computerized the process of sending delinquency letters to tenants. Coupled with this was close supervision by the SPHM of project delinquencies and instructions to the managers to process tenants more quickly for eviction. Before 1980, by the time the board heard their cases, tenants were often three months or more behind on their rent. By the mid-1980s it was not uncommon to have a tenant up for eviction six weeks after the initial default.

While the Authority was revamping its administrative structure in these ways, it was also seeking to develop an eviction board that would view cases from the new, stricter perspective it had come to prefer. One step in this direction was to fill the new slots that became available when the board was split in two with people, like private real-estate managers, likely to be sympathetic to the Authority's point of view. Also, board members were given terms, and a few members who were thought to be unduly sympathetic to tenants were not reappointed. The long-time board chair was among the first to go.

A second step the Authority took was to be more specific in its expectations about how the board should behave. A training session was held for all board members at the time the second panel was established, and another one was held several years later. The board members were told of the seriousness of the rental delinquency problem which the Authority faced, and their task was defined in neutral, judicial terms rather than from the welfare-oriented perspective of how best to help tenants. Later the chairs of the board's two panels were sent at the Authority's expense to 'judge's school' in Reno, Nevada, to encourage them further in a legalistic approach.

Complementing these formal actions were attempts at informal influence. The Authority's Executive Director and his assistant occasionally attended board parties[37] or otherwise chatted informally with board members. On these occasions they discussed the Authority's rent-delinquency problems and their expectations about how the board should act, and they complimented board members for acting in accord with their expectations.

Even after these steps had been taken, however, the Authority was not insisting on immediate evictions in all cases in which tenants owed money. Rather, board members were given discretion to allow tenants up to six months of time payments to clear accumulated debts. However, the changes in board composition and in the rigor of prosecution had their effect. The board evicted many people outright, including some who owed no rent when they appeared before the board but had histories of chronic delinquency.[38]

In about 1985, perhaps coincident with the replacement of the Authority's eviction specialist with another attorney, the Authority further toughened its policies. With the concurrence of the Executive Director, the DHM decided that the Authority should seek the immediate eviction of *all* tenants owing rent at the time of the hearing and should place all tenants who cleared their rent debts between the time they received the subpoena and the hearing on probation for six months.[39]

The Authority's Executive Director and its Director of Housing Management may have communicated these new expectations to board members, but it was largely left to the HHA's eviction specialist, whom I call C, to cement a new precedent. C may have been particularly amenable to this since he did not have experience under the former system, and the new system had its own way of allowing tenants to avoid eviction:[40] namely, by payment, after the hearing but before the time for appeal had lapsed, of all the rent that was due.[41] Moreover, since the time for appeal did not start to run until the tenant was officially notified of the board's decision, C, who handled part of the paperwork of notification, had some leeway to delay giving notice where he thought a tenant could secure money if given extra time.[42]

C recalls the process of persuading the board to change its decision-making standards as a lengthy and difficult one. It took about a year of continually pressing the board to decide cases as he wanted—which is to say always to evict when rent was owing—to persuade them that this was the right thing to do. C's strategy was to persuade the board to take a legalistic approach and to convince the board members that their vote to evict immediately did not make them responsible for a tenant's eviction. C recalls:

My argument was that the board had to make findings of fact and if the findings of fact were that the person was delinquent then they had to—they could give some kind of conditional deferment—however that was more the prerogative of the Commission than of the board members. Once they saw that these tenants were not going to get evicted for sure just because they said, 'Well you're behind and we order an eviction' . . . the board felt more comfortable in saying, 'OK we will send it on up to the Commission . . .' I remember going in there and standing up and addressing the board with what their functions were . . . the selling point was that this board wasn't going to be responsible for the people getting thrown out on the street, that there was still a safety net . . . [Once they saw this] that was probably the major reason for the change.

In addition, C tried to justify stringency by noting that the welfare of all tenants depended on the rents collected and by pointing out that evicted tenants were replaced by equally needy and presumptively more responsible tenants from the Authority's waiting list. Themes like this, along with the 'safety-net' and 'legal-duty' points, recurred in C's presentations to the board for as long as he held his office.[43]

In making his arguments and persuading the board to exercise its discretion to change the way in which it routinely decided cases, C benefited from more than the logical force of what he said. First, on each of the board's panels several members, as I have pointed out, had been chosen because they were likely to be sympathetic to the Authority's position.[44] In part for this reason, the board had since 1980 moved a substantial distance in the direction C wanted it to go.[45] Secondly, C was a lawyer officially attached to the Hawaii Attorney-General's Office and not the Housing Authority. Thus he spoke not fully as a partisan and with considerable legitimate authority. Thirdly, he was a repeat player before the board. Every week he addressed the board, and he could stress his themes without counter-argument. Tenants usually appeared before the board only once, and defense counsel, including legal-aid paralegals, were seldom present in non-payment case hearings. Thus there was no adversary knowledgeable enough to question C's characterization of the board's duties or of the tenant's situation, and no one, other than the more experienced board members, to point to the board's historic ability to set conditions. Few tenants even knew enough to plead that, while they could make time payments, they were unable to pay off their debt in one lump sum. Instead, the natural reaction of tenants was to emphasize their ability to pay, in the belief that if they could convince the board they could clear their rent debt they would stave off eviction. Thus, 'Most people', according to C 'would tell the board straight out that they are going to get the money together and would pay it'. For C, such promises made it easier to secure outright evictions, for, if such tenants did as they promised, they would be allowed to stay by the Commission. Thus the tenants' promises to pay distanced the board from responsibility for the consequences of its eviction decisions.

In sum, what we see behind the transformation I have discussed is an adjudicative body responding to various pressures to change the way in which it exercised the discretion the law accorded it. Behaviorally and phenomenologically, however, the new standard the board developed was no more discretionary than the one that existed in 1969. Whereas tenants once had to be given a chance to repay their debts over time even if they were poor risks, by 1987 tenants had to be evicted if they owed money, even if they were unlikely to meet a lump-sum demand but were a good bet to clear their debt on an instalment plan. In 1969 members who predicted tenants would never pay nevertheless voted to defer their eviction. In 1987 members whose sympathies were aroused by tenants voted to evict them.[46] The board, in short, possessed throughout the period legal discretion to change its standards and it did so, but it never developed a standard that allowed much room for the play of discretion.

I expect that the discretion of trial judges and other 'first-instance' adjudicators is often of this sort.[47] Its operational locus is not where we usually imagine it—in deciding particular cases; rather it is in deciding on a rule to apply in categorizing cases and in deciding how categories of cases should be treated (cf. Tweedie 1989). In so doing, trial judges who have discretion to evaluate facts and reach wise judgments may often be exercising a discretion more akin to that of appellate judges. They are, as a matter of practice rather than pronouncement, making law for the range of cases that come before them, and they are then acting as if they are without discretion; that is, as if they are bound by the law they have made. To understand adjudicative discretion one must understand the rules that judges make for themselves. To appreciate how discretion is, as a behavioral matter, constrained, one must understand the forces that lead judges to make particular rules. These forces may not be the same for all courts, and this study of the HHA's eviction board may provide no more than a few general clues. However, any court is likely to exist in a context of forces which systematically constrain its so-called discretionary decisions in a particular direction.

Conclusion

I said at the outset of this chapter that discretion can be a property of rules, a property of behavior, or a sense that people have of their freedom to act. Legal philosophers tell us that, when rules authorize discretion, it means that decision-makers are free to choose from a range of legally permissible options. Yet, if we look at how adjudicative discretion is actually exercised—that is, at the pattern of decisions generated—little advantage may be taken of this supposed freedom. Although the law leaves open a range of decisions, adjudicators may adopt routine ways of disposing of cases that admit of only a few outcomes within that range. Following routines, however, may be more than a matter of convenience, for routine ways of disposing of cases are easily transformed into subjectively binding precedent. The result is that a decision-maker with legally authorized discretion may lose the awareness of discretion and come to feel that in a particular situation a particular decision is required.

Thus where rules accord a range of discretion to a decision-maker, the decision-maker may be both less and more bound than he appears. The decision-maker is less bound because there is always discretion to ignore the limitations of the discretion given. Whether this occurs will depend both on the decision-maker's conception of his role and on the kinds of incentives that shape decisions to comply with any rule. For example, a police officer who is bound by law to ticket a speeding motorist may, if no one is watching, pocket a twenty-dollar bill and let the motorist go.

The decision-maker with discretion may be more bound than he appears because he may feel that he is bound. Thus a police officer authorized by law to stop any motorist going faster than sixty-five miles per hour may never stop any motorist travelling less than seventy miles per hour, and he may come to feel that he has no authority to do so, perhaps believing that motorists are entitled to a range of grace or that radar guns have a five-mile-per-hour margin of error. His beliefs, however, may have been shaped by motorists who responded with particular hostility when they were stopped for barely exceeding the speed limits, by courts that chose to believe motorists' speed estimates in close-to-the-limit cases, or by fellow officers who mocked him for 'chicken-shit' arrests.

What the law gives in discretion—that is the authorization to reach one of a number of possible decisions and the awareness of this freedom—social forces may take away. This is not surprising, for what legal discretion necessarily accords is the freedom to be influenced by factors other than the law. When the law leaves open a range of choices, unless the choice is made randomly, it must be influenced by something other than and in addition to the law. Not only is the exercise of discretion influenced by the social and psychological circumstances in which a decision-maker finds himself, but the existence of discretion invites others to try to influence its exercise. Moreover, the very act of choosing in one case affects the choice made in the next, and the experience of making a number of similar choices often leaves a decision-maker feeling that no choice exists at all.

This sense of constraint is not necessarily a bad thing, for the consistent exercise of discretion is ordinarily something to be aimed at. Problems arise, however, because the tendency to use shallow case logics in repetitive decision-making make it likely that not all the factors that might shape the wise case-by-case exercise of discretion are considered. Thus the 1987 eviction board, in routinely evicting non-payment tenants who owed rent, ignored the reasons why the family was in debt, the family's need for public housing, and the likelihood that a family that could not make a lump-sum payment was nevertheless a good prospect to repay its rent debt over time. The 1969 board behaved similarly. In regularly giving second chances, it ignored the Authority's valid interest in immediately evicting tenants who, with no prospects of meeting the terms of a conditional deferment, could only increase their debt. At both times it might be said that the board abused its discretion by not using it, for the reason the legislature granted the board discretion was, at least arguably, so that it could consider each case on its peculiar facts and reach an appropriate decision. Had the legislature wanted non-payment tenants behind on their rent to be always evicted or always given a second chance, it could have written this standard into law.[48]

Behavioral regularities do not, of course, necessarily reflect the subjective mental processes that underlie them. It is conceivable that actors conscious of their own discretion and scrupulously attending to the variety of factors they are authorized to consider might none the less generate a pattern of decisions that an observer could easily categorize knowing only a few particulars. In the case of the eviction board, however, phenotypes do not obscure genotypes. Board deliberations indicate that board members feel bound by the same factors that one with access only to decisions would identify as crucial. Yet, there are one or two cases (e.g. the case of Mrs Sua) that do not fit the mould. These exceptions suggest that board members retain some sense of having discretion, but that it takes a truly striking situation to awaken this sense. Moreover, even exceptional decisions are constrained by the usual way of dealing with cases 'of this type', with type being defined not by the family's extraordinary situation, but by the factor or factors that are ordinarily sufficient to determine outcomes. Thus the 'breaks' that the 1987 board gave a few tenants in circumstances that made an extraordinary case for leniency would have been regarded as particularly hard-hearted and narrow by the board that sat in 1969.

I expect the same is true of other decision-makers with discretion. When decisions can be consistently predicted by a case feature or two that stand out, it is likely that the decision-maker, like the observer, senses that little is left to discretion.[49] In these circumstances it takes extraordinary circumstances to awaken in the decision-maker a sense that a range of choices is open to him, and even then the range is unlikely to be co-extensive with the decision-maker's legal authority but is instead likely to be constrained by the decision-maker's sense of what is usually allowed.[50]

These conclusions are based on a study of one institution, a public housing eviction board. Thus they must be regarded as hypotheses to be tested rather than as *a priori* predictions confirmed by investigation. Nevertheless, I believe that the eviction board does not differ greatly from many other decision-makers in the way it exercises the discretion that law accords it, and I have cited studies of other decision-makers that support this claim. Notably absent from these citations have been studies of the law-making activities of appellate courts,[51] the discretionary decisions that have received the most attention from philosophers and legal scholars.

This is not an accident. Although a number of my conclusions may apply to discretionary law-making at the appellate level, there are two important reasons to expect that not all my conclusions will hold. First, the ability of most higher courts to control their dockets[52] means that such courts are less likely than lower courts to be confronted with the steady stream of similar cases that is conducive to shallow case logics and a retreat from discretion.[53] Secondly, the law-making discretion that is accorded appellate courts contemplates that they will establish rules that are not only binding on others but will also bind their own future behavior except in exceptional circumstances. Thus the substitution of rules for discretion, which can betray the authority given by law in the cases of trial courts, hearing boards, and street-level bureaucrats, may embody that authority in the case of appellate courts, particularly 'highest' ones. One reason why legal philosophers have focused as much as they do on discretion as a quality of rules may be that for the courts that most attract their attention—supreme courts—there is less of a disjunction between discretion as a quality of rules and discretion as a quality of behavior than there is when legal discretion is exercised at other levels of the system. I hope, however, to have shown in this chapter that, if we are to understand discretion in all its aspects, we must not only look at the stars—we must cast our eyes down as well.

Notes

1. Occasionally, when a quorum of the board cannot be mustered, the HHA will appoint a hearing officer to try cases. Also some tenants when threatened with eviction leave before the board can hear their cases.
2. Field-work in Hawaii is a tough assignment, but someone has to do it.
3. In the early 1960s the Hawaii Housing Authority was apparently unique among US public housing authorities in the degree to which it extended these due process protections to tenants it sought to evict. In the 1970s federal rules extended in somewhat different form similar protections to tenants in all federally aided projects across the United States.
4. These orders, called 'conditional deferrals' or simply 'conditions', were most common in cases brought for non-payment of rent, and the usual condition was that the tenant pay back the rent owing by a certain date and pay all rent when due for a certain period of time. A 1980 amendment to the statute establishing the eviction board could be read as removing the board's discretionary authority to issue conditional eviction orders where the HHA proved a lease violation, but a 1982 statement by the lawyer who then handled the authority's eviction cases did not interpret the law that way nor, in a training session held that year, was the board told that this was what it meant.

5. In both 1969 and 1987 the rent in most of the HHA's projects was set at a percentage of a family's annual income and there were income limits on eligibility for placement in the projects. In 1969 but not in 1987 there were also income limits on continued occupancy.
6. The longer cases are ordinarily trouble-behavior cases, in which the Authority presents a number of witnesses and in which tenants are disproportionately likely to be represented by attorneys (Lempert and Monsma 1988). The hearings of twenty or thirty minutes common in open-and-shut non-payment cases may seem short but are in fact longer than the typical hearing in at least some housing (Lazerson 1982), small claims (Conley and O'Barr 1990), and misdemeanor courts (Mileski 1971).
7. Cases were also more rapidly processed in 1987 because two full-time staff positions—a secretary and a lawyer—were devoted to the management of the eviction process. In 1969 the eviction process was managed by the Supervising Public Housing Manager (SPHM) and a secretary, each of whom had numerous other responsibilities that they regarded as more central to their roles.
8. The law establishing the eviction board allowed the Authority to staff it with Authority officials, and it was so staffed before 1960. The authority's central-office official decided to reconstitute it as an independent body staffed by community volunteers because, I was told, they did not want a 'kangaroo court'.
9. Project managers did not share these views. Four of the five managers felt strongly that, if they could prove a lease violation, they had the right to an eviction regardless of the credibility of a tenant's promise to reform.
10. In 1987 more board members worked in real-estate property management than in any other occupation. In 1969 a majority of board members either had a social work background or did extensive volunteer work for the poor.
11. I shall refer to each panel as the 'eviction board'. In the data I collected panel identity is not significantly related to case outcome.
12. DAG stands for Deputy Attorney-General. The Authority's prosecutor from 1982 on was a Deputy Attorney-General assigned by the Hawaii State Attorney General's Office to the HHA. Although the DAG remained technically a member of the Attorney General's Office and not of the HHA, for all practical purposes the DAG was an employee, reporting in 1987 to the SPHM and through her to the Director of Housing Management (DHM). The Authority had in 1979 appointed a full-time eviction specialist to prosecute cases and handle the other legal and quasi-legal work necessary to a smooth-running eviction process. The first such specialist was an Authority employee who was not a lawyer. His two successors were DAGs. As far as I can determine, the presence of a full-time eviction specialist was an important influence on the eviction process, but the fact that the specialist was a lawyer was not (Lempert 1989).
13. The attorney was also fond of reminding the board that evicting an apparently needy family would free an apartment for a family that would follow project rules and was presumptively just as needy.

14. Chapter 360 § 3 Hawaii Revised Statutes, as amended May 1980. The usual 'new fact or evidence' that tenants alleged on appeal was that since the hearing they had repaid all the rent that was owing. The 1980 Amendments made some other changes in the law establishing the eviction board, but these need not concern us.
15. The specific changes need not concern us here, for their relevant consequences have been described. They involved the conversion of the HHA Executive Directorship from a Civil Service to a gubernatorial appointed position and the retirement of the long-time head of HHA—who had come up through the housing-management ranks—and his replacement by a more business-oriented head who had no prior housing experience. These changes in turn reflect vast new responsibilities—including the task of building and selling middle-income housing—that had been given to the HHA in the 1970s and a local scandal that developed over the way these responsibilities were handled. The case that almost scuttled the eviction board also has little to do with this chapter. It is entitled *Tileia* v. *Chang* and is described in Lempert and Monsma (1988).
16. All names used in this chapter have been changed.
17. The jury is perhaps the best example of a legal decision-maker that gains considerable discretionary power from the fact that its decisions are not reviewable. The power is not accorded the jury by law but is rather a matter of the jury's structural position. See *Bushell's Case* 124 Eng. Rep. 1006 (1670), and *Sparf and Hanson* v. *United States* 156 US 51 (1895).
18. An example of a structural condition is the fact that in the United States there is no higher court than the Supreme Court and hence no appeal, except through a cumbersome amendment process, from its constitutional decisions. An example of a legal condition is a restriction on interlocutory appeal which allows a trial court to harm a party through a mistaken ruling in a way that a higher court cannot, even by reversing the decision made by the inferior court, fully undo.
19. Some project managers, on the other hand, knowingly denied tenants their right to an eviction hearing by bluffing them out (Lempert 1989). They knew their actions, which involved misleading notices and, on occasion, blatant lies, were unauthorized denials of rights given to tenants by Authority policy. The managers, however, saw bluffing as a way of recapturing from the board a discretion—to decide when tenants could not be 'saved'—that was rightfully theirs. The bluff system began in the mid-1960s and endured for about a decade.
20. From the face of the statute, particularly after it was amended in 1980, it is not clear that the board is authorized to exercise such discretion but the board's statute has always been interpreted by it and the Authority to confide such discretion in it. The existence of this discretion was acknowledged in a 1982 memorandum describing the powers of the eviction board that the Deputy Attorney General who ran the authority's eviction process wrote to the acting

SPHM: 'The Hawaii Housing Authority's hearing boards perform three basic functions: determining whether tenants violated provisions of the rental agreement with the Authority; determining whether the rental agreement should be terminated as a result of the violation; and determining whether tenants should be evicted for the aforementioned violations'. According to a 'script' in the Authority's files, a similar description of the board's powers, with explicit mention of the power to set conditions, was given to the board members at a training session held for them in 1982.

21. All these stages appear salient to the DAG, at least in some cases. He recognizes that he must show a lease violation, and, while he argued in 1987 that it was inappropriate for the board to refuse evictions when it found that a tenant had not fully met his or her rent payment obligations, he never argued that it was beyond the board's power to do so. Thus he was aware of the board's discretion to refuse eviction notwithstanding a lease violation and if so, he recognized, as in Mrs Park's case, that the conditions set by the board were important.

22. I was told in a letter by one board member that, several months after the case I observed, Mrs Park again let her beans burn, was brought before the board, and was this time evicted. The incident may have happened because the stove had not been removed or disconnected, or because Mrs Park had impermissibly reconnected her stove, or because Mrs Park found a way to burn her beans in a microwave. My correspondent did not tell me.

23. The major exception is in the occasional case involving troublesome behavior where different witnesses present different stories about an event (e.g. who started a fight) or the Authority has difficulty finding credible witnesses to testify to the defendant tenant's misdeeds.

24. There are exceptions such as juries that give a plaintiff greater damages than his lawyer sought or a judge who imposes a stiffer sentence than a prosecutor requested. However, these exceptions are empirically rare occurrences. The plea-bargaining system, for example, could not work if judges insisted on more severe sentences than those agreed to by prosecutors, and rejections of civil settlements are almost unheard of, even in class actions where judges have a special obligation to consider the interests of the plaintiff class as a whole.

25. It is not clear that the law was correctly interpreted for the board, since the provision in question specifically addressed the time that a tenant whose rent had been inappropriately set would have to repay the difference between the proper rent and the rent actually paid, rather than the time a tenant would have to repay a debt accumulated by defaulting on the proper rent.

26. The tendency applies to judges in general, but is not confined to them. Other examples include insurance adjusters (Ross 1970), public defenders (Sudnow 1965), private defense counsel (Skolnick 1967), and prosecutors (Maynard 1984b).

27. See the discussion of the 'representativeness heuristic' in Nisbett and Ross (1980: 24–8; cf. Fromm 1965).

28. This is not to call into question the existence of rule-granted discretion, nor is it necessarily to call into question the phenomenological reality of discretion. Rather it is to suggest that for practical purposes the adjudicator appears to be acting without discretion, and one who did not know the rule but only observed behavior might reasonably think that the law did not authorize discretion and that the adjudicator in acting had no sense of exercising any. In fact, I believe that adjudicators who act in the way described in the text often will have the sense that they lack discretion, but a sense of being without discretion is not entailed by behaviorally regular decision-making.

29. Where the wisdom of a discretionary decision will be validated by another's (or even an object's) hard-to-predict future behavior, discretion is likely to be abdicated in favor of rules of thumb or, as is the case with many college admissions officers and parole boards, in favor of mathematical formulae. I am indebted to my colleague Carl Schneider for calling my attention to the general importance of 'subject unpredictability' in his comments on an earlier version of this chapter.

30. This changed in 1975 with an informal ruling by the Attorney-General's Office that deferred tenants who did not meet their rent-payment conditions were entitled to a hearing before the eviction order could be executed.

31. Joe Sanders made this observation in a conversation many years ago. I have often been indebted to him for it. One reason for this, as Fred Schauer (1987) notes, is that the values of precedent are logically distinct from the values of a precedent.

32. The board members were correct if the criterion is the Authority's net rent-collection experience in cases where the board deferred eviction. Even allowing for tenants who did not meet the board's conditions and fell deeper in debt before they were evicted, the Authority's losses were less in cases where the board set conditions than they would have been had the board evicted immediately (to cut the possibility of further loss) in each instance. I discovered this in examining data from the 1960s, and an internal authority memorandum tells the same story based on data from the mid- to late 1970s.

33. At one point the Authority was labelled by the United States Department of Housing the Urban Development (HUD) a 'Financially Troubled Housing Authority'. HUS provided operating subsidies to the HHA and the HHA had to subject itself to HUD audits and comply with certain HUD policies in return.

34. It was, however, always recognized that this was not the primary source of the Authority's financial troubles. Rather, these troubles were due to the HHA's failure to establish adequate reserves for maintenance and renovation and a federal subsidy which, because of the formula that had been used to calculate it, was inadequate. HUD audits confirmed this diagnosis of the source of the Authority's financial troubles, even while suggesting that there was a great need to tighten the rent-collection process.

35. It appears that the HUD judgment did not reflect their auditor's independent

judgment of the situation, but instead reflected the field staff's acceptance of the manager's explanation for their rent-collection problems. The managers, at least in 1969, believed that board leniency cost the Authority substantial amounts of money, and they shared horror stories about tenants who failed to meet board conditions and eventually left or were evicted owing three or four times what they had owed at the initial hearing. As mentioned in n. 32 above, both my research and an Authority investigation reveal that the managers' views about the costs of board leniency were wrong. Although there were horror stories, the incremental losses in such cases were more than offset by cases in which rent debts were eventually paid in full or where partial repayment occurred before further default, so that the tenant when evicted owed less than at the initial hearing.

The managers also told a general deterrence story, arguing that knowledge of board leniency was common and that this prospect encouraged tenants to fall behind on their rent in the first instance. Eventually, I hope to analyse some data that may bear on this, but for the moment all I can say is that, although the argument sounds plausible, based on what I know of the eviction process and the few tenants I talked with, I would be surprised if it were true. An argument not made by the managers may, however, hold: namely, had the board been very strict and had this strictness been publicized at the project level, rent-collection patterns that were not altered by unpublicized board leniency might have been improved. Some data I saw are consistent with this hypothesis. In 1986 and 1987 the proportion of tenants behind on their rent was strikingly low at one project. Two of the tenant members of the eviction board resided at this project. They told me that one or the other goes to every tenants' union meeting at their project and reminds tenants that, if they do not pay their rent, they will be 'kicked out'.

36. The reforms were motivated entirely by a concern with non-payment cases. By 1979 there were no income limits on the federally aided projects, and the board was always more willing to evict in behavior cases than in non-payment cases, so no great problems were seen in this area. After the reforms, non-payment cases were given a special priority, so the proportion of cases brought for non-payment was somewhat higher in the 1980s than it had been in earlier decades.

37. Board members are paid a largely symbolic $10 per meeting attended. Rather than collect the money themselves, they pool it and hold parties twice a year.

38. Between October 1979 and December 1985 12.7 per cent of those owing nothing at the time they appeared before the board were evicted, as were 24.6 per cent of those owing one to three months' rent and 56.8 per cent of those owing more than three months' rent. Some of those evicted were allowed to stay by the HHA's Commissioners on appeal. The figures on board evictions suggest that the board may have been exercising genuine discretion in this period. Unfortunately, I was not in Hawaii then. My interviews suggest that board members who served at this time had more of a sense that they had real

choices to make than did the board members serving in 1987. Clearly the rent owing influenced these choices and it may be that a tenant's rent-payment history, whether good or bad, did as well. Other factors are harder to identify.

39. If such tenants did not attend the hearing or had records of chronic delinquency, they might be evicted.
40. C began practice as a legal-aid attorney and in conversation expressed sympathy for the plight of poor people.
41. Paying the debt in a lump sum became more feasible as the Authority's eviction process grew more efficient, since tenants often found themselves before the board with less than two months' rent owing. At an earlier time, when eviction actions were not so speedily commenced, many tenants who could have managed time payments to clear rent debts of three months and more would have found it difficult or impossible to come up with a lump sum to repay their debt. Of course, there are still tenants who cannot pay off everything they owe before their time for appeal has lapsed who could have paid their rent debt on an instalment plan.
42. By 1987, however, C claimed the process was so efficient and the backlog of cases so small that much of the leeway he once enjoyed was gone. C was also continually pressed to speed up his end of the process. The DHM took what he called a 'business-like attitude', which some might see as a hard line towards non-payment tenants. For example, in 1985 the DHM was apparently instrumental in getting the Authority to adopt a rule that, if a tenant was evicted for non-payment of rent, he or she would never again be admitted to an HHA project. Unless this rule has an unlikely general deterrent effect, it can only cost the Authority money, for evicted tenants seldom repay what they owe unless they seek readmission to the HHA project and are told—as they once were told—that they are ineligible until their old debts are cleared.
43. C left for another position in September 1987, shortly after I completed my second stint of fieldwork.
44. Two members of each panel were tenants who were chosen in consultation with the HHA's island-wide tenant association. For reasons that need not be explored here, tenant board members were usually disposed to deal severely with tenants brought before them.
45. The other major factor was that C's two immediate predecessors had pushed the board for a much harder line toward non-payment tenants, although they had not suggested that the board always evict those behind in their rent.
46. I witnessed several occasions on which one member, who had just voted to evict a tenant, spoke privately to her, after she had been informed of the board's decision, about ways she might acquire money to pay her debt. He even directed some tenants who were not Catholics to his Catholic church for help.
47. The discretion of law-making appellate courts obviously includes the discretion to change received standards and create precedent that is presumptively binding even on itself.

48. It is conceivable, although I do not think it is the case here, that a legislature that wanted always to give tenants a second chance nevertheless wanted that chance to appear to be a fortuitous act of grace rather than a legal entitlement.
49. By contrast, the ability to predict decisions on the basis of decision-maker characteristics like those one funds in some studies of judicial behavior (Ulmer 1973; Goldman 1975) is likely to tell us little about the decision-maker's sense of acting with discretion.
50. This may reflect a psychological phenomenon called 'anchoring and adjustment'. This phenomenon suggests that, when a right answer is *suggested* but is *known* to be wrong, final decisions are distorted in the direction of the answer originally suggested. See, e.g., Tversky and Kahneman 1974.
51. I have cited some research on appellate courts to support other points and I have suggested that an appellate court's exercise of discretion to hear cases conforms to what one would expect based on principles derived from this study of the eviction board.
52. The dockets of appellate courts are also shaped and limited by sociological factors such as the costs of appeals.
53. Where there are streams of cases that can be easily seen to be of the same type, one should expect to see appellate court decision-making that resembles the decision-making of the eviction board. Thus intermediate appellate courts seem to deal with the stream of criminal appeals they confront by broad rules of thumb (Davis 1969), and I would argue that the recent decisions of the United States Supreme Court on the administration of the death penalty has, as a result of the many such cases they confront, resulted in decisions that try to deny the relevance of distinctive facts and in this sense constitute a retreat from discretion. See, e.g., *Lockhart* v. *McCree* 476 US 162 (1986)—holding that, even if data show that death-qualified juries are more conviction prone than juries that are not death qualified, a defendant has no cause of action—and *McCleskey* v. *Kemp* 481 US 279 (1987)—holding that, even if data show that death sentences in the aggregate appear to turn in part on racial considerations, an individual defendant has no cause of action. Recently a Committee of the Judicial Conference of the United States, appointed by Chief Justice Rehnquist, recommended that the number of appeals allowed defendants sentenced to death be drastically limited, thus removing occasions on which courts could exercise discretion. When the full Judicial Conference and the Congress refused to act on this recommendation, the court managed to implement the rule by judicial decision. See *McCleskey* v. *Zant* 111 S. Ct. 1454 (1991).

References

ABEL, R. (1982a) *The Politics of Informal Justice, i The American Experience*. Academic Press, New York.

ABEL, R. (1982b) *The Politics of Informal Justice, ii Comparative Studies*. Academic Press, New York.

BARAK, A. (1989) *Judicial Discretion*. Yale University Press, New Haven.

BARDACH, E. and KAGAN, R. (1982) *Going by the Book: The Problem of Regulatory Unreasonableness*. Temple University Press, Philadelphia.

BLACK, D. (1971) 'The Social Organization of Arrest' in *Stanford Law Review* 23.

CONLEY, J. M. and O'BARR, W. M. (1990) *Rules versus Relationships: The Ethnography of Legal Discourse*. University of Chicago Press, Chicago.

DAVIS, K. C. (1969) *Discretionary Justice: A Preliminary Inquiry*. Louisiana State University Press, Baton Rouge.

DWORKIN, R. M. (1977) 'No Right Answer' in Hacker, P. M. and Raz, J. (eds.) *Law, Morality and Society*. Clarendon, Oxford.

GOLDMAN, S. (1975) 'Voting Behavior on the United States Courts of Appeals Revisited' in *American Political Science Review* 69.

LAZERSON, M. H. (1982) 'In the Halls of Justice, the Only Justice is in the Halls' in Abel, R. (ed.) *The Politics of Informal Justice i*. Academic Press, New York.

LEMPERT, R. (1989) 'The Dynamics of Informal Procedure: The Case of a Public Housing Eviction Board' in *Law and Society Review* 23.

—— (1990) 'Docket Data and "Local Knowledge": Studying the Court and Society Link over Time' in *Law and Society Review* 24.

—— and MONSMA, K. (1988) 'Lawyers and Informal Justice: The Case of a Public Housing Eviction Board' in *Law and Contemporary Problems* 51.

—— and SANDERS, J. (1986) *An Invitation to Law and Social Science*. University of Pennsylvania Press, Philadelphia.

LIPSKY, M. (1980) *Street-Level Bureaucracy: Dilemmas of the Individual in Public Services*. Russell Sage, New York.

LOFTIN, C., HEUMANN, M., and MCDOWALL, M. (1983) 'Mandatory Sentencing and Firearms Violence: Evaluating an Alternative to Gun Control' in *Law and Society Review* 17.

MAYNARD, D. W. (1984) 'The Structure of Discourse in Misdemeanor Plea Bargaining' in *Law and Society Review* 18.

MILESKI, M. (1971) 'Courtroom Encounters: An Observation Study of a Lower Criminal Court' in *Law and Society Review* 5.

NISBETT, R. and ROSS, L. (1980) *Human Inference: Strategies and Shortcomings of Social Judgment*. Prentice Hall, Englewood Cliffs, N.J.

ROSENBERG, M. (1970–1) 'Judicial Discretion of the Trial Court, Viewed From Above' in *Syracuse Law Review* 22.

ROSS, H. L. (1970) *Settled Out of Court: The Social Process of Insurance Claims Adjustment*. Aldine, Chicago.

SCHAUER, F. (1987) 'The Jurisprudence of Reasons' in *Michigan Law Review* 85(5 and 6) 847–70.

SKOLNICK, J. H. (1967) 'Social Control and the Adversary System' in *Journal of Conflict Resolution* 11.

SUDNOW, D. (1965) 'Normal Crimes: Sociological Features of the Penal Code in a Public Defender Office' in *Social Problems* 12.
TVERSKY, A. and KAHNEMAN, D. (1974) 'Judgment Under Uncertainty: Heuristics and Biases' in *Science* 185.
TWEEDIE, J. (1980) 'Discretion to Use Rules: Individual Interests and Collective Welfare in School Admissions' in *Law and Policy* 11.
ULMER, S. S. (1973) 'Social Background as an Indicator to the Votes of Supreme Court Justices in Criminal Cases: 1947–1956 Terms' in *American Journal of Political Science* 17.
WERTHMAN, C. and PILIAVIN, I. (1967) 'Gang Members and the Police' in Bordua, D. J. (ed.) *The Police: Six Sociological Essays*. John Wiley and Sons, New York.

ACCOUNTABILITY, RECOURSE, AND LEGAL CONTROL

The Administration of Benefits in Britain

N. WIKELEY AND R. YOUNG

The influence of the tribunal on local office decisions

There are a number of reasons for supposing that the effect exerted by tribunals on decision-making in local offices is slight, not least the fact that fewer than 1 per cent of decisions by adjudication officers are taken on appeal to a tribunal. On this ground alone, it could be argued that the tribunal is largely irrelevant in terms of controlling the procedures adopted within local offices. While the decisions of the tribunal are sent to local offices for implementation by the adjudication officer concerned, these decisions do not have the status of precedents and there is no obligation to regard them as having laid down any generally applicable rules or principles. In addition, appeals are *de novo* and so are not concerned with establishing whether errors were made in the initial decision-making. Where an appellant attends a hearing, further evidence is likely to be produced and the case is thus different from that which was before the adjudication officer. As a result of this, appellate hearings rarely provide any yardstick by which to assess the quality of the initial adjudication and offer little useful feedback when the adjudication officer in the case implements the tribunal's decision.[1]

On the other hand, many adjudication officers are exposed directly to the way in which the tribunals adjudicate on benefits since they appear before them as presenting officers. It follows that the procedures adopted by the tribunals and the approach which they take to adjudication may influence adjudication officers in their everyday work.[2] Moreover, the mere existence of an independent appellate forum may have a bearing on local office practice. In exploring the wider ramifications of the existence of a right of appeal, we asked adjudication officers whether, when making an initial decision, they were influenced by the prospect that their decisions could be appealed to a tribunal. Affirmative responses were categorised according to whether the respondent claimed simply that this prospect would ensure that the adjudication process would be carried out thoroughly and documented fully (a 'procedural' effect), or whether the larger claim was made that were it not for the existence of this appellate mechanism the outcome of the decision-making process might be different (a 'substantive' effect). The results are set out in Table 1.

TABLE 1 *Effect of Tribunal on Initial Decision-Making according to Type of Work*

	Department of Social Security CB AOs		IS AOs		Dept. of Employment UB AOs		Total	
	No.	%	No.	%	No.	%	No.	%
No effect	24	58.5	27	50.0	10	47.6	61	52.6
'Procedural effect'	7	17.1	12	22.2	10	47.6	29	25.0
'Substantive effect'	5	12.2	14	25.9	1	4.8	20	17.2
Other responses	5	12.2	1	1.9	0	0.0	6	5.2
	41	100.0	54	100.0	21	100.0	116	100.0

Note: Column headings refer to types of adjudication officer.
(CB = Contributory Benefit, IS = Income Support, UB = Unemployment Benefit)

It must be borne in mind in interpreting these figures that a 'procedural' effect may itself make some substantive difference to a decision. The process of justifying a decision (which may, for example, involve carefully checking the law, seeking out more facts, tapping official sources of guidance, etc.) may lead ultimately to a different decision being taken.[3] This effect must surely be seen as desirable. Different considerations apply, however, where a 'substantive' effect is produced through the adjudication officer seeking to second-guess the tribunal. Where the adjudication officer strives to give a decision in line with what she or he thinks a tribunal would do, then the cause of accurate decision-making is served only if the tribunal's predicted approach is the correct one. Similarly, it is possible to imagine adjudication officers allowing ill-founded claims in an attempt to save themselves the extra workload that is produced when claimants appeal against decisions.[4] On the other hand, one can also imagine hard-pressed officers disallowing claims without giving them proper consideration on the basis that the claimant can always appeal if something has gone wrong. It was therefore important to explore these various possibilities in our interviews.

It will immediately be seen that the most striking feature of Table 1 is that it reveals that the proportion of adjudication officers who claimed not to be influenced at all by the prospect of a decision being appealed was about one half in all three areas of work. We shall return to this finding in the course. There were, however, important differences amongst those who said that their approach was affected. For example, unemployment benefit officers were virtually as one in denying that they could be swayed into taking particular decisions by the prospect of an appeal, but nearly half said that the prospect of an appeal being made led to their taking extra care in reaching a decision, as in the following comments:

> When I make a disallowance or disqualification decision, I look at it and think, 'If this person appealed, have I got enough information to write an appeal?' So I look at it as if it was going to appeal rather than just thinking, 'Oh, I don't ike the way this man's written this, he's a bit rude, so I'll disallow or disqualify him'. On the other hand neither do I think, 'I've got a lot of appeals in, I'll allow a few [claims] and then I won't get so many appeals back.' (Interview 105, South Eastern)

> Am I influenced by the prospect of an appeal? Only insofar as when I make decisions I also have in the back of my mind, 'If I have to write an appeal on this decision, can I write that appeal?' But that, I think, only safeguards the fact that, yes, you have applied the law correctly and you can justify the decision you are making. If you can't justify the decision, it shouldn't be made. The effect is that I have to believe that it's the right decision. I don't shy away from the decision because it may be appealed, because it's perhaps a difficult decision and it would be easier to go the other way to save you the trouble of writing the appeal.. That's not what it's about. (Interview 107, South Western)

The proportion of contributory benefit adjudication officers who claimed that the prospect of an appeal led to them taking greater care in decision-making was much smaller. About one in eight such officers conceded, however, that their own determination could be affected by the tribunal's likely decision. This minority saw their approach as one of realism in the face of what they regarded as the tribunal's willingness to be influenced by sympathy for the claimant. In other words, they would sometimes allow claims, against their better judgment, in order to save the time and expense of an appeal which they felt they were bound to 'lose'. The most frequent type of claim which provoked such a response was that where the claimant was arguing that she or he was not fit to work and the adjudication officer's view was that the claimant was 'fit within limits'.[5] The following comment by one officer illustrates the point:

> I know with 'limited fit' cases that, if they go to appeal, then the claimant will stagger into the tribunal with a walking stick, with a Welfare Rights rep., and they are going to destroy the doctor's comments and destroy my job descriptions. The tribunal should address themselves to the claimant's capacity for all types of work,

but once the claimant gets a good rep. who says, 'Look, these job descriptions are just a joke, look at the state of him,' then 99 times out of 100 the tribunal loses sight of the question of incapacity for all work. When I get those cases, I wonder whether it is really worth it. This is where you shouldn't be influenced, but you are. You know they are going to appeal, which will take a lot of time. It's going to mean setting up the tribunal, and all the expense that that entails, as well as the claimant's time and the rep's time. Every now and then I'll say, 'Sod it. Carry on paying'. Once or twice a month, I'll make a decision which isn't the logical one. (Interview 31, North West)

Such sentiments as these were not expressed frequently in the interviews we conducted, but there is clearly a small minority of adjudication officers who consider themselves more likely than tribunals to reach correct decisions. These officers are in effect questioning the legitimacy of tribunal decision-making. If their criticisms of tribunals are well founded, then it would follow that the tribunal's influence in some contributory benefit matters would be a negative one, causing inaccurate decisions to be made. Yet we would argue that the tribunal's approach to these matters is more likely to be the correct one (see later) and that the 'substantive' effect of the tribunals on contributory benefit adjudication is generally to be welcomed.

Out of the three types of adjudication officer studied, one might have expected that it was those working on income support who would be the least influenced by the prospect that their decision could be appealed to a tribunal, since none of them are obliged to appear before the tribunal to explain their decisions. Our findings, however, suggest that the converse is true: close to half of the income support officers we interviewed said that the prospect of an appeal being made either caused them to take greater care in taking decisions or affected the content of those decisions. The impact of the tribunal on this type of adjudication officer was also the most variable. The largest group of those claiming to be influenced by the tribunal spoke in terms of investigating claims more carefully before reaching a decision, whilst others said that they tried to give a decision in line with a tribunal's predicted approach. Other effects could not so easily be regarded as beneficial. A small minority of income support officers admitted to treating the tribunal as a safety net which relieved them of the obligation to be meticulous in reaching a decision to disallow a benefit claimed. The attitude seemed to be that, if the decision was wrong, the claimant could always appeal. For others the temptation was to allow a claim, regardless of its merits, in order to avoid the extra work entailed by a potential appeal. Examples of these latter two approaches can be seen in the following comments by income support officers:

> Sometimes, if I'm not sure of something, I'll disallow it hoping it will go to appeal. Well, not hoping it will go to appeal, but saying to myself that I haven't got time to sort this one out, or I give up with this one, and I'll disallow it, and say no, and then hope it will go to appeal and the appeals officer will then do it. (Interview 59, London North)

> You have in the back of your mind that if you make a decision in favour of the claimant they're likely not to appeal. So I think certainly when you've got people coming in on a Friday afternoon, you know, banging on the screen saying, 'We want payment,' then you are more likely, just basically to get them out of the office, to make a payment that might be wrong because if you didn't make a payment then they probably wouldn't go away and also there'd be this right of appeal. (Interview 42, North East)

The fact that these indirect effects of providing a right of appeal to a tribunal are more apparent in relation to income support is a reflection of the stressful and urgent nature of this type of work. Our research established that income support adjudication officers are under more intense pressure than their counterparts working on contributory benefit and unemployment benefit.[6] Yet in the majority of cases where income support officers agreed with the proposition that the tribunal influenced them, the effect was beneficial in the sense discussed above. This effect was produced despite the fact that these officers have no direct experience of the tribunals. The answer to this puzzle lies in the mediating role played by the income support appeals officer between the tribunal and the adjudication officer.

Income support appeals officers enjoy a close relationship with the tribunal. Excluding those respondents who did not present cases at appeal hearings, we found that income support appeals officers appeared before tribunals twice as frequently as did contributory benefit and unemployment benefit adjudication officers. Appeals officers had appeared before a tribunal on an average of 32 times over the previous 12 months. From interviews with many of the appeals officers, we found that the experience of appearing regularly before tribunals had profoundly affected their approach to adjudication. It was not simply that they took greater care in the making of decisions, but that they had adopted an entirely different philosophy in their work. That the tribunal looms large for appeals officers is well illustrated by the following comments:

> I don't think monitoring affects my work on appeals because I find that appeals is a very high profile aspect of our work anyway. And I think the fact that your

submission goes to an independent body, and that the claimant, the claimant's adviser, the chairman, they all see your submission and your decisions, I think that in itself makes you want to get it right. You don't feel any additional pressure with someone from Regional Office or whatever coming to look at it (Interview 120, North East)

It seems that there's almost a natural view that, when you're sitting in this seat, you start seeing things differently from an adjudication officer. I think that adjudication officers tend to think that, so far as claimants are concerned, they've got their hands on the purse strings and they are perhaps not as objective as they should be in interpreting their instructions and regulations. From this seat you know full well that you've got to put a case in front of a tribunal and, although it's not your decision, the tribunal does view it as your decision.' (Interview 133, London North)

What really made me diligent of course was getting it in the neck from the tribunals. The chickens really come home to roost then. I found that embarrassing. It's not really internal monitoring that makes you careful so much as tribunals. If the chairman says, 'You don't seem to know the law here,' it only needs to happen once. (Interview 135, London North)

The strong influence of the tribunal on income support appeals officers can also be shown in quantitative form. We asked adjudication officers and appeals officers whether, at the stage of an internal review of a decision that had been appealed, they considered a tribunal's likely approach to the appeal. The results, categorised in the same way as in the previous table, are set out in Table 2.

It is interesting to compare the findings presented in Tables 1 and 2. At the initial decision-making stage (Table 1) there is only a possibility that a

TABLE 2 *Effect of Tribunal on Internal Reviews of Decisions according to Type of Work*

	\multicolumn{2}{c}{}	\multicolumn{4}{c}{Department of Social Security}	\multicolumn{2}{c}{Dept. of Employment}	\multicolumn{2}{c}{}
	CB AOs		IS AOs		IS Appeals		UB AOs		Total	
	No.	%	No.	%	No.	%	No.	%	No.	%
No effect	29	76.3	29	54.7	14	34.1	10	47.6	82	53.6
'Procedural effect'	4	10.5	5	9.4	12	29.3	10	47.6	31	20.3
'Substantive effect'	3	7.9	19	35.9	15	36.6	1	4.8	38	24.8
Other responses	2	5.3	0	0.0	0	0.0	0	0.0	2	1.3
	38	100.0	53	100.0	41	100.0	21	100.0	153	100.0

Note: see Table 1 for explanation of column headings. (IS Appeals = Income Support Appeals Officers)

tribunal will be called upon to review a decision, whereas at the internal review stage (Table 2) it is highly likely that the decision will be subjected to external scrutiny unless it is superseded internally. It might have been predicted, therefore, that the tribunal's approach to a case would have a stronger influence on adjudication officers once a decision had been appealed. Yet from Tables 1 and 2 it can be seen that at the review stage the tribunal's influence is stronger only in the case of income support officers, despite the fact that they do not attend tribunal hearings. The explanation for this lies in our finding that they usually take the advice of appeals officers at the review stage (rarely disagreeing when they are advised to supersede an initial decision)[7] and, as Table 2 demonstrates, appeals officers generally pay close attention to how a tribunal will approach a case. Appeals officers make income support adjudication officers aware of how the tribunal is likely to react to a decision, and this awareness seems to have a bearing on the way in which these officers take decisions, as well as on their attitude towards reviews.

By contrast, unemployment benefit officers appear to be no more influenced at the review stage than they are when making initial decisions, while contributory benefit officers are, if anything, less likely to take into account the tribunal's approach at the review stage. The explanation here lies, we suggest, in the fact that these adjudication officers are responsible for reviewing their own decisions when appealed. We detected a certain unwillingness on the part of these officers to countenance the possibility that they took a wrong decision in the first place.[8] The department's own regional monitoring teams have reported that contributory benefit adjudicators 'show a reluctance to revise decisions, even where the decision is based on incorrect or incomplete evidence'.[9] The reluctance of officers to supersede their own decisions is reflected in their indifference as to how a tribunal might deal with appeals against those decisions. The disadvantages of such a system of self-review seem to be almost self-evident.[10] Certainly, the results produced by this form of review when compared with review by a specialist appeals officer (as happens in income support cases) suggest that the balance of advantage lies in specialisation. This can be seen by examining the figures for 1989 (the year in which the bulk of our fieldwork was conducted), as set out in Table 3.[11]

These figure several considerable disparities in the processing of appeals according to the type of benefit in question. For example, whereas nearly three-quarters of successful income support appeals are disposed of at the internal review stage, over two-thirds of successful contributory benefit appeals are upheld only when the tribunal has held a full

TABLE 3 *Stage at which Successful Appeals were upheld in 1989 by Type of Benefit*

	Department of Social Security Contributory Benefit		Income Support		Dept. of Employment Unemployment Benefit	
	No.	%	No.	%	No.	%
Appeals superseded by internal review	1,756	28.9	16,259	72.5	6,387	47.4
Appeal upheld by tribunal	4,328	71.1	6,170	27.5	7,085	52.6
	6,084	100.0	22,429	100.0	13,472	100.0

Source: *Social Security Statistics 1990* (HMSO), Table H6.01.

hearing on the case. In other words, a much greater proportion of appeals is upheld at an earlier stage of the adjudication process for income support cases than for other cases, with consequent savings of time, money and inconvenience for all concerned.[12] The conclusion must be that the review process on income support is more effective at filtering out meritorious appeals than that employed for unemployment benefit and contributory benefit appeals.

To summarise, we found that the existence of the appeal tribunal and its distinctive approach to adjudication appeared to have a greater impact on initial decision-making and the review process within local offices than might have been expected, given the proportionally low number of appeals overall. The strength of this 'tribunal effect' cannot be precisely measured and appears to differ (both in strength and nature) according to the type of benefit and the particular administrative structures within the departments for making decisions and processing appeals. However, the tribunal's influence, such as it is, may not be operating in every case dealt with by adjudication officers, and this *caveat* is particularly needed in relation to what we have termed a 'substantive effect.

It is important to remember that about one-half of all adjudication officers interviewed claimed not to be influenced by the tribunal at all. We have our doubts about such claims, however, and suspect that the tribunal exerts a generalised influence on most adjudication officers, even if it is not always acknowledged as operating. It is plausible to suppose that any adjudication officer who appears before a tribunal to present cases will become aware of the need to substantiate decisions more carefully. Moreover, the local office culture, nurtured by frequent consultations amongst frontline adjudication officers, has its positive as well as negative side. For if it acts as a conduit for negative perceptions of claimants to flow around local offices, it also helps spread working practices and attitudes which are informed by an awareness of the way the tribunal approaches cases.

Such an awareness would, however, be unlikely to make any significant difference to the behaviour of adjudication officers in the absence of an acceptance by them that tribunals were operating in a fair and even-handed manner. We have already noted how some contributory benefit officers harbour doubts about tribunal decision-making, but such reservations were not widely shared. When we asked presenting officers for their views about particular hearings which we had observed, the responses were overwhelmingly positive. In 87 per cent of cases, for example, they said that they had been given a fair hearing and full opportunity to present their case; in 86 per cent, the tribunal was thought to have grasped the relevant issues in the case; in 80 per cent, chairmen were viewed as handling the hearing in a helpful manner, and in 83 per cent no criticisms whatever were raised about the procedures adopted at the hearing. Our interviews with appellants confirmed that those who appear before tribunals are generally satisfied with the standards of procedural fairness maintained there, even though their own appeals may be unsuccessful. These findings represent impressive testimony to the efforts made to create a more professional approach in tribunal adjudication, and they indicate that the legal chairmen have made considerable headway in ensuring that participants perceive hearings to be fair.

So what is distinctive about the tribunal's approach to decision-making?[14] It is not just that appeal hearings are conducted in a fair and open manner, but that the tribunals, notwithstanding the relative informality of the proceedings when compared to courts, act in a proper judicial fashion. The legal chairmen on these tribunals are well used to such concepts as burdens and standards of proof, and are experienced in weighing evidence and remaining impartial in determining claims. Many of these ideas are alien to adjudication officers who, after all, are not legally trained and are fairly low-level civil servants. Internal monitoring has not proved effective in bringing home to them the essentials of their quasi-judicial task. It is rather through exposure to the tribunal, and through being asked to play an *amicus curiae* role, that they come over time to realise the desirability of collecting sufficient evidence, of weighing that evidence objectively and of applying the relevant law impartially.[15] They accept—with evident reluctance in some cases—that they ought not to be swayed by gut feelings about a case, and that they should ignore any prejudicial comments made within benefit offices about claimants. Finally, we return to our argument that where adjudication officers seek to give decisions in line with their expectations of how a tribunal would decide the matter, the outcome is generally a desirable one. Doubtless the tribunals make mistakes, as is indicated by successful appeals made from their decisions to the Social Security Commissioners, but the procedures followed make it more likely that the right results emerge.[16] Moreover, tribunal hearings are characterised by a quality which so much first-tier adjudication lacks—sufficient time to consider a case properly.

In his celebrated critique of the Franks report, Griffith argued over 30 years ago that tribunal procedures have always given much less cause for concern than 'the closed, dark and windowless procedures' which precede them.[17] Our own study suggests that first-tier adjudication in social security is not completely insulated from the outside world. To some extent at least, social security tribunals have opened up procedures to critical scrutiny, shed light on the work of adjudication officers, and blown fresh 'judicial' air into the anonymous corridors of local offices.

Conclusion and implications

These findings suggest that the different levels of the social security adjudication hierarchy should not be viewed in isolation from each other but rather be seen as intermeshing parts of a process. Social security appeal tribunals, though fully independent of the Departments of Social Security and Employment, cast a long shadow over the work carried out in local offices. It cannot be denied that the strength of that shadow must at times be very pale in view of the enormous pressure on adjudication officers to clear cases quickly, nor that the standards of first-tier decision-making remain unacceptably low, but at least the tribunals provide a partial corrective to this situation. Notwithstanding the cautionary note we have sounded concerning the occasional unintended effect of providing a right of appeal to claimants, the system of social security adjudication would be much the poorer without the educative and supervisory contribution of the appeal tribunal.

We are aware that these findings may be seen as running contrary to the view expressed by Mashaw (1983) in his influential book, *Bureaucratic Justice*.[18] He concluded, in the context of a study of a disability benefit scheme operating in the United States, that 'If appeals make any contribution to the pursuit of high quality adjudication . . . it is very limited' (p.

149) and argued that it was much more important to develop effective systems of internal quality control. We would certainly agree that greater attention should be given to ensuring that decisions are made accurately in the first place. We suggest, however, that Mashaw underestimated the possible contribution that appeals could make to improving standards of adjudication. In the British context, where adjudication officers have to present cases before the appeal tribunal, the indirect influence of the tribunal extends over a much greater range of cases than those brought directly before it. This influence is all the more important, given the evidence that the internal monitoring of adjudication presently carried out by the Departments of Social Security and Employment (not to mention the Office of the Chief Adjudication Officer) has little impact, and that internal quality control remains a low priority for the administrators of the social security system.[19]

The implications of these findings warrant further consideration. Over recent years the government has shown a preference for establishing systems of internal review for handling grievances against decisions taken by civil servants, rather than providing a right of appeal to a tribunal.[20] Indeed, two of the most important examples are to be found in the social security field. Claims for housing benefit and most claims against the Social Fund are administered entirely separately from the mainstream system of social security adjudication we have described.[21] Decisions on housing benefit are made by local authority staff while decisions on the Social Fund are made by 'social fund officers' working within the Department of Social Security. In neither case is there a right of appeal to a tribunal, and all the claimant can do is initiate a process of internal review.[22] These systems lack the combination of a fresh internal look at the case within the responsible bureaucracy, informed by the prospect and experience of re-examination by a genuinely independent external agency—features identified in our research as crucial. We would further argue that if the staff administering housing benefit and the Social Fund had to appear before the tribunal as presenting officers, the standards of initial decision-making in these areas would be likely to improve.

Our findings are also significant in the context of other aspects of social security adjudication. For example, decisions on certain types of benefit, or for certain geographical areas, are concentrated in central offices. Thus all family credit decisions are taken in North Fylde and all child benefit decisions in Newcastle. Income support decisions for certain inner city offices with the worst performance ratings are now taken centrally at Belfast, Wigan and Glasgow.[23] The problem is that whilst these offices also carry out internal reviews and prepare appeals submissions in those cases where appeals are lodged, they do not provide the presenting officer at the tribunal hearing.[24] This is left to the local office's presenting officers. The essential link between the tribunal and the decision-makers is thus broken. As we have argued, it is not tribunal decisions *per se* that influence adjudication officers, but the prospect and experience of appearing in person before an independent body.

The value of such external accountability does not appear to be fully recognised. It is not just a matter of providing an open and manifestly independent grievance mechanisms to satisfy the minority of people with sufficient determination both to lodge and pursue an appeal, but of making accountable those who operate the 'closed, dark and windowless' procedures of first-tier adjudication. Even under the tightly drawn regulations which are characteristic of the social security system, the determination of claims by adjudication officers is not just a matter of routine and automatic application of rules to a set of self-evident facts, such that only one decision is possible. There is still much room for sensitive judgment in the process, and a need for objectivity and impartiality if a fair decision is to be reached. The social security appeal tribunal has an important part to play in promoting these values.

Notes

1. J. F. Handler, *Protecting the Social Service Client—Legal and Structural Controls on Official Discretion* (1979), p. 60. See also Mashaw, *supra*, n. 13, pp. 148–149, and R. Sainsbury, *Deciding Social Security Claims: A Study in the Theory and Practice of Administrative Justice* (1988, unpublished Ph.D. thesis, University of Edinburgh).
2. See further Sainsbury, *supra*, n. 1.
3. 'While process never wholly determines outcome, attention given to the information collection and processing aspects of decision-making should certainly foster the pursuit of correct adjudication': Mashaw (1983), p. 150.
4. See further Mashaw (1983), p. 74.
5. Entitlement to invalidity benefit depends on the claimant being 'incapable of work,' and 'work' is defined as 'work which the person can reasonably be expected to do': Social Security Contributions and Benefits Act 1992, s. 57(1)(a). For analysis of this requirement, see D. Bonner *et al.*, *Non-Means Tested Benefits: The Legislation* (1992).
6. On the stressful nature of income support adjudication, see Baldwin, Wikeley, and Young, (1992) at pp. 44–45.
7. *Ibid.*, at pp. 78–80.
8. *Ibid.*, at pp. 81–82.

9. DHSS, *Appeals Handling in Local and Central Offices* (1987), para. 3.21.
10. In its consultative document, *Disability Allowance: Detailed proposals for Assessment and Adjudication* (August 1990) the DSS asserted that 'We think it is important that any review should be conducted by a different AO' (para. 2.22). See now Social Security Administration Act 1992, s. 30(11).
11. The contributory benefit figures in Table 3 encompass those benefits dealt with by contributory benefit adjudication officers in local offices: disablement benefit, guardian's allowance, industrial death benefit, invalidity benefit, maternity allowance, maternity benefit, retirement pension, severe disablement allowance, sickness benefit, statutory maternity pay, statutory sick pay, and widow's benefit.
12. The figures for 1990 confirm this pattern: 71.5 per cent of successful income support appeals were 'upheld' at the review stage compared with 43.1 per cent of unemployment benefit appeals and 30.5 per cent of contributory benefit appeals. See Table H6.01, *Social Security Statistics 1991* (1992).
13. For views of appellants and presenting officers on the appeal tribunals, see Baldwin, Wikeley, and Young (1992), Chaps. 6 and 7.
14. *Ibid.*, Chap. 4. The most obvious defect of the social security tribunals remains the unavailability of legal aid. See *ibid.*, at pp. 124, 204 and 211–212, and H. Genn and Y. Genn, *The Effectiveness of Representation at Tribunals* (Lord Chancellor's Department, 1989).
15. This is not to imply that presenting officers are generally successful in playing the role of *amicus curiae*: see N. Wikeley and R. Young 'Presenting Officers in Social Security Tribunals: the Theory and Practice of the Curious *Amici*' (1991) 18 J.L.S. 464.
16. See Baldwin, Wikeley, and Young (1992), Chaps 4, 5.
17. J. Griffith, 'Tribunals and Inquiries' (1959) 22 M.L.R. 125, 127.
18. Mashaw (1983).
19. *Annual Report of the Chief Administration Officer (C.A.O.) for 1989/90 on Adjudication Standards* (1991).
20. See *Annual Report, Council on Tribunals 1989–90* (H.C. 64, 1990–91). Bradley (1991), and T. Moloney and R. Young, 'Community Charge Appeals: A Major Problem' (1991) 10 C.J.Q. 206. See also R. Sainsbury, 'Social Security Appeals: in Need of Review?' in W. Finnie, C. Himsworth, and N. Walker (eds.), *Edinburgh Essays in Public Law* (1991), p. 335.
21. Claims from the Social Fund in respect of cold weather, funeral and maternity payments carry a right of appeal to a social security appeal tribunal.
22. See R. Drabble and T. Lynes, 'The Social Fund: Discretion or Control?' [1989] P.L. 297, N. J. Wikeley, 'Reviewing Social Fund Decisions' (1991) 10 C.J.Q. 15 and R. Sainsbury and T. Eardley, *Housing Benefit Reviews* (DSS Research Report Series No. 3, 1991).
23. See the National Audit Office Report by the Comptroller and Auditor General, *Support for Low Income Families* (H.C. 153, 1990–91) and the Committee on Public Accounts, *Twenty Sixth Report* (H.C. 216, 1990–91). See

further H.C. Social Security Committee, *The Organisation and Administration of the DSS: Minutes of Evidence* (H.C. 550–i, 550–ii, 1990/91).
24. See the concerns of the C.A.O. on the presentation of family credit appeals, *supra*, n. 19, paras. 5.42 and 6.7.

References

BALDWIN, J., WIKELEY, N., and YOUNG, R. (1992) *Judging Social Security: the Adjudication of Claims for Benefit in Britain*.

BRADLEY, A. W. (1991) article in [1991] P.L. 6.

Mashaw, J. L. (1983) *Bureaucratic Justice: Managing Social Security Disability Claims*. Yale University Press, New Haven.

Praying Patience: The Patients' Perspectives The Responsible Medical Officers

J. PEAY

Patients' views about the tribunals

New and improved?

Nineteen of the patients had experienced tribunal review before and had a reasonable idea of what to expect. Some had had as many as ten previous hearings, although two or three was the more common figure. The time period over which these had taken place meant that a number of patients had experiences under both the 1959 and the 1983 Acts.

Virtually all the patients able to make the comparison said that the new tribunals were an improvement. Only one cynically remarked, 'the Act is the same as before, they've just changed the numbers of the sections'. Patients noted that tribunals were less formal and they believed that more care was taken in reviewing their cases. Also mentioned as positive features were the opportunity to see the reports and attend throughout the hearing, as well as being able to question the Responsible Medical Officer ('RMO'). Finally, patients noted the rapidity with which they received the decision after the tribunal and the more frequent opportunities to reapply.

Curiously, the patients did not speak of these procedural improvements merely as basic rights, which they might reasonably have expected to be fulfilled. Rather, they expressed considerable gratitude. Moreover, patients especially welcomed the tribunal clerks' reassurances before their hearing and the tribunals' efforts to make them feel at ease.

As patients knew that tribunals did not readily make discharge decisions, most of them pitched their requests at a comparatively low level or, as Patient 9 remarked of those who didn't, 'they put it out of their reach and ask for too much'. Sometimes they would only seek a recommendation for transfer to another secure hospital (even Bendene was regarded as a 'way out'). Of the sample, eighteen patients were seeking transfer, usually to their catchment area hospital. Understandably, some expressed their desires simply in terms of going out—either on trial leave (one patient remarked 'going five miles up the road and back would suit me'), or, frequently unrealistically, to be back with their families. Two patients were trying for conditional discharge; one because he knew his RMO was opposed even to his transfer; as a last resort, he felt he had to persuade the tribunal to reach a decision they could put into effect. The other patient contested his guilt; from his perspective, transfer to another hospital would be inappropriate. If he had not committed the offence he was not ill. The RMO shared this view—provided the patient could establish his innocence.[1]

Thus, patients were both reasonably attuned to and realistic about their positions. They accepted that an application to a tribunal could help to change a RMO's mind, or might act as a catalyst and spur along plans that had been verbally agreed with the patient but not put into effect. But they also recognized that a tribunal recommendation for transfer was meaningless without the support of their RMOs, since they might frustrate or merely ignore those recommendations with which they disagreed. Some patients therefore argued that tribunals should ensure that their recommendations were implemented.

In eleven of the cases which went to a tribunal the RMOs were opposed to any change in a patient's status. In nine cases the RMOs were reasonably supportive. Undoubtedly the best way to achieve a positive result from the tribunal was for the patient and the RMO to present a mutual package for action; but this occurred only infrequently. Where patients had their RMOs' active support and yet still nothing changed, alternative barriers were identified. Often the Home Office or receiving hospitals would be criticised for blocking a deserved move.

Patients were aware that the RMOs were capable of influencing the tribunals' decisions in ways more subtle than simply opposing discharge. Patient 4 noted that where the RMO had an untried treatment, the tribunal would delay making any decision to discharge or transfer until its outcome could be assessed. This viewpoint is amply illustrated in Chapter 6 and 7 [see *Tribunals on Trial*, Chs. 6, 7].

The decisions did not surprise them. Patients were grateful for the feedback even where the decision was not to discharge or recommend transfer. As one, Patient 25, remarked 'It did not set me back at all. It made me think what I've got to do'. Indeed, they said they prepared themselves for turn-downs and that it was only the 'really mad' who expected to get out. Some mentioned that they would have preferred fuller reasons—'where you are in the dark, how can you work through your problems?' (Patient 11)—but the following reflection on a decision not to discharge from Patient 8 was not atypical: 'I got the result. I wasn't disappointed. But then I've been a bit disappointed since I came back [to Acheland]'

Why patients apply

In essence, patients did not apply to the tribunal because they expected the hearing to result in their release. The outcome was not the crucial criterion for assessing whether the process was worthwhile: 'having a tribunal is a good thing, even if you get a turndown'. Tribunals gave you a chance to discuss your case, to say that you disagreed with your RMO and to vent your feelings and frustrations.[2]

Even those exceptional patients who believed that they had a good relationship with their RMO did not necessarily fully understand his viewpoint. Patient 13 said he learnt things from his doctor at the hearing; in his view, the tribunal was an invaluable mechanism for improving communication—notoriously problematic even within general medicine. For those patients who found it difficult to talk to their RMOs, receiving their medical reports provided a necessary source of information.

However, one unexpected explanation was volunteered for applying—namely, that to have on your record that you had not exercised your rights 'made you look like a real screw-ball'. This reflected a general anxiety amongst the patients that they, as individuals, should be distinguished from those in Acheland who were 'really crazy'.

Finally, patients had more confidence that a tribunal hearing would facilitate an independent assessment of their circumstances. Under the 1959 Act it was believed that in 99 per cent of the cases the tribunal went with the RMO; patients had had few expectations. But by and large they believed that the tribunals were prepared to use their new powers, even if they did so infrequently. Many of the patients knew of others who had received conditional discharges (presumably the same handful of cases) and this gave them some confidene.

Admittedly, dissatisfaction was still expressed when the tribunal's decision merely 'repeated word for word what the RMO had said'. Or, as Patient 8 put it: 'very few get out on tribunals; unless the psychiatrist agrees to back you, you don't stand a chance; if the doctor is going to disagree with you why waste the time of the tribunal'. Tribunals, he believed, were only 'a straw to clutch at'. Others claimed that it was only the more disturbed patients who saw tribunals as a 'toothless formality, with no willingness to go against the doctors'. Patient 11, for example, believed that the approach of the tribunal depended upon its constituent members. He said that his first tribunal under the new Act had been probing towards absolute discharge (they had in fact recommended that he should be moved on from Acheland as he would not benefit from any further treatment there and that he was no longer a danger to himself or the public). His second tribunal, on virtually the same information, found him to be 'ill and dangerous'. He felt that this decision was attributable to the impact of the views of a 'hard-line judge' on the other two members. He described this particular tribunal as being like 'a puppeteer and two marionettes'.[3]

Preparing for the tribunal: the role of the representative

How did patients set about preparing themselves for a tribunal hearing? An obvious, though not necessarily reliable source of information, would have been to discuss the matter with fellow patients. However, virtually all of those questioned said that their tribunal experiences wee not a topic of conversation. The reason was simple; those who had 'failed' did not want to talk about it and those who had 'passed' refrained for fear of 'rubbing it in'. Tribunal applications were nobody's business but their own.

Most patients found out about the tribunal system by obtaining a representative. Which representative they chose to instruct varied: some did follow a recommendation by a fellow patient; some were advised by members of staff; and some chose a legal representative from the lists of solicitors posted on the ward notice-boards.[4] Others simply returned to the solicitors who had dealt with their case at court. But this was less usual, since a proportion blamed these solicitors for their original confinement in Acheland.[5]

Of the 26 patients, 19 were legally represented at their current hearing.[6] Six of these had kept the same solicitor from their previous tribunal hearing, 3 had changed their solicitors and 10 had not been previously represented. By and large, patients were impressed by the efforts their solicitors made; and, despite remaining in Acheland, they believed they benefited from the continuity of retaining the same representative. Or, as one patient who had been referred automatically and had an approved solicitor from the Law Society's list remarked, 'he did not know me well enough to do a good job'.[7]

Of the 6 patients who had chosen not to be represented, 4 had never been represented. One of these claimed she had never thought of having a solicitor, but, if this application was not successful (and it was not), she would get one for the next. Another said that he wanted to speak for himself and was confident that he would be able to do so.

It is difficult to assess how many benefits representation brings. Clearly, cases are better prepared, independent psychiatric reports sought, and inconsistencies in the medical evidence and other reports explored. Also as a result of this contact, patients do gain some insight into the operation of the tribunal system and the law. However, it must be noted that even represented patients could be surprisingly vague about the tribunal's powers and practices.

As important as the quality of preparation is the proper presentation of the case at the hearing.[8] One of the most articulate patients, who had long experience of the tribunal process, said that even he 'felt inhibited, and didn't speak well' at his most recent hearing. Other patients felt that they had been rushed at the tribunal and didn't remember to say all they wanted to. No matter how hard the tribunal try to put them at their ease, patients understandably remain apprehensive. Challenging the RMO, so obviously a figure of authority, was not undertaken lightly. As one remarked, since 'I'm only a patient, whose word will be taken?'

Several patients were disturbed about the financial barrier for those patients who did not qualify for legal aid because they had some savings.[9] The question was asked, 'Should you have to buy your way out of Acheland?' Patients who were both ineligible for legal aid and not confident of getting a recommendation for discharge or transfer were doubly reluctant to pay for legal assistance—regarding it as a 'waste of money'.[10] Patients were saving money for when they *left* Acheland, not to enable them to leave; those who did save were doubly penalized.

One further inequity emerged. One patient who had only been in Acheland for a short period (two years) following a serious offence was denied legal aid because, he was informed, he stood no chance of getting out. These limited expectations about both the outcome and the purpose of tribunal review were, it was claimed, shared by some other patients. Where it was believed that there was no chance of getting out, patients might refrain from exercising their rights to apply.

One final point to which patients brought attention should be stressed. Acheland, like many institutions, has a significant ethnic minority population, which fluctuates around 15 per cent. Although only a tiny proportion of patients (approximately 3 per cent) fail to cite English as their mother tongue, where language difficulties did arise, patients could be particularly disadvantaged. Not only might they have problems in relieving their stresses by talking to the staff or other patients, but also their limited resources could hinder them finding out how to set about applying to a tribunal. Thus, even this safety-valve might be denied to them.

Patients and their reports: is seeing believing?

Under rule 12(1) of the Tribunal Rules 1983, the tribunal is obliged to send 'a copy of every document it receives which is relevant to the application to the applicant, and (where he is not the applicant) the patient'. Although rule 10(5) obviates the need to send documents to a patient where these have been provided to his or her representative, in practice chairmen in some regions require papers to go to patients in *all* cases. Whichever of these procedures is adopted, the outcome is that patients who apply for a tribunal hearing will have the opportunity, for probably the first time, to read or be made familiar with the contents of all the reports written about them—including the RMOs' reports. Although subsection 2 of rule 12 can prevent the disclosure of documents that would 'adversely affect the health or welfare of the patient or others', this is no longer routinely invoked by the RMOs.

Where a patient is represented, the documents would normally be sent to the representative. Representatives may convey the information contained in the report either orally to the patient or by showing or sending him or her copies of the reports. For patients who are unrepresented, the reports will be sent to them direct. A potential difficulty may thereby arise. It is not unusual for unrepresented patients to be among the more inadequate and vulnerable; indeed, they are often the illest. It is these patients about whom the greatest concern arises when reports are disclosed. Yet, it is this very group who may not enjoy the protection which representation can bring. They will not necessarily have anyone with whom to discuss the content of the reports; whatever damaging effects these may have could be enhanced.

A number of patients were surprised by the contents of their reports. They were also upset that there were, in their view, so many inaccuracies. These could either be factual, for example, about personal details or incidents which had occurred in Acheland; or matters of interpretation, for example, describing a man as a 'habitual alcoholic' when he claimed to have drunk only two pints a night; or matters which were, according to the patients, opinions dressed up as facts.

The RMOs cannot be blamed for all of these. Patients were very retentive about things that had been said to them, possibly only in casual conversation, which the RMOs' reports may have subsequently contradicted. Any inconsistencies in the reports were quickly identified by the patients.

They, in turn, cannot be blamed for adopting this attitude. Patients believed that these comparatively minor matters could be very influential in the tribunal's decision-making.

Perhaps the most hurtful criticisms were those maligning their characters. For this they blamed the RMOs for not knowing them better. For example, one patient, who had both visitors and contact with the outside world through letters, was disturbed when his RMO described him as solitary and not wanting to see anyone. Clearly the RMO was misinformed; yet this was the information which the tribunal would initially receive. Another patient resented being described as paranoid when at his previous tribunal his RMO had admitted that he had been the subject of false allegations. One patient confronted his RMO about the factual inaccuracies *before* the tribunal and the report was amended, but this happened rarely. Thus, the RMOs were right to say that it sometimes upset patients to see their reports, but the patients would assert that they were appropriately upset. Finally, some patients accepted that, even if the reports made uncomfortable reading, they did consider them fair appraisals and frequently found them 'helpful'. No patient said that they would have preferred not not have seen their reports.

Patients were also concerned about the disparity between their admittedly unfavourable histories prior to coming to Acheland and their subsequent behaviour in hospital. The written reports, which could be 'deadly', made great play of their early history, whilst patients claimed that little attention was paid to the ways in which they felt they had made progress. For example, 'My records are twenty years old but they talk as if it were yesterday' (Patient 17). Since a number of patients had experienced tribunals of quite short duration—under half an hour—they did not believe there was much scope for 'getting at the truth'. Furthermore, at the tribunal, the RMOs, according to Patient 17, relied on their credibility as doctors—the quality of their reports was necessarily assumed to be good.

To redress any possible imbalance one point is worth making. The failure to communicate did not lie solely with the RMOs. Patient 15 readily admitted that his feelings had been hurt by what he considered to be derogatory statements made about him by the RMO. He said, 'but I wouldn't let them see I was needled about it'. In the report his RMO had written: 'He is somewhat solitary in the ward setting and generally uncommunicative with nursing staff'. Qr, as the charge nurse put it 'he hides his feelings'.[11]

* * * * *

RMOs and the tribunals

Given these somewhat reserved attitudes to the 1983 Act, the RMOs' assessments of the tribunals are understandable. They were regarded as at best a mixed blessing, and at worst an intrusion and distraction to their medical responsibilities. At the time the research was conducted Acheland was in a state of some disarray, and this may have exacerbated these negative appraisals. The recent death of one of the RMOs and an extensive rebuilding programme had resulted in disorganization and reorganization within the hospital. One house had been closed entirely; many patients had been transferred to Bendene; a new RMO had recently been appointed and another, semi-retired, RMO brought back part-time. The consequence was some reallocation of patients. But overall, there was a substantial reduction of the numbers of patients in each RMO's care. The RMOs' patient allocation was approximately 70 for those working full-time and 35 for those part-time, making a total in-patient population of approximately 520 patients.[12] Previous work-loads had varied—between 100 and 200 patients per RMO—as did attitudes to the appropriate work-load. Or, as the director of one Special Hospital has noted

All patients in special hospitals deserve the best of medical attention with regular review of their cases, and this cannot be provided properly when each consultant has over 100 patients to treat. One full-time consultant for sixty-five patients should prove a maximum case-load

One RMO interviewed felt there should be a maximum case-load of 30—yet another felt that he provided better individual care when responsible for nearing 200 patients. How can these views be reconciled?

Tribunals: administrative burden or therapeutic tool?

(i) Contract. Despite the assorted benefits which tribunals could bring, these were not considered to outweigh what were seen as numerous disadvantages. Some of the RMOs resented the amount of extra work placed upon them by the demands of the tribunals under the new Mental Health Act.[13] This was regarded as divisive both of their time ('more spent in office writing reports') and of a proper therapeutic relationship with and treatment programme for their patients. Indeed, one RMO forcibly maintained that the 'pressures and demands' generated by the tribunals were 'destroying [any] therapeutic relationship', and that he was no longer seeing patients on a treatment basis, but only on a review basis. Another argued that where patients believed that their 'way out' of the hospital

would come via the law, rather than through treatment and change, they were not motivated to co-operate with therapy; where a tribunal was pending planned psychotherapy might be stopped.[14] Since patients had the right to apply every year and tribunals could take many months to arrange, it was sometimes the case that patients were only available for treatment for about a third of the year. The process was regarded as a vicious roundabout; patients would only 'come round' after two or three turn-downs by the tribunal, but up until that point they would not be co-operative. Thus, tribunals 'interrupt beneficial progress'.[15]

The frequency of tribunal review could also disrupt the RMOs' independent plans for patients. First, the Home Secretary might delay transfer on the grounds that a tribunal hearing was pending and the Home Office would like the benefit of the tribunal's advice. Secondly, the tribunal might not favour the transfer or conditional discharge of a patient which the RMO sought; hence, the tribunal could act as a stumbling-block rather than a stepping-stone. Third, one RMO, again pursuing the distracting impact of tribunal hearings, noted that at least three of his patients would have been moved on 'long ago' had he had the time to write recommendations for the Home Secretary, but with all the tribunal work and tribunal reports he had not even had time to get their files out.

Not all the RMOs regarded the redistribution of their time as retrograde, since it forced them both to review patients' cases and to muster their arguments cogently for the tribunal (as opposed to knowing that the Home Secretary would simply adhere to any recommendation they made for *no* change the 1959 Act). Secondly, the opportunity for face-to-face explanations was particularly welcome. For example, one RMO said he went to the tribunal as the patient's ally; making it clear that *he* was not their custodian improved his future relationship with them. Third, some of the RMOs valued the opportunity to observe patients at hearings, under stress, in novel situations, and with their families. Others took the opposite view that they would be very surprised (and disappointed in themselves as RMOs) if anything new emerged at the tribunal hearing—whatever is learnt is at great cost'. If their patients needed to be placed in stressful situations (a dubious proposition in itself) they, as RMOs, should already be doing it. Curiously, they also emphasized that the doctor–patient relationship needed to continue *after* the tribunal. A presumption that patients would not be discharged was evident.

Finally, one group of patients were identified as benefiting specifically from tribunal turn-downs, namely the recently admitted patients. In those instances where patients claimed to have been unjustly sent to Acheland, this independent review of the necessity for their detention could help to settle them into a treatment programme.

(ii) Disclosure. The RMOs were also split over the question of whether they should object to psychiatric reports being shown to patients. Initially, the tribunal staff had experienced real problems with the Acheland RMOs as they would routinely object to patients seeing reports. Rule (6(4) of the Tribunal Rules allows RMOs to recommend to the tribunal that patients should not see parts, or indeed all, of their reports—usually on the grounds that it would be harmful to the doctor–patient relationship. Clearly, there always are patients whose particular circumstances made such provisions necessary.[16] The question becomes how much information and of what nature is it harmful for patients to know? Furthermore, does this include their present medical state, their progress or lack of it within hospital, and the psychiatrists' opinions as to what is the most appropriate course for them?

After some early rancour the Acheland RMOs had mostly decided not to pursue their objections to disclosure. Indeed, one of the most vocal opponents ultimately came to value disclosure. Another, although he remained unpersuaded, ceased objecting on pragmatic grounds. One or two RMOs objected more frequently than others, regarding specific patients as more vulnerable or less able to handle the *content* of the reports. Yet others believed that it was the *format* of the information which was most damaging; written reports could be handed around, reanalysed, misinterpreted, and brooded upon—their impact could thus be long-standing.17 The remaining RMOs almost never objected, taking the view that it was preferable to have an open relationship with patients where information was freely exchanged; in this context the tribunal hearing was regarded as a 'tremendous bonus' for the doctor–patient relationship.

One point of likely consensus emerged; none of the RMOs had control over the content of the social-work report, which would also be disclosed to patients. Yet this information was often the most potentially damaging (since it might include, for example, families' attitudes to a patient's possible return home). Although social workers could object to disclosure, there was concern about whether they would necessarily know which parts of their reports the RMOs considered the most sensitive. This concern was heightened when administrative errors, described by one RMO as 'careless not malevolent procedures', at the tribunal office (or alternatively by representatives) led to patients seeing information where the tribunal had not advised disclosure. Such slip-ups were felt to make the RMOs' role significantly harder.

(iii) Professional relationships. There was also some dispute amongst the RMOs about whether other members of staff should present their views about the patient directly to the tribunal. They may well have had much greater contact than the RMO with the patient in question. None the less, such an approach was widely considered divisive, almost to the point of being regarded as disloyal to the RMO; one RMO firmly believed that he should represent the views of the staff and that a consensus would emerge *before* the tribunal. An increasing reliance on multi-disciplinary case conferences is, of course, likely to facilitate this. However, where members of staff had reached alternative views and submitted a separate report to the tribunal, considerable hostility had arisen between them and individual RMOs.

(iv) Decision-making with responsibility. Although some of the RMOs criticized the tribunals for not being prepared to recommend for transfer those patients who remained ill, but did not require Special Hospital care,[18] they also feared that the tribunals made recommendations where the RMOs were opposed to any change in a patient's status. Thus, in cases where patients' illnesses had stabilized or receded, but where the RMOs believed they remained dangerous (or not *yet* ready to be moved on), the tribunals, being guided by their supposedly civil libertarian instincts, might recommend transfer or order a discharge. The empirical research suggests that these fears are groundless, vindicating the view of one RMO who noted 'all the tribunal is really concerned about is whether the patient is safe' and who further asserted that 'the judges were appointed simply to block the discharge of patients'. Whether these generalizations about the tribunal resulted from a lack of familiarity with the tribunal's criteria[19] or from having observed tribunal decisions is unclear. Alternatively, perhaps the tribunal merely reflected the advice they received from the RMOs; as one RMO stated 'for the very dangerous I don't mind bending the law'.

RMOs had all experienced decisions or recommendations by the tribunal with which they disagreed—usually in respect of the latter, because there was only a handful of cases where the tribunal had discharged against the RMO's advice. Further difficulties had arisen where the tribunal had given short notice of their intentions and there was little time for the RMOs to make the necessary supervisory arrangements. The

RMOs were sensitive to the possibility that the tribunal might have assumed that they had tried actively to frustrate the decision. But they did accept that in the vast majority of cases the tribunal did not go against their advice. In this respect tribunals were characterized as 'a shop window [for the public] and a safety-valve [for the patient]'. Although the *process* of tribunal review had altered with the 1983 Act, the *outcome* was not much different from that under the 1959 Act. The RMOs seemed to regard this as for their patients' benefit and not to their benefit.

Clearly locating the responsibility for decisions, particularly those to which they were opposed, was identified by the RMOs as a matter of some importance; if things subsequently went wrong they did not want the accusatorial finger pointed at them. Decisions contrary to their advice might be made, for example, where an independent psychiatrist's advice was preferred by the tribunal. Their approach was to distance themselves: 'where tribunals discharge against my advice it's not my concern'.

The dilemma which arose when the tribunal made a recommendation for transfer with which the RMO disagreed was comparatively easily resolved. The view was put that it was no part of the tribunal's function to consider whether a (restricted) patient was well enough to be recommended for transfer. Some RMOs therefore said they would do nothing (although they accepted that the judges were increasingly alive to this tactic and were making further inquiries about the responses to their recommendations). Others said they would make an approach to the appropriate receiving hospital, but make their view clear—thus, 'going through the motions'. Only one said, 'I am happy to look for an alternative placement'.

Preferably, RMOs would try to bring the tribunal round to their point of view and ensure that inappropriate decisions were not made in the first instance. Two less reputable strategies emerged, employed by two different RMOs. One claimed that he told the tribunal directly that he would not comply with their recommendations. Another, it was suggested, would withdraw a patient's medication prior to the hearing so that the tribunal would be convinced that the patient was ill and in need of treatment and/or detention.

Similarly, the RMOs objected to 'independent' (or 'private' as some preferred them to be known) psychiatrists making recommendations about a patient's suitability for transfer without offering a bed in their hospital. Discussing the case with the RMOs or the staff was seen as a prerequisite to ensuring that these 'independent' reports were not written 'in ignorance of the facts'. Those who failed to engage in these preliminary inquiries were considered, at the very least, discourteous. Only one RMO said this discussion should not occur outside the open setting of tribunal. Finally, representatives were considered irresponsible when an independent psychiatric report which did not favour any change in the patient's status, was withheld from the tribunal. This was felt to epitomize the adversarial approach so denigrated by the RMOs, rather than an investigatorial approach which tried to cope with real problems in the light of all the evidence. Where independent reports had been prepared, but not presented to the tribunal, the RMOs had, on occasions, felt it necessary to inform the tribunal of the existence of this 'alternative' view.[20]

(v) Reflective v. reflexive decision-making. Tribunal decisions were characterized by the RMOs, as essentially precipitative (alternatively, provocative). They viewed their own decisions as being made in the context of a longitudinal knowledge of the patient and a planned programme for moving a patient on. Conflict was inevitable.

This conflict arises first in the apparent gap between RMOs' recommendations in their reports to the tribunal (see Schedule 1 MHRT Rules) and those made to the Home Secretary in their annual reports, as required under S41(6) of the 1983 Act. As one RMO remarked, it was because tribunal decisions were black and white—and there was no black and white way out of Acheland—that the conclusions you were forced to draw at a given point in response to the tribunal would not necessarily tally with those you might simultaneously be sending to the Home Office in a report geared to urging transfer. Another, not dissimilar, perspective was advanced by the RMO who noted that the same information need not necessarily impel one inevitably to the same conclusion. He explained that under pressure he had changed the conclusion to one of his reports (in this instance originally being *against* admission to a Special Hospital) without changing any of the preceding content. Having the opportunity to reflect on a decision may produce a rather different outcome to that of the tribunal where a decision is required at a specific point in time.

Most of the RMOs agreed that it was appropriate for patients to have the right to have the necessity for their continued detention reviewed by an independent body. But, at the same time, they felt that it was not essentially their business. One RMO expressed the view that it was right that a patient should be able to question the authority for his detention, but wrong that he (as RMO) should be cross-examined. His report represented his view, so why was it also necessary to 'put him on trial'? And if it was, shouldn't he also be legally represented? He felt that he should not have to sacrifice his clinical time to argue with lawyers about psychiatry.

The majority of RMOs neither encouraged nor discouraged patients in their applications (one exceptionally said he did encourage patients to apply). Another said that he liked to share his difficult decisions with the tribunal, whilst others agreed that a tribunal application could act as a catalyst in formulating their own views. One even accepted that the tribunal could alter his point of view. Finally, it was implied that the tribunal might have discouraged RMOs from actively pursuing their statutory role since, as one RMO put it, he 'allowed the tribunal to push where previously he pushed'. Although sometimes in conflict, the relationship between the RMOs and the tribunals was more complex; clearly, it was dynamic and occasionally symbiotic.

Notes

1. This case (Patient 12) was a classic illustration of the difficulties which arise when a diagnosis, that of psychopathic disorder, is based on evidence of a single offence or series of similar offences. If the patient is innocent, there is no need for treatment; if he is guilty, but continues to protest his innocence, this is a symptom of his disorder demonstrating a lack of insight and no motivation to co-operation in therapy to promote change. If he is guilty and disordered and not treated, he is too great a risk to release.
2. In this respect patients' views concurred with those of the RMO who described the tribunal as a safety-valve.
3. Certainly the two decisions make an interesting contrast. It should also be noted that patients frequently remembered their tribunals mainly in terms of the judge sitting and not by the other two members. But then, as the president of the tribunal, the judicial member invariably dominates the proceedings.
4. The local solicitors benefited from this because patients believed that being close to the hospital they would receive more preparatory visits. Similarly, once a solicitor had an interest in a patient, he or she could often be prevailed upon to undertake other legal work, for example writing letters, for which the patient would otherwise have had to pay directly.
5. Patients occasionally felt that they had been badly advised or misled about the merits of a hospital order rather than arguing for a sentence of imprisonment.
6. One of these had instructed a solicitor who did not turn up at the hearing.
7. The Law Society maintains a list of approved representatives from whom patients may seek assistance. These are solicitors who have met certain minimum requirements, namely (i) they have represented 5 patients or attended 3 hearings as an observer; (ii) they have attended a MIND training course; (iii) they pass a Law Society interview.
8. Patient 13 said 'my representative put my case so much better than I ever

could have, everyone should be represented'. Another said that 'representatives help you to know what you are up against—you need all the information that you can get'.
9. This previously stood at £750, but has now been raised to £3,000 so virtually all patients would qualify for legal aid.
10. If it could be established that representation made a *real* difference to the tribunal's decisions, these patients may have been denying themselves the route out of Acheland. For those patients who chose to pay regardless, tribunals' decisions to defer for further information (often a prerequisite to a conditional discharge) result in more cost being incurred by the patient.
11. It was to emerge in another forum where patients discussed their detention that some patients believed that you couldn't afford to admit to being ill in Acheland or 'you'd never get out'. Patients' preoccupations with leaving hospital, and the fear of remaining indefinitely, could lead to some of them having a less than open relationship with those seen as detaining them.
12. Only one RMO had a mixture of male and female wards; the other RMOs confined their responsibilities primarily to one house. In addition to the RMOs there were eleven associate specialists, senior registrars, and psychotherapists variously involved with the patients and numerous other qualified individuals—for example, psychologists, social workers, and occupational therapists.
13. One RMO was an exception—he said that there was no extra work because patients were regularly under review anyway.
14. On the grounds that where resources were scarce it was better to direct them to those patients not seeking immediate discharge.
15. This impelled one RMO to recommend a return to two-yearly applications to tribunals and not, as at present, annual.
16. For example, where patients confide in the staff and not the RMO, but that information is none the less passed on by the staff to the RMO, problems can arise over its inclusion in reports. Since the need for patients to continue to confide in someone, for example about sadistic fantasies, is considered paramount, it may be thought inadvisable that patients should learn that both their RMOs and the tribunal are, in fact, fully informed of these matters.
17. One RMO made a similar point about *written* tribunal decisions.
18. Although the RMOs said *they* would be prepared to make such recommendations in the presence of illness, implying the tribunals were intimidated by psychiatric illness.
19. S72(b)(iii) and associated sections. The preceding subsections give pre-eminence to illness criteria—'dangerousness' appears rarely in the Act; its use in b(iii) above almost never applies in the Special Hospitals.
20. The empirical research clearly indicates that 'negative' reports or reports concurring with the views of the RMO are routinely submitted to the tribunal. See also *W.* v. *Egdell and others* Times Law Report 14 Dec. 1988.

The Social Psychology of Making and Responding to Hospital Complaints: An Account Model of Complaint Processes

S. LLOYD-BOSTOCK AND L. MULCAHY

I. Introduction

Complaints processed through the National Health Service (NHS) formal complaints machinery form an important part of the broader picture of expression of grievance and the emergence and transformation of disputes in a health care setting. This paper concentrates on the first stage of the procedures, that is complaints made to a local hospital and handled internally by the Health Service Unit (usually one hospital) to whom the complaint is made. Drawing on a study of 399 complaints entering the procedure, the paper seeks to further our understanding of the social psychology of making and responding to complaints.

Drawing on a study of 399 hospital complaints entering the National Health Service formal complaints procedure, this paper analyzes the interaction between complainant and hospital as a social episode in which the hospital is called to account for violation of the complaint's normative expectations and makes its response. The non-instrumental and uncrystallized character of many complaints is emphasized. Letters of complaint and replies from the hospital were readily analyzed in terms of the proposed model, providing insights into the social psychology of complaining, the goals of complainants, and the elements of successful apologies. Factors correlating with complainants' satisfaction further support the model and confirm the importance of a socially appropriate response to complaints. The implications of the study are discussed both in relation to hospital complaints and in the context of the literature on disputing more broadly.

SALLY LLOYD-BOSTOCK *is a Research Fellow in Psychology at the Centre for Socio-Legal Studies in Oxford, England, where she has held posts since 1973. Her current research concerns complainants' perspectives on National Health hospital complaints procedures, and the impact of medical negligence claims on patients and doctors. Her previous research at the Centre has included psychological aspects of personal injury claims, and responses to accidents by the Health and Safety Executive in Britain and OSHA in the U.S.*

LINDA MULCAHY *received her first degree in law from Southampton University in 1984 and her Masters in Legal Theory from the London School of Economics in 1993. She has held posts at Bristol University, the Law Commission, and the Centre for Socio-Legal Studies at Oxford University. She is currently a Senior Research Fellow in the newly established Social Sciences Research Centre at South Bank University.*

Studies of litigation and other forms of disputing have placed great emphasis on the origins and early emergence of disputes, leading in turn to theoretical perspectives that locate the initial expression of grievance as an early or low-level stage in a broader model. It is a truism that only a tiny proportion of potential legal disputes ever reach the courts or other tribunal. In order to understand the factors governing what appears to be a massive filtering of potential cases, scholars have sought to go further and further 'back' or 'down' and to trace the development and transformation of disputes as they emerge and progress towards formal legal action and the courts, or alternatively, wither on the vine. Models portraying the process in terms of a pathway towards 'claiming' (Felstiner, Abel & Sarat 1980–81) or pyramid with litigation at its top (Miller & Sarat 1980–81) have been highly influential. The structure of complaints procedures, and the flow of complaints through those procedures, can usually similarly be conveniently described in terms of levels, or a pyramid, in which rapidly decreasing numbers of complaints progress upwards from the broad base to further stages in the procedure. Concern with the efficiency or 'justice' of the operation of complaints procedures again leads to interest in complaints entering the system as informative about the origins of complaints that proceed (or could in principle proceed) to further stages (see, e.g. Rawlings 1987).

In terms of such models, complaints entering internal, NHS hospital complaints procedures are almost always at an early, low-level stage. However, this paper suggests that a richer understanding of complaints, and of their potential to evolve further, can be achieved if they are analyzed in their own terms rather than primarily as early or embryonic versions of something else. (Lloyd-Bostock (1991) discusses further some limitations of 'pathway' models in capturing the reality of the emergence and pursuit of grievances and disputes.) The approach taken in the present paper focuses attention on the nature of complaining itself. The paper proposes a theoretical framework which conceptualizes complaints quite broadly as a social process of calling the hospital to account. The interaction between the complainant and the hospital is analyzed as an 'account episode' (cf. Schönbach 1992; Lloyd-Bostock 1992), in which the hospital is called to account for violation of the complainant's normative expectations, and in which it makes its response.

Although a very extensive and theoretically rich literature now exists on disputing (see e.g. Smith & Lloyd-Bostock 1991), and there is a growing literature on complaints (see Mulcahy 1994 forthcoming), social psychological theory specifically relating to the processes of making and responding to complaints has not been developed. Previous studies of hospital complaints tend to be descriptive and policy oriented (e.g. Great Britain Committee on Hospital Complaints Procedures 1973 (the Davies Report). Besides descriptive and evaluative studies of particular complaints procedures, such as Maguire and Corbett's study of police complaints (1991), the literature on complaining has focused on individual and cultural differences in propensity to complain (e.g. Vidmar & Schuller 1987), on the informal and professional networks that channel an individual towards making and pursuing complaints and claims (May & Stengel 1990), and on the disadvantaged position of complainants in asserting their legal rights (Nader 1980). The social psychology of the act of complaining itself adds a further dimension to these analyses. In order to understand who complains and why, what 'adverse events' are likely to give rise to complaints, what factors affect complainants' satisfaction or dissatisfaction with the response to their initial complaint, and their choices to pursue their complaint or 'lump it,' it is necessary to understand what complainants are doing when they complain—what social act or acts they are performing.

Understanding the social psychology of complaints at this level is of practical as well as academic interest. With increased emphasis on cost effectiveness and on the consumer of health services, the NHS itself is attaching growing significance to complaints, both as indicating grievances that should be redressed or that might escalate, and as a form of free feedback to management and medical staff (cf. Mulcahy & Lloyd-Bostock 1992). Where a public service, such as the National Health Service, is concerned, complaints have a further significance as 'citizen grievances'[1] which goes beyond that of complaints in a commercial setting, and which subjects complaints procedures to evaluation within a framework of administrative law according to established criteria, such as openness, speed, and impartiality (Mulcahy & Lloyd-Bostock 1992; Rawlings 1987). Concern that health service complaints procedures are unsatisfactory is widely expressed (Robinson 1988; National Association of Health Authorities and Trusts 1993; Consumer's Association 1993).

The paper first outlines NHS hospital complaints procedures and the methodology of the study; and presents background data showing the broad pattern of complaints found in the study. The paper then moves on to develop further the account model indicated above, and to present data on what complainants want when they complain, the social dynamics of the process of making and responding to complaints, and factors affecting satisfaction or dissatisfaction of complainants.

A. NHS hospital complaints procedures

The study examined formal complaints about care and services at acute hospitals under the procedure set up under the Hospital Complaints Procedure Act of 1985. The relevant circular issued under the Act requires that each unit (usually consisting of one hospital) must have a 'designated officer' who is required to conduct an enquiry and coordinate the hospital's response to complaints by consumers. Complaints processed as formal are not filtered according to subject matter and include both clinical and non-clinical complaints. However, the medical profession has retained a degree of control over clinical complaints, and further stages in hospital complaints procedures differ between clinical and nonclinical complaints. Clinical complaints are being handled through the Independent Professional Review, and nonclinical are handled ultimately by the Health Service Commissioner. These appeal stages handle only a few complaints nationwide each year and were beyond the scope of the present project. For further details of the operation of formal hospital complaints procedures in practice, see Mulcahy and Lloyd-Bostock (1994, this issue [see (1994) 16 *Law and Policy*]).

B. Sample and method

The main source of data was hospital files on formal complaints held at district level. Files on complaints received in one calendar year (1 July 1989 to 30 June 1990) were studied in two health service districts giving a total sample of 399 complaint files. In addition to background data on the complaint and complainant, the present paper draws on content analysis of letters exchanged between the complainant and the hospital in a subsample of 218 cases. Because formal complaints are usually made and responded to in writing, it is possible in many cases to study the interaction between the hospital and the complainant by analyzing letters of complaint and replies. Files on written complaints do not merely contain records of complaints and responses. The letter of complaint *is* the complaint, while the letter of response is (very often) what the complainant gets in response. The methodology is detailed further below.

In depth, face-to-face interviews were conducted with a subsample of seventy-four complainants in the main files sample, whose complaints had been received by the hospital during the second six months of the study period. The interviews covered many aspects of the complainant's experiences and views, some lasting several hours. Selected data relating to complainants' satisfaction with the handling and outcome of their complaints are drawn on for the present paper.

II. Background data on complaints processed through the procedure

A. Channel of complaint

Most complaints treated as formal by the hospital are made in writing, but oral complaints can also be processed as formal. Comparison of the frequency of oral complaints in the two districts in our sample provides an interesting illustration of the scope for arbitrariness in complaints statistics. Table 1 shows that oral complaints were comparatively common in District B. This almost certainly has nothing to do with the frequency of oral complaints in the two districts, but rather reflects District B's policy of having complaints-recording forms available in outpatients departments, such as accident and emergency. Thus, whether or not an expression of grievance was responded to as a 'formal complaint' and entered complaints statistics was sensitive to local difference of procedure.

TABLE 1. *Written and Oral Complaints by District*

	District A	District B	Total
Written only	247	76	323
Oral only	15	40	55
Both oral and written	10	9	19
Totals where channel known	272	125	397

B. Who makes the complaint

By no means all complaints were received from patients. Almost exactly half, from both sexes, concerned the experiences of someone other than the complainant, usually a relative (see Tables 2 and 3). More complaints were received from women than from men, although the difference is not very large. More complaints might be expected to be from women as a reflection of the sex breakdown of the patient population. However, Table 2 shows that the difference is not fully accounted for in this way, because 60 per cent of those complaining on behalf of others are women. This may be explained by the fact that women more often than men are involved in the care of a patient-relative. Of those complaining on behalf of another, 68 per cent were performing a specific caring or visiting role in relation to a patient-relative at the time of the events at issue. It may also

TABLE 2 *Sex of Complainant by Whether on Another's Behalf*

	Male	Female	Row Total
Own behalf	79	127	206 (49.1%)
Other's behalf	85	128	213 (50.8%)
Column Total	164 (39.1%)	255 (60.9%)	419 (100%)

(Missing observations, 12)

TABLE 3 *Relationship Between Complainant and Person on Whose Behalf Complaint Made*

The complainant was the:

Mother	39 (18%)	Father	19 (9%)
Wife	22 (10%)	Husband	26 (12%)
Sister	4 (2%)	Brother	2 (1%)
Daughter	39 (18%)	Son	14 (7%)
Grandmother	3 (1%)	Grandfather	1
Other Relative	9 (4%)		
Friend	6 (3%)		
Employer	3 (1%)		
G.P.	11 (6%)		
Other/unknown	15		
Total where 'on behalf'	213		

be that women more often see it as their role within the family to deal with (and complain about) health care and hospital services on behalf of members of their family.

In 113 (53 per cent) of complaints on another's behalf, there was an apparent age or health reason why the person concerned would not have complained him or herself: forty-six were children; nineteen had died; twenty-two were in poor health; and twenty-five were elderly. The relatively high number made on behalf of a patient who had died suggests that complaints were sometimes precipitated by the stress and anxiety caused by serious illness or may be part of a process of grieving. While this possibility may help explain why dissatisfaction evolves into grievance in some cases, it does not 'explain away' the complaint.

It should be noted that the description 'on another's behalf,' while it is convenient shorthand for the purpose of analysis, is an oversimplification. It is clear that complainants whose complaints concerned the experiences of another were not merely acting as a conduit for the expression of another's grievance. Rather, they were expressing their *own* dissatisfaction with the way their relative or field had been treated. This is evident, for example, in the results of analysis of expressions of emotion in the sample of 218 letters of complaint analyzed in detail. Fifty-six per cent of letters analyzed explicitly mentioned a negative emotion experienced by the complainant as a result of the event or circumstances complained of ('upset', 'disgusted', 'embarrassed', 'worried', 'disappointed', etc.). Complainants writing 'on another's behalf' were just as likely to attribute to such emotions to themselves as were complainants complaining on their own behalf (56 per cent compared with 55 per cent). Moreover, examples of complaints on another's behalf occurred in which it was apparent that the person who had undergone the experience complained of did not wish a complaint to be made.[2]

C. Allegations made

In practice, any letter or oral comment addressed to or referred to the designated officer is handled as 'formal', irrespective of the allegations made. As well as being unrepresentative of grievances, the range of complaints handled in the procedure is therefore very wide. Allegations were classified according to thirty-five main categories developed for the purposes of the study (see Table 4). Up to seven allegations were coded for each complaint, the average per complaint being 1.8.

The classification of complaints as 'clinical' or 'nonclinical' is potentially important in determining the extent to which the medical profession controls the complaints procedure (see Mulcahy & Lloyd-Bostock 1994, this issue [see (1994) 16 *Law and Policy*]). Thirty-four per cent of complaints in the sample were wholly or partly 'clinical'—that is, they involved at least one allegation concerning an aspect of medical care, excluding communication problems, such as rudeness.

D. Taking the complaint further

After receiving a reply to their initial complaint, rather few complainants pursued their complaints any further. (That is not to say that they were satisfied by the response, see below.) A number wrote a second letter to the hospital (sixty-nine (16 per cent) complainants), and eighteen wrote further letters; but the complaints were not taken beyond district level. In the sample there was no evidence that any of the cases studied were taken

TABLE 4 *Allegations Made (column percentages in parenthesis)*

	Number of Allegations
'Nonclinical' allegations	
Waiting on hospital premises	41 (5.8%)
Waiting lists/waiting for appointment	40 (5.6%)
Other 'waiting' problems	12 (1.7%)
Cancellations e.g., operation, appointment	14 (2.0%)
Admission refused on arrival	5 (0.7%)
Hospital arranged transport e.g., late	9 (1.3%)
Difficulties parking on hospital grounds e.g., lack of space	30 (4.2%)
Reception services e.g., lack of information	7 (1.0%)
Appointment system defects and mistakes e.g., double booking	15 (2.1%)
Food e.g., cold, unappetizing	17 (2.4%)
Linens and blankets e.g., availability	13 (1.8%)
Environment e.g., noise, temperature	25 (3.5%)
Cleanliness e.g., laundry, ward	25 (3.5%)
Loss and theft e.g., clothes	17 (2.4%)
Ward Allocation e.g., mixed ward	3 (0.4%)
Discharge e.g., premature	28 (3.9%)
Treatment of corpse e.g., requests not respected	2 (0.3%)
General communication e.g., failure to answer questions	43 (6.0%)
Explanations to patients e.g., risks of treatment	20 (2.8%)
Explanations to complainant(s)/friends/relatives/carers	9 (1.3%)
Confidentiality e.g., medical details	3 (0.4%)
Behaviour and attitude problems e.g., rude, lazy	91 (12.8%)
Discrimination e.g., because of sex, race	2 (0.3%)
Staffing e.g., inadequate levels	19 (2.7%)
'Political' e.g., allocation of funds	14 (2.0%)
'Clinical' allegations	
Tests e.g., improper performance	15 (2.1%)
Diagnosis e.g., delay, inaccurate	26 (3.6%)
Treatment e.g., improper choice	61 (8.6%)
Care of patient during treatment/stay/visit	56 (7.9%)
Medication e.g., wrong dosage	7 (1.0%)
Ante-natal treatment e.g., scan procedures	4 (0.6%)
Obstetrics e.g., delay in delivery	3 (0.4%)
Anaesthesia e.g., improper technique	2 (0.3%)
Surgery e.g., foreign body retained	11 (1.5%)
Equipment e.g., failure	4 (0.6%)
Other allegations (clinical and nonclinical)	26 (3.6%)
Total allegations	713

further through the formal complaints procedure to an Independent Professional Review or the Health Service Commissioner. The possibility of referral to the Health Service Commissioner was raised in one case by the Community Health Council, but there was no evidence that it was pursued.

Three out of the 399 complaints were passed to the department responsible for handling legal claims, apparently because a solicitor had been consulted. However, the probability that a complaint will in practice evolve into a negligence claim appears to be minimal. Although a substantial number of complaints did concern matters that were prima facie actionable, making a complaint and making a legal claim appeared to be two separate activities. (See Lloyd-Bostock and Mulcahy (1993) for further discussion of the relationship between complaints and legal claims.) The question of what it is to make a complaint rather than a claim is pursued next.

III. An account model of complaining

This section returns to the broader theoretical questions tackled in the paper and discusses in further detail the question of why people complain and the social psychological dynamics of the complaint process. Hospital complaints handled at this level are very diverse, not only in subject matter, but also in the apparent goals of the complainant. In order to take further the question of why people complain, it becomes necessary to enquire more closely into the nature of complaining itself, and what it is that complainants want.

A. Instrumental and non-instrumental complaints

Before the account model is presented in more detail, this section draws attention to the non-instrumental and open-ended character of many complaints. Complaints are usually thought of as attempts to obtain a remedy of some kind. In relation to some forms of complaint, this will indeed be the norm, and the complaint will have an obvious purpose. If someone complains that their rubbish bins have not been emptied, that the local factory is polluting a stream, or that their neighbour is causing disturbances, then it is probably safe to assume that they are complaining in order to get the situation remedied. Similarly, where hospital complaints are concerned, in relation to some allegations it is possible to assume that the complainant wants some matter put right (e.g., to have an awaited operation, to have lost possessions restored to them). But many hospital complaints are not obviously instrumental in this sense, and the goal or goals are not necessarily obvious from the expression of the complaint. The event complained of is often over and done. For example, a complaint may be made abut nursing care made after the patient has returned home from the hospital stay in question. Only 26 per cent of the allegations made in complaints related to events or circumstances that were continuing or might be expected to recur at the time the complaint was made. Hospital complaints are by no means unique in this respect. Complaints against the police (Maguire & Corbett 1991) for example, are usually made after the event, with no necessary expectation of a repetition that the complaint might aim to prevent.

This raises the question of what hospital complaints do actually hope the response to their complaint will be. Are they hoping for compensation, or for someone to be disciplined? Is there any clear goal at all, and, if not, what response from the hospital will satisfy them? To what extent are complainants themselves clear about their own goals and wishes in complaining? Are some complaints ends in themselves? Some complainants may simply be voicing dissatisfaction to those they think should hear. In 38 per cent of letters of complaint analyzed in the present study, complainants stated that they were writing 'to complain', or a similar phrase.

What complainants ask for in letters of complaint. An analysis of 342 letters of complaint was conducted to clarify the extent to which complaints are instrumental in the sense of aiming to achieve some clear further goal beyond the expression of dissatisfaction. Any statement of what the complainant wanted the hospital to do, however vague or general, was coded and categorized, for example, 'I look forward to hearing your explanation', 'I hope you will give much consideration to these problems', 'I do feel strongly that more care needs to be given to out-patients'. The results are summarized in Table 5.

In 98 cases (28.7 per cent), there was no statement at all, however general, indicating what the complainant wished the hospital to do or hoped to achieve by the complaint. For the remaining cases, statements of the response or action requested were grouped into four main categories:

1. Specific requests that something should be done to put matters right *for this complainant* (or the person on whose behalf the complaint was made). This included requests that someone should be disciplined, or an apology should be made, or a meeting arranged, as well as

requests for appointments, operations, transport arrangements, etc. Fifty-three letters (15.5 per cent) included at least one statement in this category.

2. Requests that steps should be taken to put matters right *for others* in the future. As Table 5 shows, this was more frequent than the first category. Sixty-eight letters (19.9 per cent) contained a statement in this category. Some were quite general, for example, 'Please, please don't let any more people suffer in this manner'. Sometimes a specific change to procedures and practices might be suggested for the benefit of others in the future, for example, improvements to appointments systems, providing separate-sex wards. Also included in this category are statements expressing the hope that the hospital will make use of the information that the complainant has supplied, without specifying particular steps that should be taken.

TABLE 5 *Statement of What Complainant Wants*

Specific remedy for this complainant/patient	
Arrange/help arrange treatment	10 (2.9%)
Expenses/charges paid/compensation	9 (2.6%)
Other specific request e.g., to see notes, transport, letter written to GP	11 (3.2%)
Punish/reprimand/sack/discipline	7 (2.0%)
Apology	6 (1.6%)
Ensure will not recur for this complainant/patient	5 (1.5%)
Meeting arranged/interview	5 (1.5%)
Remedy for others/future	
Suggested specific change in policy/procedure	17 (5.0%)
Ensure will not recur/improve/stop it happening (for the future)	25 (7.3%)
Ensure others will not suffer in future	13 (3.8%)
Use the feedback/pass on information to someone who can do something	15 (4.4%)
Investigation/Explanation	
Answer questions/provide information	38 (11.1%)
Investigate/look into/find out	32 (9.3%)
Treat as formal complaint	8 (2.3%)
'Vague'	
Put right/resolve matter/sort out	32 (9.3%)
Let me have your comments/reply	45 (13.1%)
'Help' (unspecified)	13 (3.8%)
Give attention to problem	9 (2.6%)
No statement	98 (28.7%)
Total No. of letters analyzed	342

3. Requests relating to the provision of information about what has gone wrong, including asking the hospital to conduct an investigation. Seventy-one letters (20.8 per cent) contained statements in this category.
4. 'Vague' requests that the hospital should 'resolve the matter', 'take some action', or 'let me have your comments'. Seventy-three letters (21.3 per cent) contained statements in this category; and in sixty-seven of these (19.6 per cent of all letters analyzed), there was no other codeable statement of what the complainant wanted. Thus, in 165 letters (48.2 per cent) there was either no statement at all or only a 'vague' statement of what the complainant wanted.

The results confirm that many complaints are not clearly instrumental. The frequency with which the complainant asks for an explanation, or makes a 'vague' request for a response is, however, consistent with the suggestion that the complainant is calling on the hospital to provide an adequate account, as discussed next.

B. *Making and responding to complaints as 'account episodes'*

It is clear that any model of complaining behaviour must accommodate not only a wide variety of allegations, but also a variety of goals in complaining. As indicated in the Introduction, the approach developed therefore looks at both instrumental and non-instrumental complaints and responses to them quite broadly as social sequences in which the hospital is called to account for violation of the complainant's normative expectations, and in which the hospital makes its response. The appropriate response from the complainant's perspective may or may not include taking some particular action.

Taken together, the interchanges between the complainant and the hospital might be described in Schönbach's terminology as an 'account episode' (Schönbach 1990, 1992; Lloyd-Bostock 1992). Schönbach proposes a model of account episodes as social sequences that are initiated by a 'failure event' and which serve a mitigating function, containing conflict. Escalation of conflict indicates failure of the account episode and is likely to occur when a reproach for the failure event is strongly phrased, provoking a defensive rather than conciliatory response.

Because the interaction between the complainant and the hospital responding to the complainant so often takes place entirely in writing, it is possible to explore the process by analyzing what is said in letters. No previous research has examined complaints in this way, and the model does not yield detailed hypotheses but rather provides a broad framework within which the content of letters can be analyzed. The framework used in the analysis was developed during preliminary phases of the research, and draws on concepts in attribution theory (e.g. Kelley 1967; Hewstone 1989), the social psychology of explanation and accounting (e.g. Scott & Lyman 1968; Antaki 1988; Schönbach 1990, 1992; Harvey, et al. 1992) and Goffman's work on apologies (1971).

Broadly, the approach views the complainant as putting forward a negative evaluation of some event or circumstance for which he or she holds the hospital responsible and calls on the hospital to respond. The complaint can be expected to be supported by, for example, statements (or statements implying) that the hospital has violated a normative expectation, evidence that the matter complained of is worthy of complaint, and suggestions as to what action the hospital should take to rectify matters. Conversely, the response from the hospital can be expected to acknowledge or deny the violation, offer justifications and excuses, express remorse or regret, and perhaps offer to make amends. Like making a legal claim (cf. Lloyd-Bostock 1984), making a complaint involves making an accusation, if only implicitly, that there has been a failure to meet expected standards, and perhaps that someone is at fault. It is thus in itself a somewhat hostile act, and the interaction between the complainant and the hospital has potential to become emotionally charged. The complaint is likely to provoke a defensive response, although in terms of Schönbach's model, the function of an account episode is conciliatory.

The analysis pays particular attention to apologies. It is frequently said that complainants 'want an apology': the analysis develops further what more exactly this might mean. Rather than examining whether the complainant succeeds in obtaining a remedy, the analysis concentrates on the extent to which the response amounts to a full apology. Goffman (1971) suggests that full apologies have several elements including that the person making the apology must acknowledge that something blameworthy has occurred, must sympathize with the censure of others, and must evidence this repentant attitude by making amends or indicating that he or she will do better in future. In this sense, the response as a whole can be evaluated as an apology; and it can be predicted that the satisfaction of the complainant will relate to the extent to which the response amounts to a full apology by acknowledging that the complaint is justified, showing regret, and indicating intention to improve.

Letters of complaint and replies from the hospital were analyzed in a subsample of 218 written complaints. The analysis consists of categorizing phrases or passages from letters, and it draws on the literature on accounting and attribution mentioned earlier. The categories used in the analysis were derived from many hours spent reading complaints files in five hospitals, and from an independent exercise carried out by a member of the research team (Ann Bullen). She was given twenty letters of complaint and their replies, and asked to split the content into different things that it seemed to her the complainant or respondent was doing, continuing until she had categorized everything in the letters.

The main sample of letters and replies was then analyzed according to a schedule based on the categories thus derived. The main focus is the complainant's perspective and the factors likely to affect complainant's satisfaction with the hospital's response. It is perhaps worth emphasizing that the aim is to further our understanding of the social psychology of making complaints. Whether complaints are justified, or should be upheld, are not questions of concern.

1. *Letters of complaint.* In spite of the great variety of complaints, the content of letters fell into clear patterns that were readily analyzed in terms of an 'account' model of complaining. As well as describing the matter complained of, letters comprised several categories of statement making the case (implicitly as well as explicitly) that something worthy of complaint had occurred; recognizing that making a complaint is a hostile act that may need justifying, and may evoke a defensive response; preempting the hospital from dismissing the complaint as, for example, unjustified, mistaken, or trivial; and calling on the hospital to make an appropriate response.

The procedure outlined above yielded seven main categories of statement, together with further less frequently occurring categories. First, (a) by definition, every letter of complaint included a statement of the substance of the complaint. This varied from a short sentence to several pages setting out details of events with dates and times. Beyond that, the major categories were statements; (b) that what had occurred caused harm; (c) that what had occurred was in other ways significant and complaintworthy; (d) that what had occurred failed to meet acceptable standards; (e) calling on the hospital to make a response; (f) indicating that the complaint is not the result of the complainant's own complaining nature; and (g) anticipating excuses or justifications that the hospital might offer. Other less frequently occurring analytical categories were mention of actions taken by the complainant to correct the situation (e.g., 'I mentioned it several times to the nurse'), statements about misfortunes or difficulties not part of the substance of the complaint (e.g., about continued ill-health), statements about the complainant's status or medical knowledge, and threats as to action the complainant might take (such as seeing a solicitor, or going to the press). Table 6 gives frequencies for categories occurring in over 25 per cent of the 218 letters analyzed (a subset of the sample of 342 letters analyzed in Table 5).

(a) Substance of the complaint. As well as the more specific allegations, analyzed earlier (Table 4), Table 6 shows that a number of statements of the substance of the complaint included a general statement, such as 'the whole visit was a misery' or 'the nurses were uncaring'. In these cases, more specific allegations (e.g., that the patient was left in a wet bed for several hours), rather than being in themselves the substance of the complaint, became illustrations of a more general or global complaint.

(b) Effects of the event/circumstances complained of. Statements categorized as showing that the event was significant and worthy of complaint took

TABLE 6 *Frequencies of Main Categories of Statement in 218 Letters of Complaint*

Category	Frequency	% of letters with statement in this category
1. Statement of substance of complaint		
Specific allegation	218	100%
General statement (e.g. 'the ward was in chaos')	73	33%
2. Effects of event/circumstances		
Physical harm	76	35%
Negative emotion/other effects on complainant or patient	156	72%
3. Reasons event/circumstances significant	163	75%
4. Appeals to standards	90	41%
5. Asking for response	163	75%
6. Consensus/distinctiveness		
Consensus (other(s) agreed there was a problem)	50	23%
Praise for other visits/aspects of visit	76	35%
Complainant not prone to making complaints	43	20%
7. Anticipating hospital's justifications (e.g. I know the clinic was busy, but . . .)	61	28%
8. Attempts to put matters right	63	29%
9. Difficulties not part of the complaint (e.g. continued ill-health of patient; difficulties coping at home)	54	25%

various forms. The first group shown in Table 6 is statements of harm caused, negative feelings, or other effects of the matter complained of, such as interference with plans, or expenses. Thirty-five per cent of complaints explicitly alleged that physical harm had been caused, ranging from temporary cold and discomfort to severe and lasting pain or permanent harm. One hundred and twenty (56 per cent) mentioned a negative emotion experienced by the complainant or, in thirty-six cases, the patient was, for example, 'worried', 'embarrassed', 'upset'.

(c) Significance of the event/circumstances complained of. The second group of reasons offered as to why the event is significant appeared to anticipate that the hospital's response might be to view the matter as trivial, and suggested that what to the hospital might seem an unimportant event is in fact important. Seventy-five per cent of letters contained a statement or passage indicating why the matter complained of is important or significant. For example, complainants pointed out that the vulnerable or frail state of the patient magnified the importance of an incident (sixty-nine letters); that although no serious outcome had occurred on this occasion, the consequences could have been far more serious (twenty-six letters); and that the incident, though not serious in itself, indicated a more general problem (twenty-three letters). Thus, one complainant wrote that, while it may seem unimportant, the offhand manner of a member of staff is significant:

because my mother is 80 years old, frail and suffering from angina and osteoporosis . . . I wonder what effect such a man has on people visiting patients about whom they are desperately worried.

Sixteen of the statements in this category remarked that non-medical considerations such as dignity, politeness and comfort are important to patients. For example:

I had to use the toilet in the same room. Again another person came in and out of the room part of the time and though I felt almost too ill to care I found it very humiliating.
   I realize that to hospital staff this is a familiar routine but to the patient it is not. . . . I left the hospital angry at being degraded.

(d) Appeals to standards. Statements referring to failure to meet standards, which appear in 41 per cent of letters, are of particular interest since they support the proposition that complainants hold the hospital responsible for violating a normative expectation. In forty-seven letters

(23 per cent) there was an appeal to professional standards (e.g. that doctors or nurses should be more caring; should protect confidentiality); in forty-one (19 per cent) an appeal to moral standards or standards of humanity (e.g., 'how can you let a child suffer in this way'); and in twenty-seven (12 per cent) an appeal to standards of fairness (why should others be given priority?).

(e) Calls for a response. Analysis of 'what the complainant wants', already described above in connection with discussion of instrumental and non-instrumental complaints (see Table 5), showed that rather few complainants actually expressed a clear and specific goal. However, the majority made some form of statement calling for a response, including general statements asking for (or 'looking forward to') the hospital's explanation, comments, reply, etc. It is questionable whether such statements as 'I look forward to your reply' are sensibly counted as statements of the complainant's goals, but they are readily seen as a general statement that an appropriate reply is expected, in the context of an account episode. Thus, even where there is no obvious material goal, complainants, having initiated a social sequence, may be indicating that they wish for a socially appropriate response to their complaint.

The remaining categories in Table 6 consist of statements not about the complaint worthiness of the event or circumstance that is the subject of the complaint, but rather explaining and justifying the complaint itself, and preempting excuses or justifications the hospital might offer.

(f) Consensus and distinctiveness information. Complainants sometimes seem to anticipate (quite rightly) that the complaint might be seen as arising from their complaining nature rather than more objectively from the event or circumstances. Letters often include statements that might preempt that conclusion, providing what in attribution theory terms might be described as consensus or distinctiveness information, for example, statements to the effect that 'I don't usually complain' (sixteen letters), 'I don't like to complain' (fourteen letters), or 'I felt I had to write' (twenty-five letters); mentioning that others also felt that something worthy of complaint had happened (fifty letters); or praising other visits or other aspects of the hospital stay (seventy-six letters).

(g) Anticipating hospital's justifications. Statements in this category can again be understood as preempting the hospital from dismissing the complaint by displaying that the complainant has considered the hospital's

point of view and possible excuses, and still regards the matter as complaintworthy.

Some of the further, less frequent categories (not shown in Table 6 unless occurring in over 25 per cent of letters) can similarly be seen as serving to justify the fact that the complainant has complained, and to preempt excuses or dismissal of the complaint as arising from the complainant's ignorance, unreasonableness, or complaining nature. Thus, mention of attempts to put matters right imply that the complainant has not complained lightly, but tried to remedy matters before resorting to complaining. Statements about the complainant's status (thirty-two letters) or medical knowledge or contacts (twenty-six letters) can also be seen as attempts to preempt reasons the hospital might not take the complaint seriously. In addition, many (25 per cent) mentioned difficulties and problems that were not part of the substance of the complaint, such as the continued poor health of the patient, or difficulties coping at home. These statements might again tend to explain or justify the complaint, as well as to throw an obligation on the hospital to reply sympathetically.

2. *Letters of response.* The analytical framework developed for letters of response was largely the converse of that used for complainants' letters. A natural response to being called to account is to offer a defence. Whereas complainants were putting forward a case for complaint worthiness, letters of response commonly corrected (17 per cent) or cast doubt on details of the complainant's account (15 per cent of replies); included comments to the effect that if the event did occur, it was not complaintworthy (18 per cent of replies), or in some other way undermined the complainant's account or competence. Only 26 per cent of replies clearly accepted that all the alleged events occurred as described by the complainant. For example (correcting the complainant's account):

I don't think the bath would actually have been uncleaned for the whole week your mother was in hospital—though it is possible the bath wasn't clean whenever she went . . .

Or, questioning complaint worthiness:

I know no-one likes having a barium enema done, but we do try to be as nice as we can . . . One has to expect a little bit of non-privacy in an X-ray department. As for the room opening from the room where you are examined, this is the main dark room and film sorting area and we cannot expect people there to be entirely quiet.

(a) Apologies. Sixty-three per cent of letters of response contained some form of expression of regret (excluding condolences on a bereavement). However, as indicated above, Goffman (1971) suggests that full apologies have several elements including that the person making the apology must acknowledge that something blameworthy has occurred, must sympathize with the censure of others, and must evidence this repentant attitude by making amends or indicating that he or she will do better in future. The mere use of words or phrases such as 'I am sorry' or 'I apologize' is not by itself a full apology in this sense. Many of the apologies offered to complainants might on analysis be called incomplete or 'pseudoapologies' (Lloyd-Bostock 1992). They contained the word 'apology', 'apologize', 'sorry', or 'regret' but did not acknowledge that anything complaintworthy had happened nor indicate a willingness to improve matters. If an apology states I apologize *if you felt that* [x occurred]' it does not actually concede that x did occur.

Thus, one real apology read (at the end of a letter acknowledging that the complainant is quite right):

Can I offer you and your sister my apologies and assurances that we will do everything we can to avoid inflicting such discomfort on our patients in the future.

Another, concerning a rude remark by a nurse:

We have identified the person concerned . . . May I once again offer you my apologies on behalf of the hospital and nursing staff, and thank you for drawing the problem to my attention, as without important feedback we may never have known.

However, some 'pseudoapologies' read:

I am sorry that you felt as you did, but we try to be as kind as we can and I think that our staff go out of their way to do this.

I apologise for any misunderstanding that may have arisen over the reason for her visit.

What at first glance may seem to be an apology may not relate to the substance of the complaint at all, but rather be an expression of regret about something else, for example, 'I was sorry to learn of your husband's continued ill-health', 'I was sorry to get your letter', 'I am sorry that you felt you had to complain'.

3. *Complainant's views of the hospital's response: interview data.* Interviews with complainants confirmed that hospitals' replies frequently appeared

defensive; and that the elements of a full apology related strongly to their satisfaction with the response. Thirty-six per cent rated the hospital as 'not at all' accepting responsibility for the event complained of; 57 per cent rated the hospital as 'very much' or 'rather' trying to defend itself. Forty-one per cent said that they had been given an 'unsatisfactory explanation'. Complainants interviewed were asked to rate their satisfaction overall with the handling of their complaint on a ten-point scale. It was explained that a score of five or less indicated that they were more dissatisfied than satisfied, and six or over that they were more satisfied than dissatisfied. The average overall satisfaction rating was 4.8, with 58 per cent giving a rating of 5 or less, that is, more dissatisfied than satisfied. Twenty per cent rated their dissatisfaction at the extreme of the scale.

Table 7 shows the correlations (Pearson's r) between the satisfaction measure and other aspects of the complainants' views about the response to their complaint, rated on four-point scales. The results support a theoretical perspective that emphasizes the importance of an appropriate social response to the complaint. It is evidently important to complainants that their complaint be acknowledged and taken seriously, and that the hospital accept responsibility. It is particularly interesting to note the strong positive correlation between satisfaction and the complainant's belief that the hospital intended to 'improve things for the future'. This finding suggests that when complainants state that they are complaining in order to prevent others suffering in future, they genuinely wish for this and are not merely justifying their complaint with reference to altruistic goals. The finding also supports the proposition underlying the analysis of apologies offered here, that if an apology is to fulfil its mitigating social function, it is important that the hospital give substance to its

TABLE 7 *Correlations between complainant satisfaction and ratings of hospital's responses*

Statement rated	Correlation with Satisfaction
The hospital understood the complaint	.41***
The hospital accepted responsibility for what had happened	.59***
The hospital was trying to defend itself	−.34***
The hospital took the complaint seriously	.53***
The hospital saw that there was a problem	.39***
The hospital doubted that events had occurred as described	−.39***
The hospital felt the complaint was a nuisance	−.43***
The hospital regretted what happened	.45***
The hospital intended to improve matters for the future	.67***

*** $p<.001$ (one-tailed)

explanations and statements of remorse by conceding that matters are (or were) unsatisfactory and indicating its intention to take remedial action.

IV. Conclusion

This paper has proposed an account model of making and responding to complaints, and explored its usefulness in the context of hospital complaints. As the study shows, complaints entering the NHS formal hospital complaints machinery are an extremely varied, fluid, and often emotionally charged form of expression of grievance. Some are comparatively clear-cut, some are more diffuse. The vast majority 'go away' after the first response from the hospital, although most complainants remain dissatisfied with the response they receive. A few become continuing disputes.

Despite their variety, the study confirms that hospital complaints can be readily analyzed within a single 'account' model. The analysis focuses on the social processes of complaining and responding to complaints, looking in-depth at the interaction between the complainant and the hospital and seeking principally to understand complaints as social episodes. It therefore complements other approaches that examine, for example, the possible role of individual differences in litigiousness, the social and professional networks that channel individuals towards making and pursuing complaints, the experiences of complainants who pursue their complaint further, and the extent to which complaints procedures achieve their stated goals.

The approach promises to prove fruitful in furthering our understanding of the nature of complaints, their usefulness and limitations as feedback, why they can provoke a defensive response, and the ways in which hospitals' responses may mitigate or aggravate the complainant's sense of grievance. The analysis treats potential for defensiveness and dissatisfaction as endemic to complaint processes and indicates the elements of responses likely to affect complainants' satisfaction or dissatisfaction with the response to their complaint. Further study and analysis to explore more closely the relationship between actual responses and satisfaction would be valuable. The present study relates satisfaction only to complainants' *perceptions* of the responses they have received.

The study has concentrated on the complainants' perspective, but the approach has potential to be extended to explore further the organization's perspective and the aspects of complaints that provoke defensiveness in response. Responding to hospital complaints is felt to be a difficult and often unwelcome task. Obviously it will not be appropriate to respond to every complaint with an apology in the full sense as developed in this paper. However, even in those cases where the hospital feels that a complaint is totally unjustified, understanding of the complainant's perspective and wishes may nonetheless help the organization to respond to complaints in ways that will maximize the complainant's satisfaction and avoid aggravating his or her sense of grievance.

Making a complaint has been analyzed quite broadly as calling the hospital to account for violation of the complainant's normative expectations. Some theorists (notably Lempert 1980–81; Griffiths 1983) similarly conceive of the initiation of a dispute as a normative claim, but with the crucial difference that the claim is a claim *to resources*. Under such approaches, an expression of grievance is conceived of as centrally involving seeking redress, virtually by definition. The present study has emphasized the non-instrumental and uncrystallized character of many complaints and demonstrated the difficulty of analyzing complaints as attempts to obtain some form of remedy or redress. In contrast, a broader account model that emphasizes social rather than material goals readily embraced the range of vague requests for an explanation, that something should be done and 'looking forward' to the hospital's reply; and the model accommodates both clearly instrumental complaints and complaints with no clear further goal beyond a satisfactory reply.

Complaints cannot be neatly categorized as 'instrumental' or 'non-instrumental', but the absence of an evident material goal in many complaints serves to highlight the essentially social character of complaining. Many complaints appear to be non-instrumental, not in the sense that there is *no* further goal or desired outcome,[3] but rather in the sense that part if not all of the desired outcome is a satisfactory *social* response to the complaint. This proposition is well supported by the finding that complainants' satisfaction ratings were highly correlated with their sense of having been understood, taken seriously, and offered a satisfactory explanation, and with whether the hospital evidenced its repentant attitude by showing that it intended to improve matters for the future. To the extent that the complainant is seeking something specific in response to the complaint, the paper has argued that it will include, and may sometimes consist primarily in, an appropriate social response.

This conclusion brings an interesting perspective to research on satisfaction with procedures and with outcomes (e.g. Lind & Tyler 1988). It has seemed somewhat counterintuitive that people's satisfaction with legal processes should relate at least as much to the procedure as to the outcome. On the present analysis, at least where complaints are concerned, procedure cannot be neatly distinguished from outcome. If (as the study finds) complainants' satisfaction with the hospital's response relates to their sense of having been understood, believed, and taken seriously, is that satisfaction to be regarded as relating to procedure or outcome? Such factors are part of the way the complaint was handled, and therefore in one sense part of the procedure, but it has been argued in this paper that 'what the complainant wants' may well be primarily to be taken seriously, etc. These aspects of procedure are therefore part of the socially appropriate response or 'outcome' sought. On the other hand, the question of whether or not in a more material sense a complainant achieved redress or a desired 'outcome' is often uncertain because it is often not clear what if any further goal or redress the complainant is seeking. If what the complainant wants is an apology, on the present analysis that too will involve the hospital displaying that it takes the matter seriously otherwise the apology will be a 'pseudo-apology'. Especially in respect of those complaints characterized as 'non-instrumental', aspects of procedure blur into outcome.

It is not a new suggestion that many complainants do not want, for example, compensation, but rather want an explanation, a sincere apology, and/or to improve things for others in the future. Risk managers and those involved in handling complaints know that an apologetic attitude and an immediate explanation work wonders, raising the question of whether it is ethically desirable to 'cool out' complainants in this way. Techniques of handling consumer complaints in more commercial settings recognize that a complaint may be a 'blasting off' with no specific goal and that the most important part of satisfying the complainant is to agree that something worthy of complaint has occurred and to indicate that the matter is taken seriously (Finkelman & Goland 1990). But in the health services, the view that complainants may want an explanation, apology, or to help others, has sometimes been met with scepticism and reluctance to risk negligence claims by admitting responsibility or giving a full explanation. The analysis presented here suggests that where NHS hospital complaints are concerned, scepticism about complainants' motives is likely to be unfounded, and possibly counterproductive, leading at best to dissatisfaction and possibly provoking a continuation of the dispute.

This brings us to the question of how complaints fit into the broader picture of the initiation and pursuit of grievances about health care and, in particular, to what extent they are a pool from which claims may develop.

As discussed in the Introduction, studies of litigation and other forms of disputing have placed great emphasis on the origins, early emergence, and transformation of disputes. Concern with the transformation and progression of disputes to further stages, and the filtering or dropping out of potential 'cases', has led to models of disputing dominated by the theoretical potential for disputes to continue and transform, perhaps becoming a legal claim or reaching the courts. In the context of that literature, the present study goes 'back' or 'down' towards the initial expression of grievance (although in search of 'initial' expressions of grievance one may need to look further back than complaints responded to as formal by the hospital). Complaints to hospitals are, as the study shows, extremely varied and fluid. The complainant frequently appears not to have formed clear goals, and even the substance of the complaint may remain unclear. The study has sought to develop a theoretical framework for examining these often uncrystallized, heterogeneous types of dispute, and for understanding why a small subset continue. To what extent then are complaints appropriately viewed as an early or low-level stage in a larger dispute process? Do complaints belong towards the bottom of a pyramid with courts at the top, or the beginning of a pathway that may end in legal action?

The notion of a dispute that is a 'potential claim' is as problematic as that of a set of circumstances that is a 'potential dispute'. It was clear that in terms of the *substance* of the allegations made, complaints contained substantial numbers of potential negligence claims against the hospital or medical staff (cf. Lloyd-Bostock & Mulcahy 1993). But grievances about health care evolve and progress in a rather more complex way than a pyramid or pathway model might suggest. In particular, complaining seems to be a different avenue and form of expression of grievance from claiming. Despite high levels of dissatisfaction, the number of complainants who turned to the legal system was negligible, while people who brought claims did not usually complain first and then move on to making a claim. Only about 17 per cent of claims made against the hospitals passed through complaints procedures at any stage (for fuller analysis see Lloyd-Bostock & Mulcahy 1993). We argued (Lloyd-Bostock & Mulcahy 1993) that complainants are engaged in a different activity from claiming. Indeed, complainants sometimes view a claim as a hostile or greedy action that they positively do not want to take. Rather than a logical 'next step' or endpoint for a complainant wishing to pursue their grievance further, embarking on a claim is to initiate a different process.

A small number of complaints are taken further within the complaints procedures, and a small number (not necessarily the same complaints) turn into claims. Some of the above discussion has suggested ways in which responses to complaints may mitigate or aggravate the complainant's sense of grievance, and settle or provoke a continuation of a dispute. Understanding complaints is an important part of understanding the evolution and filtering of potential disputes. However, on the approach taken in this study, the theoretical potential for some complaints to evolve into a claim is incidental to understanding the process of making and responding to complaints.

Where complainants go on to make claims, it may sometimes be because their purposes in complaining are frustrated, and they may use the legal system to achieve other goals than compensation. Some, for example, initiate legal action in order to obtain information or to force an admission that a mistake has been made—goals that many complainants share. One patient (in a related study) began by trying to find out what had happened during an operation that had resulted in a far lengthier and more serious episode of treatment than anticipated. She described how she met with defensiveness and refusal to give a full explanation and ended wanting to 'pin the consultant to the wall'. She was making a legal claim.

Studying hospital complaints in detail has highlighted the non-material goals of those voicing a grievance, because complaints are so often non-instrumental in the sense described here, throwing into sharper relief other aspects of complaining. Much complaining makes little sense in terms of the pursuit of redress. Less obviously, those nonmaterial goals may apply also to those bringing legal claims or in other ways expressing more clearly instrumental grievances. Understanding complaints processes can illuminate not only why some complaints transform into claims, but also why for the overwhelming majority of complainants the possibility of a claim is virtually irrelevant, and why some who make claims insist that they are not primarily interested in compensation, but in an explanation, apology, or preventing the same thing happening to others in the future.

Notes

1. The research reported in this paper was supported by ESRC Award number YE1325007 under an ESRC Initiative on Citizens Grievances, and by the ESRC Centre for Socio-Legal Studies, Oxford. A group of research projects examined grievance procedures in different settings: NHS hospital services, housing, local authority services, customs and immigration, and appeals over decisions in the administration of the government's social fund. The research team included Ann Bullen, Tracey Hunt, Gail Eaton, Samantha Jones, and Susan

Clay. Interviews with complainants were conducted by professional, in-depth interviewers Faith Barbour and Jenny Green. We are also extremely grateful to the complainants interviewed, and to the health service staff without those cooperation the research would not have been possible.

2. A category of complaints occurred that were more literally 'on another's behalf', and which we categorized as complaints via an agent. Sixty-five complaints were forwarded to the hospital by agents such as an MP, Community Health Council GP or patient support group acting solely as a conduit in the way suggested. In these cases, the person whose experiences the complaint related to was coded as the 'complainant' and the person forwarding the complaint as the 'agent'.

3. Though it is probable that making a complaint serves therapeutic goals, and in this sense a complaint may sometimes be an end in itself.

References

ANTAKI, CHARLES (1988) *Analyzing Everyday Explanation: A Casebook of Methods*. Sage, London.

CONSUMERS' ASSOCIATION (1993) *NHS Complaints Procedures: The Way Forward*. Policy Report, London.

DAVIES REPORT. See GREAT BRITAIN. COMMITTEE ON HOSPITAL COMPLAINTS PROCEDURES (1973).

FELSTINER, WILLIAM L. F., ABEL, RICHARD, and SARAT, AUSTIN (1980–81) 'The Emergence and Transformation of Disputes: Naming, Blaming, Claiming . . .', *Law & Society Review* 15: 631–54.

GOFFMAN, ERVING (1971) *Relations in Public: Microstudies of the Public Order*. Penguin, Harmondsworth, England.

GREAT BRITAIN. COMMITTEE ON HOSPITAL COMPLAINTS PROCEDURES (1973) *Report of the Committee on Hospital Complaints Procedure*. Chaired by Sir Michael Davies. HMSO, London.

GRIFFITHS, JOHN (1983) 'The General Theory of Litigation—A First Step', *Zeitschrift für Rechtssoziologie* 2: 145.

HARVEY, JOHN H., ORBUCH, TERRI L., and WEBER, ANN L. (eds.) (1992) *Attributions, Accounts, and Close Relationships*. Springer-Verlag, New York.

—— (1992) 'Convergence of the Attribution and Accounts Concepts in the Study of Close Relationships' in J. H. Harvey, T. Orbuch, & A. L. Weber, 1992.

HEWSTONE, MILES (1989) *Casual Attribution: From Cognitive Processes to Collective Beliefs*. Basil Blackwell, Oxford.

KELLEY, HAROLD H. (1967) 'Attribution Theory in Social Psychology' in *Nebraska Symposium on Motivation*, vol. 15, edited by D. Levine. Univ. of Nebraska Press, Lincoln.

LEMPERT, RICHARD O. (1980–81) 'Grievances and Legitimacy: The Beginnings and End of Dispute Settlement', *Law & Society Review* 15: 707–15.

LIND, E. ALLAN, and TYLER, TOM R. (1988) *The Social Psychology of Procedural Justice*. Plenum Press, New York.

LLOYD-BOSTOCK, SALLY (1983) 'Attributions of Cause and Responsibility as Social Phenomena' in *Attribution Theory and Research: Conceptual Developmental and Social Dimensions*, edited by J. Jaspars, F. Fincham, and M. Hewstone. Academic Press, New York.

—— (1984) 'Fault and Liability for Accidents: The Accident Victim's Perspective' in *Compensation and Support for Illness and Injury*, by D. Harris, M. Maclean, H. Genn, S. Lloyd-Bostock, P. Fenn, P. Corfield, and Y. Brittan. Clarendon Press, Oxford.

—— (1991) 'Propensity to Sue in England and the United States of America: The Role of Attribution Processes: A Comment on Kritzer', *Journal of Law & Society* 18: 428–30.

—— (1992) 'Attributions and Apologies in Letters of Complaint to Hospitals and Letters of Response' in J. H. Harvey, T. Orbuch & A. L. Weber, 1992.

LLOYD-BOSTOCK, SALLY, and MULCAHY, LINDA (1993) 'Hospital Complaints: A Reservoir of Potential Claims'. Unpublished paper.

MAGUIRE, M., and CORBETT, C. (1991) *A Study of the Police Complaints System*. HMSO, London.

MAY, MARLYNN, and STENGEL, DANIEL B. (1990) 'Who Sues Their Doctors? How Patients Handle Medical Grievances', *Law & Society Review* 24: 105–20.

MILLER, RICHARD E., and SARAT, AUSTIN (1980–81) 'Grievances, Claims and Disputes: Assessing the Adversary Culture', *Law & Society Review* 15: 526–66.

MULCAHY, LINDA (1994 forthcoming) *Redress in the Public Sector*. National Consumer Council, London.

MULCAHY, LINDA, and LLOYD-BOSTOCK, SALLY (1992) 'Complaining—What's The Use?' in *Quality and Regulation in Health Care: International Experiences*, edited by R. Dingwall and P. Fenn. Routledge, London.

—— (1994) 'Managers as Third-Party Dispute Handlers in Complaints About Hospitals', *Law & Policy* 16: 185–208.

NADER, LAURA (ed.) (1980) *No Access to Law: Alternatives to the American Judicial System*. Academic Press, New York.

NATIONAL ASSOCIATION OF HEALTH AUTHORITIES AND TRUSTS (NAHAT) (1993) *Complaints Do Matter*. NAHAT, Birmingham.

RAWLINGS, RICHARD (1987) *Grievances Procedures and Administrative Justice: A Review of Socio-Legal Research*. Economic and Social Research Council, London.

ROBINSON, JEAN (1988) *A Patient Voice at the GMC*. Health Rights, London.

SCHÖNBACH, PETER (1990) *Account Episodes: The Management or Escalation of Conflict*. Cambridge Univ. Press, Cambridge.

—— (1992) 'Interactions of Process and Moderator Variables in Account Episodes' in J. H. Harvey, T. Orbuch & A. L. Weber, 1992.

SCOTT, MARVIN B., and LYMAN, STANFORD M. (1968) 'Accounts', *American Sociological Review* 33: 46–62.

SMITH, RUSSELL, and LLOYD-BOSTOCK, SALLY (1991) *Why People Go to Law: An Annotated Bibliography of Social Science Research.* Centre for Socio-Legal Studies, Oxford.

VIDMAR, NEIL, and SCHULLER, REGINA A. (1987) 'Individual Differences and the Pursuit of Legal Rights: A Preliminary Inquiry', *Law and Human Behavior* 11: 299–317.

Expectations and Experiences of Tribunal Hearings

H. GENN AND Y. GENN

The knowledge and expectations that appellants and applicants have of the tribunal process has an influence on whether they seek advice about their tribunal case, and conditions their response to what occurs at the hearing itself. Appellants were therefore asked what they thought would happen when they decided to apply for a tribunal hearing, what they thought the meaning of 'appealing' was, and what they thought would happen as a result of their appeal.

The process of appealing

The majority of those interviewed at all of the three tribunals were experiencing their first tribunal hearing. As a result, they could not draw upon past personal experiences. Instead, appellants and applicants relied upon information that had been given to them by friends or relatives or from 'general knowledge'. Those interviewed often found it difficult to say what they had thought was going to happen when they lodged their appeal or application.

Interviewees had had little or no experience of 'legal' processes which accounted for the lack of knowledge or awareness about appealing. Appellants at social security appeals are, by definition, among the most disadvantaged groups in society and often have experience of being passed back and forth between different bureaucratic departments. They tend to have low expectations of bureaucracy.

Immigration appellants are nationals of countries other than the United Kingdom. They often lack knowledge about the British legal system and they believe that there may be prejudice and discrimination against foreigners. Many appellants and sponsors also experience cross-cultural problems because in other countries immigrants have no right of redress. They also, therefore, may have low expectations of the immigration appeals process.

Applicants before industrial tribunals will have been in employment. On the whole, industrial tribunal applicants tend to have a greater awareness of 'rights' in general, and often appeared to be more articulate than appellants at the other two tribunals.

Among all three groups of appellants and applicants, many had little idea about what 'appealing' or making an application actually meant. They certainly had little accurate knowledge about the powers of tribunals or what the possible outcome of their hearing could be.

When appellants at social security appeals tribunals and immigration hearings were asked how they knew that they could have the decision of the Department looked at again, most replied that they learnt about this from the letter sent to them by the relevant Department advising them that their original application had been refused. Many did not know what would happen next; they did not seem to know what asking for an appeal would lead to. Applicants to industrial tribunals, on the other hand, had a clearer idea of the consequences of applying to the tribunal, but many said that they had not thought the case would get to the stage of a hearing. They had hoped that the case would settle at an earlier stage, and that they would not have had to go through the ordeal of facing their former employer at a tribunal hearing.

In spite of the unclear, and often unrealistic expectations of what appealing actually involved, appellants and applicants were usually very clear about why they were appealing and what they wanted from the tribunal. Appellants and applicants described this, for example as follows:

Having another go at trying to get what the DSS have refused. (Social Security Appellant)

To clear my name. (Industrial Tribunal Applicant)

This time they will grant my [relative] to come here. (Immigration appellant)

Appellants received many documents from the various tribunal offices in connection with the hearing. These usually described the tribunal and informed the appellant or applicant of the date, time and place of the hearing. Unfortunately, the recipients of these documents often found them confusing and although they had, in principle, been advised what to expect at their hearing, this frequently had little meaning for appellants. It is difficult for people to appreciate and anticipate what is going to take place in a situation totally outside of their experience. Appellants often did not realise that an appeal would lead to a hearing. As a result, appellants often appeared at the hearings in a state of confusion, having little idea about what they were doing there. For example:

I never thought it would get to this stage.

I thought this would just be another interview.

I don't know what happens here.

This confusion is sometimes the result of the 'appeals conveyor-belt' which can operate in social security and immigration hearings. Once the appeal is lodged, the appeal process takes over. The appellant receives various pieces of paper which may mean nothing to them or confirm in their mind that the claim has been rejected a second time. At some point, the appellant will be notified about the hearing and they will be requested to attend. By this time circumstances may have changed:

I appealed back in December last year, since then I'd forgotten what the problem was all about. [Social Security Appellant]

I just got this letter saying that I had to be at [this place] at 10.00 and my case would be heard. I don't know what is going to happen. [Immigration Appellant]

These situations occur less frequently at industrial tribunals, where applicants exhibit a greater degree of 'participation' in the process. They are more involved and more aware of the possibility that they may have to attend a hearing where they will be asked questions. Nonetheless, many admit that they had hoped at the outset that the matter would not have to come to a hearing.

This relatively high level of confusion about what is involved in bringing a case to a tribunal provides a partial explanation for failure to seek advice about tribunal hearings.

Advice and representation

There are two stages in the process of obtaining advice and representation about tribunal hearings. First, appellants and applicants must perceive a need for advice and/or representation, and secondly such advice or representation must be available.

(a) Perceiving the need for representation

Social security appellants. The majority of those who appeal to social security appeals tribunals are informed of their right to appeal by the DSS. The appeal forms sent out by the tribunal list the types of agencies an appellant might approach, should they wish to obtain advice about their appeal. This information is sent to appellants after they have lodged their appeal.

Perceiving the need for advice depends on the claimant understanding what making an appeal is about and what the consequences of that appeal are likely to be. Lack of knowledge results in appellants being unaware that they may have to do more than simply re-state their case. Many of those interviewed regarded the appeal as an opportunity to 'have another go'; they did not have any appreciation of the fact that they would be required to provide a persuasive argument explaining why the DSS decision was incorrect.

At the outset at least, lack of knowledge about the process of appealing and appellants' inability to anticipate what they must do in order to have the original decision revised, often means that they do not appreciate that they many need help to do this. The belief, or assumption, that tribunals can simply decide appeals without reference to rules or regulations is reinforced by the description of the tribunal as 'independent' and 'informal', which leads appellants to believe that it is a simple procedure for which no special expertise or knowledge is required. As will be discussed later, this misconception often led appellants to be stunned by the formality and complexity of the proceedings.

A minority of appellants recognised that the appeal would involve a formal hearing which might require a representative, but felt that they were able to present their own case without representation. This was usually because appellants were concerned that if they had a representative, they would not be allowed to have their say, or because they simply felt that they were able to put their case themselves. Some claimants felt strongly that they did not wish to appear helpless or incapable:

I've always been very independent.

I've always done this sort of thing myself. At the end of the day, you've only yourself to rely on.

[A representative] couldn't tell me more than I already know . . . I can manage without help.

Some appellants were also afraid that attempting to obtain any kind of assistance with their hearing would involve expense which they could not afford.

Applicants to industrial tribunals. The route by which industrial tribunal applicants find out about their right to apply to the tribunal for compensation is more complicated and more haphazard than that for social security appellants. Applications to industrial tribunals are dependent on information from Job Centres, Unemployment Benefit Offices, union officials, ex-colleagues and friends. Many industrial tribunal applicants are advised by their trade union, who then go on to represent the applicant at the tribunal hearing.

Among those applicants who were interviewed, only a tiny minority attended their hearings without a representative through choice. Two or three of those interviewed appeared in person because they felt that they were the best person to present their case, since they believed that only they knew the circumstances. The overwhelming majority of applicants interviewed, however, perceived clearly the need for advice and representation.

Appellants before immigration adjudicators. Those wishing to appeal against decisions made by the Home Office, immigration officers, visa officers, or entry clearance officers are informed of their right of appeal when their request is refused. The appeal form has printed on it the address of the United Kingdom Immigration Advisory Service, the free representation service funded by Government. Many appellants therefore nominate the UKIAS as their representative. Nonetheless, some immigration appellants do not perceive the need for advice and representation, particularly those who imagine that they are coming to a form of interview. Some immigration appellants associated the need for representation with wrong-doing. They felt that people only needed to have a representative to speak for them if they had done something wrong, or committed a crime, or if they intended to lie. Some found it difficult to accept the need for representation since they thought that they were only coming to the hearing to explain their situation.

(b) Obtaining advice and representation

Appellants at social security appeals tribunals. Many of those who attended their hearings unrepresented felt that they needed advice and assistance with their appeal, especially with the hearing itself. They had often, however, experienced difficulty in obtaining advice and representation.

The distribution of advice agencies is very uneven throughout England and Wales, and the ease with which people can obtain advice is dependent upon where they live. In London, for example, some boroughs have many different types of specialist advice agency whereas others may rely on one overstretched Citizens Advice Bureau. The disparity in distribution of advice centres is reflected to some degree in the differing knowledge and expectations which people have about appealing and obtaining assistance with their appeal. However, even where people are aware of existing advice agencies, they may be unable to obtain help as a result of lack of resources. For example:

I suppose I might have gone to the CAB. It's good that those people give up their time, but ours has only got one person there at a time and it's not a proper place. It's in a room at the back of the Church. There is no sign or anything. You have got to know it's there.

I've spent whole days on that phone just trying to get through to the CAB. In the end you just give up.

My husband did go down to the Citizen's Advice to ask what they thought we would get, but unfortunately the man there said that he didn't know because the law had changed and he wasn't up on the new rules. (November following the April 1988 changes)

The problems of increased demands (particularly since the changes to the social security legislation since April 1988) and low funding of advice agencies, means that claimants often arrive at advice agencies and find a long queue of people before them. Many agencies do not have the facility to open during the evening, making it difficult for people who cannot get to them during the day. Those appellants with mobility problems find difficulty in getting through to CABx on the telephone.

Among those interviewed, many had used the CAB as their source of advice. Unfortunately, the restricted opening hours of many Bureaux presented difficulties:

The CAB's difficult because they are only open three days a week for two hours, and they are so crowded.

It's not easy to find a CAB that is open.

It said in the letter (from the DSS) to go to the CAB. Ours has a staff shortage, and it's closed at all different times. I couldn't get anything from them and I didn't know anyone else to get in touch with.

In addition, people are often unclear about what advice agencies can do. For example:

I didn't bother with Citizens Advice. I didn't think that they did this sort of thing. I thought they were more for sorting out arguments with your neighbours.

They are only ordinary people, just volunteers. I didn't think that they would know much more about it than me.

Claimants experience a variety of difficulties in trying to get assistance with their appeals and the pressure of trying to get advice, and failing, leads people to drop their appeals or fail to turn up on the day.

There were a few social security appellants who consulted solicitors about their appeals. These were most often solicitors who were already acting for the appellant on other matters, for example, divorce or custody of children. The majority of appellants who considered the possibility of obtaining legal advice simply felt that they would not be able to afford to consult a solicitor.

I just couldn't afford a solicitor. Mind you, they would probably consider this sort of thing to be beneath them.

Those who appeal to social security appeals tribunals are therefore dependent on whatever advice and assistance is provided in their local area as a result of local authority funding. The experience of those interviewed is of long queues, full waiting rooms and hours spent waiting to be seen. Even when they have succeeded in seeing an advice worker, the agency may not be able to provide representation as a result of lack of trained staff, or staff shortages. If an appellant does, therefore, attend their hearing with a representative, it must be considered an achievement.

Industrial tribunal applicants. Industrial tribunal applicants are often given initial advice by a Job Centre or their local Unemployment Benefit Office when they sign on. ACAS also advise clients about where to get advice. If an individual either fails to sign on within the time limit for applying to the tribunal, or, as in the case of many married women, fails to sign on at all, they are unlikely to receive information about the tribunal. In spite of this, industrial tribunal applicants are likely to attend their hearing having received some advice, most frequently from a solicitor. The cost of having a solicitor represent them at the hearing, however, is often prohibitive in their circumstances. There are a number of applicants who obtain advice about their application under the Green Form Scheme, but then appear before the tribunal unrepresented. For example:

The solicitor wanted £250 before he would do anything. It doesn't make sense to me this Legal Aid. If I was a murderer or a sex maniac, you would represent me in court. But because I am a working class man, worked all my life, and never had a penny from the State, why do they want £250 to represent me in court? They go on about closed shops, but if you think of justice, it's a closed shop, and the only freedom you have got is what you have got in the bank. (Unrepresented applicant)

The Citizens Advice Bureau gave me a list of solicitors, but I just knew that I wouldn't be able to afford it so I didn't bother. (Unrepresented applicant)

Appellants at hearings before immigration adjudicators. Appellants who bring cases before immigration adjudicators have the greatest ease in being directed to an advisory body, since they are usually advised to go to UKIAS, whose name and address are printed on the appeal application form. Many appellants nominate UKIAS at the outset. UKIAS do not represent all of those who nominate them, since some cases are without merit, and other appellants subsequently seek private legal advice, or go to other advice agencies. Appellants who come before immigration adjudicators span a wide range of social class and background. Many are able to afford the services of a private solicitor and in spite of the existence of UKIAS, a number of appellants prefer to pay for representation. Some consider that the representative will work harder on the case if they are being paid. Some believe that there is a tactical advantage in paying for representation, since it might be evidence of the seriousness of their intention. Some appellants are directed by friends and relatives to a number of independent advice agencies who are well-known within the established ethnic communities, and who advise on a range of related areas, including immigration appeals.

Summary

It is evident from case files that in all tribunals, except social security, a majority of appellants and applicants feel a need to obtain advice about their cases, or are made aware by someone else, of a need for advice about appealing to a tribunal. Advice is most often obtained by appellants to immigration adjudicators, applicants to industrial tribunals, and patients applying for mental health review tribunals. In social security cases, industrial tribunal cases and immigration cases, advice is being provided by a relatively wide range of individuals and agencies. Legal advice is more frequently obtained by those appealing to immigration adjudicators and applicants to industrial tribunals than those appealing to social security appeals tribunals. In mental health review tribunals advice and representation is almost exclusively provided by solicitors and barristers.

In all of the four tribunals studied there were significant regional variations in the extent and source of advice and representation. This was equally true of the two tribunals in which free representation is available. Outside of urban centres the availability of specialist lay advice is virtually non-existent. CABx are geographically widely spread and provide generalist advice in all areas. In the absence of specialist advice agencies, solicitors provide advice about social security appeals.

The likelihood that advice and representation would be obtained was also related, in all four tribunals, to the type of case under appeal or review. Thus where cases are more serious or more difficult to pursue, appellants are more likely to obtain advice and representation.

The provision of advice has an important effect on the manner in which cases are ultimately decided. In social security appeals tribunals those appellants who had obtained advice were more likely to attend their tribunal hearing. In immigration appeals, those appellants who had obtained advice were more likely to have their cases determined on the basis of a hearing, rather than on the papers. In industrial tribunals, applicants who obtained advice were less likely to withdraw their applications than those who had not been advised, and more likely to settle their application.

Obtaining advice is the first step to obtaining representation at tribunal hearings. Although a majority of those obtaining advice were represented at their hearing, in many cases appellants and applicants who had obtained advice nonetheless were unrepresented at their hearing. Interviews with appellants and applicants who attended their hearings indicate that a small proportion of those who attend their hearings unrepresented do so either from choice or from ignorance of the availability of advice. Among the majority who had sought advice and attempted to obtain representation, the failure to obtain representation was most often the result of lack of resources on their own part, or on the part of advice agencies.

Summary of main findings relating to outcome of mental health review tribunal hearings

1. Those patients detained under Section 2 of the Mental Health Act 1983 were the most likely to be discharged following a hearing. Restricted patients were those most likely to obtain some change in their situation, but this was most often limited to a recommendation for transfer.
2. Patients detained in special hospitals were less likely than patients detained in district hospitals to obtain a favourable outcome at their review hearing.
3. Patients with a criminal record were less likely to obtain a favourable outcome to their hearing, irrespective of the section under which they were detained.
4. The recommendation of the responsible medical officer ('RMO')

significantly affects the probability that a patient will receive a favourable decision at a hearing.
5. A report from an independent psychiatrist which disagrees with the recommendation of the RMO can substantially mitigate the effect of the RMO's recommendation, holding constant other factors.
6. The probability of a patient obtaining a favourable outcome at a review hearing is significantly increased where the patient is represented, holding constant factors such as the section under which the patient is detained and the recommendation of the RMO.

General summary and conclusion: the effect of representation on outcome of hearings

1. The information presented in this chapter has demonstrated conclusively that the presence of a representative significantly increases the probability that social security appellants, appellants before immigration adjudicators, industrial tribunal applicants and patients detained in mental hospitals will succeed with their cases at a tribunal hearing. This finding holds true when other measurable factors related to outcome are held constant.
2. In social security appeals the presence of a representative will increase the probability of success from 30% to 48%. In hearings before immigration adjudicators the presence of a representative will increase the probability of success from 20% to 38%. In mental health review tribunal hearings the presence of a representative will increase the probability of success from 20% to 35%. In industrial tribunal hearings, where the representation of both sides must be taken into account, the presence of a legal representative will increase the applicant's chance of success where the respondent is not represented from 30% to 48%. Where the respondent is legally-represented and the applicant is unrepresented, the applicant's probability of success is reduced to 10%. These relative increases in the probability of success have been calculated after taking into account all other observable influences on outcome.
3. The type of representation obtained by appellants and applicants has an effect on the probability of success. In social security appeals specialist representatives, such as welfare rights centres, tribunal units and law centres have the greatest effect on the probability of success. In immigration hearings, those represented by UKIAS, solicitors and barristers have a greater probability of success than those represented

by other advice agencies. In industrial tribunals, legal representation is of the greatest benefit to both applicants and respondents; and representation by a barrister results in the highest probability of success for either an applicant or a respondent.
4. Other factors independently associated with success were the type of case, number of witnesses, and in social security appeals, geographical location. In social security appeals, immigration hearings and industrial tribunals the identity of the chair or adjudicator was found to have a significant and independent effect which could either increase or reduce the probability of success.

Representation evidently increases the probability of a favourable outcome to a tribunal hearing. Assuming that, in the vast majority of cases, the favourable decision reached by the tribunal is correct in the light of the law and facts of the case, then representation can be said to be increasing the accuracy of tribunal decision-making processes.

If the objectives of tribunals are not simply to provide a quick cheap and accessible forum for the resolution of disputes, but include accurate and fair decision-making, the evidence of this chapter suggests that representation may be both desirable and necessary.

The ways in which representation contributes to the accuracy of decision-making, through preparation and presentation of tribunal cases, is the subject-matter of Part II of this report [see *The Effectiveness of Representation Before Tribunals*, Pt II].

Changing Patterns in Use of Judicial Review

M. SUNKIN, L. BRIDGES AND G. MAZEROS

It has become commonplace to observe that use of judicial review, as measured in case numbers, has grown considerably ever since the reforms of the late 1970s. The official statistics on judicial review (see Appendix 2 [see *Judicial Review in Perspective*, App. 2]) show that, between 1981 and 1992, the number of applications for leave made in each year increased over four-fold, from 558 to 2439. Growth has not been constant throughout this period, however. There were two years (1986 and 1988) when the number of applications for leave declined significantly (by 30% and 20% respectively) from the previous year. On the other hand, in 1987 the rate of growth, at 87%, was very considerably above the average for the period as a whole, and this was also the case in 1990 (35% growth). The most recent figures for 1992 show that the number of leave applications increased by 17% over 1991.[1]

Statistics relating to the number of applications for leave to apply for judicial review, however, can be a somewhat misleading indication both of the way judicial review is being used and the nature of the caseload. Research in the mid-1980s showed that large numbers of applications in particular subject areas tended to obscure the low levels of use of judicial review in many other fields.[2] Also, all judicial review cases begin with an application for *leave*, a procedure under which each case is subject to an initial vetting by a High Court judge to determine whether it is arguable and should be allowed to proceed to a full hearing and determination by the court.[3] Data relating solely to initial applications do not take account of the effects of the leave stage itself in regulating (whether by design or not) the overall flow of cases through the procedure. Such data, therefore, do not tell us very much about the nature of issues that are actually being reviewed by the courts.

The official statistics show that although the number of applications has risen over the past decade there has been a decline in the proportion of applications that have been granted leave to proceed to a full hearing.

During each of six years up to 1986, over two-thirds for applications for leave were granted. However, in 1987, the year of the dramatic 87% increase in applications, the rate of grant of leave fell below 60%, and with the exception of 1989 it has continued to decline since then. In the most recent year, the rate of grant of leave has fallen to 47%. As a result, while the number of applications for leave in 1992 was over 4.3 times higher than in 1981, the number of cases granted leave was only 3.1 times higher in 1992 than in 1981.

The official statistics also indicate that the number of judicial review cases going to a substantive hearing in each year has failed to keep pace even with the rate of increase in grants of leave. Following what appears to have been an exceptional year in 1991, when over 600 judicial review cases reached a final hearing (an increase of over 40% from the previous year), the number of cases heard in 1992 dropped back to just over 500. This latter figure was only 1.6 times greater than the number of cases determined in 1981. In other words, the rate of increase in the number of substantive hearings between 1981 and 1992 has been approximately half the rate of growth in cases granted leave and less than two-fifths the rate of increase in the number of initial applications for leave.

This widening gap between the number of cases determined each year and the numbers applying for and granted leave helps to explain the lengthening delays in the judicial review procedure. However, other factors may also be significant in this respect. For example, the official statistics for the last five years show that a significant number of applications for judicial review are being withdrawn, apparently even once leave has been granted. Indeed, in 1992 such withdrawals constituted no less than 42% of the judicial review cases disposed of from the Crown Office list. The issue of withdrawals will be considered further in Chapter 4 [see *Judicial Review in Perspective*, Ch. 4].

Subject areas of judicial review

As we have seen, the seemingly rapid growth in the number of applications for judicial review has attracted a good deal of judicial comment. However, such judicial comment, especially when addressed to applications in specific fields, has sometimes been inaccurate in the light of the empirical evidence. Sunkin, in his analysis of the pre-1987 judicial review caseload, was particularly critical of Lord Brightman's dicta in the 1986 case of *Puhlhoffer* v. *London Borough of Hillingdon*.[4] In delivering the unanimous judgment of the House, Lord Brightman said that the Housing

(Homeless Persons) Act 1977 had resulted in a 'mass of litigation' and in 'the prolific use of judicial review . . . [to challenge] . . . the performance by local authorities of their functions under the [Act]'. He expressed hope that 'there will be a lessening in the number of challenges . . . mounted against local authorities who are endeavouring in extremely difficult circumstances, to perform their duties' and called upon the High Court to grant leave in homelessness applications only if there were exceptional circumstances.[5]

Sunkin showed that despite these references to prolific use of judicial review the number of homeless persons' applications had never exceeded 75 in any one year (1983) and that in the year prior to *Puhlhoffer* there had only been 66 applications. Moreover, of these only six had been refused leave. The low refusal rate implied that the vast majority of these applications raised an arguable case of illegality. The numbers of applications may be compared with the approximately 100,000 unsuccessful applications to local authorities each year for accommodation under the relevant statute.

The *Puhlhoffer* decision had two immediate effects on the caseload. First, it reduced the known number of homelessness judicial review applications. These fell from 66 in 1985 to 32 in 1986. The second was to increase the failure rate at the leave stage, from less than 10% to over 30%.

Sunkin's earlier analysis also showed that the number of judicial review cases relating to immigration had a far more significant impact on the overall caseload during the early 1980s than did homeless person cases. Between 1931 and 1985 the number of immigration judicial review applications increased more than three-fold, to a point where such cases represented nearly three-fifths of the total civil judicial review caseload. In fact, once immigration cases were excluded from the analysis, Sunkin's figures indicated that use of judicial review in all other civil fields actually declined by nearly a third between 1984 and 1985, when statistics indicated an increase of 28% in the overall caseload. Equally, the fall off in the number of immigration judicial review applications between 1985 and 1986, from 516 to 409 (or 43.5% of all civil cases), helped to contribute to the general drop in applications shown in the official statistics for this latter year (see Appendix 2, Table A [see *Judicial Review in Perspective*, App. 2].

Turning to data drawn from our current research, Table 1.1 shows the number of applications for leave in each subject area in which there were 10 or more applications in any one full year. As will be seen, the 23 subject areas listed in the table account for between 91.3% and 94% of all judicial review applications in the periods covered by our data. More significantly, applications relating to just three areas—crime, immigration and housing—have dominated the use of judicial review throughout this time, accounting together for between 57% and 68% of all leave applications.

There have, nevertheless, been some significant shifts in the pattern of applications even in these major areas of use. These can best be seen in Table 1.2, which provides a simplified summary of the number of applications for leave over the study period. The number of applications for leave relating to immigration, which had previously declined between 1985 and 1986 (see above), rose sharply again in 1987, when such cases accounted for over 44% of all applications and over half of those relating to civil matters. In fact, 1987 represents the high point for the number of immigration judicial review applications, with a further sharp fall seen in

TABLE 1.1 *Applications for Leave to Seek Judicial Review by Subject Areas, 1987–1989 and 1st quarter of 1991*

	1987		1988		1989		1991 (Jan–Mar)	
	No	%	No	%	No	%	No	%
Criminal:	214	14.2	164	13.4	219	14.2	71	15.6
Civil:								
Immigration	671	44.4	356	29.1	430	27.7	103	22.7
Housing	141	9.3	161	13.2	232	15.0	108	23.8
Planning	59	3.9	84	6.9	132	8.5	17	3.7
Family	42	2.8	44	3.6	26	1.7	7	1.5
Discipline	36	2.3	27	2.2	18	1.2	6	1.3
Tax	29	1.9	17	1.4	41	2.6	6	1.3
Education	27	1.8	24	2.0	46	3.0	27	5.9
Legal Process	25	1.7	43	3.5	36	2.3	20	4.4
Local Govt. Affairs	25	1.7	37	3.0	38	2.5	17	3.7
Prisoners	17	1.1	25	2.0	16	1.0	5	1.1
Health	17	1.1	24	2.0	34	2.2	3	0.7
Environment	15	1.0	11	0.9	21	1.4	3	0.7
Employment	15	1.0	16	1.3	18	1.2	2	0.4
Rates	14	0.9	9	0.7	13	0.8	2	0.4
Agriculture	12	0.8	7	0.6	7	0.5	-	-
Transport	12	0.8	7	0.6	14	0.9	-	-
Legal aid	10	0.7	20	1.6	32	2.1	5	1.1
Coroners	10	0.7	9	0.7	15	1.0	3	0.7
Benefits	10	0.7	15	1.2	29	1.9	4	0.9
Trade	9	0.5	7	0.4	19	1.2	4	0.9
Compensation	6	0.4	6	0.8	11	0.9	9	2.0
Travellers	6	0.4	4	0.8	16	1.0	3	0.6
Other	90	6.0	107	8.7	123	7.9	29	6.4

TABLE 1.2 *Applications for Leave to Seek Judicial Review by Major Subject Areas, 1987–1989 and 1st quarter of 1991*

	1987		1988		1989		1991 (Jan–Mar)	
	No	%	No	%	No	%	No	%
Criminal:	214	14.2	164	13.4	219	14.2	71	15.6
Civil:								
Immigration	671	44.4	356	29.1	430	27.7	103	22.7
		(51.7)		(33.6)		(32.3)		(26.9)
Housing	141	9.3	161	13.2	232	15.0	108	23.8
		(10.9)		(15.2)		(17.4)		(28.2)
Other	486	32.1	543	44.4	669	43.2	172	37.9
		(37.4)		(51.2)		(50.3)		(44.9)

Figures in brackets show percentage of total civil applications in year represented by particular subject areas.

1988, and smaller increases in subsequent years. The official published statistics show that for the most recent three years, 1990–1992, the number of immigration leave applications has been fairly steady at between 500 and 570 per year.

While the number of immigration applications has therefore remained at a fairly even level, the proportion of the total judicial review caseload represented by immigration declined to below 30% in 1988 and 1989 and to less than a quarter in the first three months of 1991. The most recent official statistics for 1992 show that immigration applications constituted just 22.3% of the total judicial review caseload and 25.8% of all civil applications.

At the same time, the share of judicial review applications represented by housing has increased steadily throughout this period, rising from 9.3% in 1987 to 23.8% in the first three months of 1991. Again, this trend is confirmed by the latest official statistics which distinguish housing cases from other types of judicial review application for the final six months of 1992. In this period there were 239 housing applications or an estimated 20% of the total.

The third major area of judicial review litigation is applications arising from criminal court proceedings. Our data show that criminal matters account for a remarkably steady percentage of all leave applications, at between 13.4% and 15.6%. In fact, the official statistics do indicate that in 1990, a year not covered by our own data, there was a sharp rise in the number of applications relating to criminal matters, with such cases accounting for 29% of all applications in that one year. This was apparently the result of a surge of challenges relating to the legality of certain 'drink-drive' prosecutions. This appears to have been an exceptional occurrence however, with the overall proportion of leave applications relating to crime returning to 15% in 1991 and 13% in 1992.

Table 1.2 indicates that between 1987 and 1989 there was a 38% increase in the use of judicial review in areas other than immigration, housing and crime. Indeed, the 20% decline in the number of applications for leave between 1987 and 1988 noted earlier was entirely attributable to the fall in applications relating to immigration (–47%) and crime (–23%), while applications in housing rose by 14% and those for all other subject areas by 12%. The overall share of judicial review applications represented by subjects other than immigration, housing and crime rose from 32% in 1987 to 44% in 1988, and this share was maintained in 1989. However, by the first quarter of 1991 such applications accounted for just 38% of all cases, even though the actual number of 'other' applications in this period was roughly comparable to the number in 1989 (i.e. 172 applications in three months representing a rate of 688 per year, as compared to 669 in 1989).

Returning to the various individual subject areas listed in Table 1.1, it will be seen that there were only two other than immigration, housing or crime that accounted for more than 5% of leave applications in any of the periods covered. These were planning (including land) and education. The former represented 6.9% of applications in 1988 and 8.5% in 1989, but fell back to only 3.7% in the first quarter of 1991. Education accounted for 5.9% of applications in the first quarter of 1991. It might be speculated that the rate of applications for judicial review in planning would be sensitive to shifts in the housing and land development markets. On the other hand, the rise in judicial review applications in the field of education is likely to be explained by the 'politicisation' and 'judicialisation' of this field since the Education Reform Act 1988, with greater emphasis on schools 'opting out' of local authority control and on parental choice of schools for their children, and to particular activity in challenging local authorities over their assessments of children with special educational needs (or their failure to carry out such assessments).

Other less significant growth areas in terms of the rate of applications for judicial review since 1987 are 'legal process', local government affairs, legal aid (up to 1989) and travellers cases (especially in 1988). Applications relating to family law matters showed an increase in 1987 and 1988 (42 and 44 applications respectively) from earlier data recorded by Sunkin[6], but have fallen off since then. By contrast, prisoners' applications for judicial review, which accounted for 32, 42 and 64 cases in three six month samples in 1984, 1985 and 1986 respectively[7], declined sharply to only 17 applications throughout the whole of 1987 and have remained at a similar level since.

The 23 subject areas listed in Table 1.1 (and the more than 160 included in the coding scheme covering the total sample of applications) are testimony to the potential breath of judicial review. Yet, this presents one of the main contradictions to be explained about judicial review: given the potential scope of litigation and that the procedure is perceived as being reasonably accessible and the law dynamic, why are there so few challenges in many key areas of governmental activity affecting vital rights and interests of individuals and groups? The decline in challenges by prisoners since the mid-1980s has already been noted, but equally significant is the fact that the whole structure of welfare benefits, encompassing millions of individual decisions about citizens' entitlements each year, produced at most 29 applications for judicial review in any of the years covered by our current research. Similarly, while the 'politicisation' and 'judicialisation' of education in recent years may have led to a significant increase in judicial reviews in this area, the fields of health and environment, which have also been subject to major legislative activity and public controversy, also appear to have been hardly touched by judicial review. This suggests that however accessible judicial review may seem in purely legal and procedural terms, other factors such as the availability of funding or of expert legal advice and assistance (see Chapter 3 [see *Judicial Review in Perspective*, Ch. 3]) or the perceived formality of the process continue to provide major barriers to its use.

It is important to stress that the data presented relate solely to applications for leave. As already indicated, the judicial review 'funnel' narrows sharply at the point of leave, and there are interesting contrasts between the nature of the caseload in terms of subject areas before leave and the nature of the caseload after leave. This is an issue we will examine further in Chapter 4 [see *Judicial Review in Perspective*, Ch. 4]. Before that, however, we look in detail at the make-up of applications in the two main fields of judicial review activity associated with administrative law.[8]

Judicial review and immigration

As we have seen, immigration has remained the largest single area of use of judicial review throughout most of the past decade. However, in this period there have been some important shifts in the composition of the immigration caseload.

Judicial review usually comes into play in immigration cases in one of three ways. First, there are those situations in which individuals have rights to appeal from within the United Kingdom to an adjudicator and, with leave, to the Immigration Appeal Tribunal ('IAT'). These include cases in which someone is refused entry despite having obtained entry clearance and decisions affecting rights to remain in the United Kingdom after lawful entry has been obtained.[9] In this type of case judicial review may be used either to challenge the IAT's refusal to allow leave to appeal from an adjudicator or, where an appeal has been made, to challenge the ultimate decision of the IAT.

Secondly, there are those decisions in which there is no right of appeal from within the United Kingdom. These include most cases of refusal of entry clearance and refusal to grant entry to those who arrive without entry clearance, as well as appeals against removal of alleged illegal entrants.[10] Unlike the first category, individuals in this group can only appeal from outside the United Kingdom and in some circumstances the right of appeal can be 'virtually useless'.[11] Here judicial review is used either to challenge the immigration officer's decision prior to (or instead of) appealing or to challenge the IAT after an appeal from abroad has been made.

Use of judicial review prior to appealing became very common during the mid 1980s, particularly in 'genuine visitor' cases.[12] Indeed, 'genuine visitor' applications became numerically the most important single type of judicial review case during 1985, accounting for approximately 20% all civil judicial reviews during that year.[13] Mounting judicial concern that the judicial review procedure was being abused by applicants seeking to avoid the appeal procedure culminated in the Court of Appeal's decision in *ex parte Butt and Swati*.[4] In this the Court of Appeal said that however inconvenient it was to appeal, leave to seek judicial review prior to appealing would only be granted in exceptional cases. The *Swati* decision, as we shall see, had an immediate and expected impact on the use of judicial review in this area.

The third type of decision is that against which there is no right of appeal, either from within or beyond the United Kingdom. This includes refusals of entry (and entry clearance) on grounds that the Secretary of State is of the view that the exclusion is conducive to the public good, and deportations of those otherwise lawfully here on the ground that the deportation is in the interests of national security, diplomatic relations or for reasons of a political nature. Judicial review may be the only way of challenging these decisions.[15]

Unfortunately, the Crown Office records do not distinguish clearly between these three categories but simply refer to immigration applications by reference to their subject matter, for example, 'asylum', 'entry clearance', 'leave to remain', etc. It is these headings that we have adopted in Table 1.3. It will be seen from this table that asylum decisions generated the single largest area of immigration judicial review in three of the four periods covered by our research, although their numbers declined sharply from a peak in 1987. Up until 1985 there were approximately 20 asylum applications for leave each year, but in 1987 this jumped to over 300. This increase in litigation involving asylum in 1987 was largely due to the number of applications brought by Tamils seeking sanctuary in the United Kingdom from the civil war in Sri Lanka.[16] The known numbers of Tamil asylum applications in our sample were 258 in 1987, 23 in 1988, 4 in 1989 and 1 in the first three months of 1991. The scale of judicial review applications from this group no doubt reflected delays in the processing of applications for refugee status coupled with the high rates of refusal of this status by the Home Office.[17]

A similar, although numerically less significant, upsurge in litigation occurred in the late 1980s when large numbers of Kurdish refugees sought sanctuary in this country from repression in Iraq. The known numbers of judicial review applications from Kurds in our sample were 11 in 1987, 6 in 1988 and 55 in 1989.

The tremendous increase in asylum litigation echoes the growth in 'genuine visitor' cases during the mid-1980s. This is illustrated in diagram 1A, showing trends in both asylum and entry cases during the 1980s.[18] The peaks that can be seen in these two areas highlight the degree to

TABLE 1.3 *Immigration Applications for Leave to Seek Judicial Review by Type, 1987–1989 and 1st quarter of 1991*

Subject	1987	1988	1989	1991 (Jan–Mar)
Asylum	305	85	81	34
Entry clearance	107	94	57	19
Leave to enter	44	29	60	19
Leave to remain	18	24	13	5
Illegal entry	46	42	42	6
Deportation	24	20	35	11
Removals	24	7	6	1
Others and unclassified	103	55	136	8
Totals	671	356	430	13

which judicial review is susceptible to surges in applications involving very specific issues and indicates some of the difficulties faced by managers in predicting and reacting to caseload trends. At the same time, the figures also provide a warning against either the judiciary or the Government taking precipitate action to limit access to judicial review in specific fields, since 'bulges' in applications arising from particular issues may be relatively short-lived.

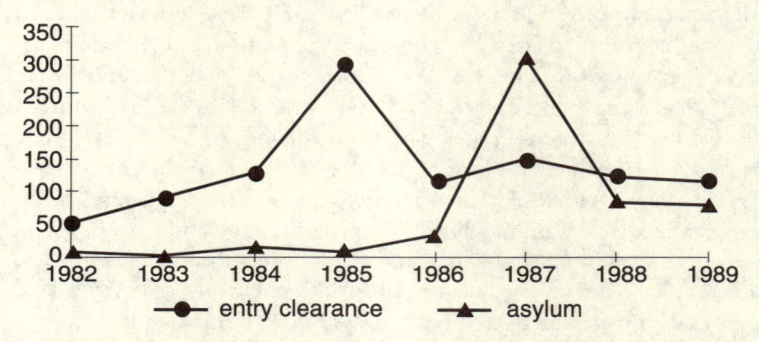

DIAGRAM 1A *Trends in Number of Applications for Leave to Seek Judicial Review in Immigration Entry Clearance and Asylum Cases, 1982–1989*

The Asylum Bill now before Parliament will give asylum-seekers new rights of appeal to adjudicators, a step that may divert cases out of judicial review. However, the extent to which the provision of alternative mechanisms for appeal will in practice lessen the number of applications for judicial review very much depends on the quality of these appeal rights and of the body adjudicating them. In this respect, the proposed 'fast-track' appeal which it is intended to be used in many asylum cases is bound to raise concern and may lead to an even greater resort to judicial review in this area.

It is already the case that in a considerable number of immigration judicial reviews the respondent is the Immigration Appeal Tribunal rather than the Home Office, thereby indicating that judicial review follows an administrative appeal. This is the case in the majority of 'entry clearance' judicial reviews, as shown in Table 1.4.

The 'leave to enter' cases concern decisions made in the United Kingdom by immigration officers. In some of these the individual will already have obtained entry clearance and will have exercised rights to appeal within the United Kingdom prior to seeking judicial review. In other cases the individual will have arrived at a port of entry without

TABLE 1.4 *Applications for Leave to Seek Judicial Review in Immigration 'Entry Clearance' Cases by Respondent, 1987–1989 and 1st quarter of 1991*

	1987		1988		1989		1991 (Jan–Mar)	
Respondent	No	%	No	%	No	%	No	%
Home Office	46	43.0	18	19.1	14	24.6	5	50.0
Immigration Appeal Tribunal	61	57.0	75	79.8	43	75.4	5	50.0
Others	–	–	1	1.1	–	–	–	–
Total	107		94		57		10	

entry clearance and will either have been refused entry and left to appeal from abroad prior to seeking judicial review, or they will have sought judicial review immediately instead of leaving the country (the *Swati* type situation). How many applications fall within these classes is difficult to determine from the records alone. If we look at the respondents we find that a relatively small (and declining) proportion of 'leave to enter' cases were known to be challenging decisions of the IAT. In 1987 there were 15 known cases of this type, but only four in 1988, five in 1989 and two in the first three months of 1991. By contrast, there were seven, none, eleven and five applications in each of these periods respectively involving challenges to refusals of 'leave to enter' where the respondent was the Home Office, i.e. the challenge was directly against an immigration officer's decision without an intervening appeal. In other words, there is some evidence of a small increase in the number of *Swati* type challenges during the study period, but the numbers were clearly very much lower than they were during the mid-1980s.[19]

As with the asylum cases, future trends in the use of judicial review in relation to 'entry clearance' and 'leave to enter' cases may well depend on the outcome of the current Asylum Bill. In this respect, the legislation was recently subject to powerful judicial criticism, not least over its proposal to deny overseas visitor applicants refused entry their present direct right of appeal to an adjudicator. This provision, which is intended to make more adjudicator time available for consideration of asylum cases (see above), is seen by judicial and legal commentators as risking a revival of judicial review as a 'mass remedy' in visitor cases.

Judicial review and homelessness

The inter-relationship between appeal rights, or the lack of them, and resort to judicial review also lays at the heart of the rapid increase in applications relating to housing over recent years. Table 1.5 shows the detailed make up of judicial review housing applications over the study periods.

TABLE 1.5 *Applications for Leave to Seek Judicial Review in Housing Cases by Type, 1987–1989 and 1st quarter of 1991*

	1987 No	1987 %	1988 No	1988 %	1989 No	1989 %	1991 (Jan–Mar) No	1991 (Jan–Mar) %
Homeless persons	84	59.6	105	65.2	176	75.9	77	72.6
Housing benefits	9	6.4	11	6.8	13	5.6	15	14.2
Compulsory purchase order	6	4.3	5	3.1	4	1.7	2	1.9
Rates	7	5.0	5	3.1	3	1.3	2	1.9
Repairs	8	5.7	2	1.2	1	0.4	–	–
Possession	–	–	1	0.6	5	2.2	3	2.8
Grants	2	1.4	1	0.6	8	3.4	1	0.9
Transfers	1	0.7	2	1.2	4	1.7	3	2.8
Squatting	1	0.7	12	7.5	6	2.6	-	
Other/unclassified	23	16.3	17	10.6	12	5.2	3	2.8
Totals	141		161		232		106	

% figures relate to total number of housing applications in the relevant year

This table clearly shows the extent to which use of judicial review in relation to housing issues is dominated by homeless person cases. Homelessness applications accounted for an increasing proportion of housing judicial reviews over the study period. At least up to 1989 the growing significance of housing in the overall judicial review caseload was entirely a function of the increase in homeless persons' applications.

Looking over a longer period, Diagram 1B shows the incidence of homelessness judicial review applications between 1981 and 1989.

The projected number of homelessness applications in 1991, based on our data for the first quarter of the year, would have been over 300, i.e. some 75% higher than in 1989. These figures clearly demonstrate that the House of Lord's decision in the *Puhlhoffer* case in 1986 (discussed above) had only a very limited and short-term impact in restricting the use of judicial review in this field.[20]

Despite these trends, it remains the case that the number of homeless-

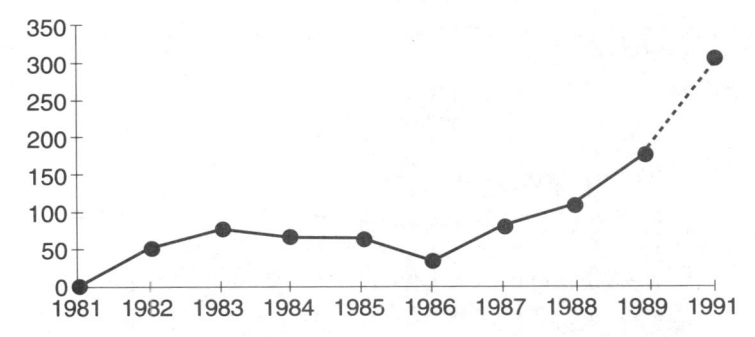

[Chart showing values approximately: 1981: 0, 1982: 50, 1983: 80, 1984: 70, 1985: 65, 1986: 40, 1987: 85, 1988: 110, 1989: 180, 1991: 310]

1991 figure based on projection of number of applications in January to March multiplied by four

DIAGRAM 1B *Trends in Number of Applicants for Leave to Seek Judicial Review in Homelessness Cases, 1981–1989 and 1st quarter of 1991*

ness judicial review applications is minute when placed against the background of the growing social crisis in this field and the very large numbers refused access to housing by local authorities under the legislation. Given the lack of an independent appeal right against local authority decisions on homelessness, applicants denied accommodation must rely on local authorities' own internal review procedures, or have immediate resort to judicial review. While the frequent use of judicial review in this area is perfectly understandable from the perspective of the applicants, the way in which some local authorities use the procedure as a substitute for tighter internal scrutiny of decisions and as an additional hurdle into accommodation requires much more critical attention from judicial and legal commentators than it has so far received.

The relatively low level of use of judicial review is also highlighted by the infrequency of challenges to public authorities' decisions across a wide range of housing rights and benefits. It may be noted that between 1987 and 1989 the number of housing judicial review applications relating to non-homelessness matters remained almost completely constant at 57 or 56 each year. Research has shown that a significant proportion of housing enquires handled by advice organisations such as Citizens' Advice Bureaux contain issues that are open to public law challenges but are not recognised as such.[21] A particular area where a higher rate of applications for judicial review might have been anticipated is in relation to housing benefit. Even more people are affected by decisions on housing benefit than on homelessness. Although a form of appeal is available to local councillors from decisions by local authorities to deny housing benefit, the independence and quality of these appeal rights have been questioned.[22]

Summary and conclusions

Much of the current drive toward the reform of judicial review appears to be based on a perceived overload of the procedure arising, it is said, from the vast increase in numbers of applications. Our data call into question, not so much the fact that the machinery of judicial review is under pressure, but rather the precise causes of these strains. The official statistics reveal a significant differential between the very rapid rate of growth over the past decade in the number of initial applications for judicial review and the much more modest increase in the number of cases reaching a final hearing. This suggests that intervening variables have an important role in regulating the flow of cases through the system.

Use of judicial review, at least in terms of numbers of applications, is dominated by two civil areas of use: immigration and housing. Together they account for around half of all applications. The rate of applications in respect of immigration has been subject to wide fluctuations and to surges in litigation around particular issues, such as 'entry clearance' or, more recently, asylum seeking. These surges in litigation in specific fields are difficult to predict, and for this reason they do not constitute a sensible basis on which to plan reform of the overall procedure. This is indicated by the fact that, during the period of our current research, immigration has actually been of declining significance as a source of judicial review applications.

Nor have past efforts by the judiciary to stem the flow of litigation in specific areas proved to have lasting effect. Despite the judicial sentiment expressed in *Puhlhoffer* in 1986, urging stricter control over the use of judicial review in relation to homelessness, this has in fact been the most significant growth area for judicial review over recent years. In this respect, judicial review provides a 'safety net' for asserting basic rights to fair treatment within the system of administrative decision-making where no other appeal mechanisms is available. At the same time, the continued frequent use of judicial review to challenge decisions of the immigration adjudicators and Immigration Appeal Tribunal may serve as a reminder that it is as much the perceived inadequacy of alternative appeal rights, and not just their lack of availability, that may produce a demand for judicial review.

Despite its growing importance in the overall judicial review caseload, the rate of homelessness applications remains low when compared with the much wider potential to challenge decisions in this and related housing fields. This is even more true of the other subject areas covered by judicial review which, while they theoretically span the whole scope of governmental and other public decision-making, tend to produce relatively few applications. In the following two chapters [see *Judicial Review in Perspective*] we examine in more detail precisely who it is that uses judicial review at present and against which public bodies; and how patterns of use may be affected by the availability of different forms of legal assistance.

* * * * *

Our findings on the progress of applications through the judicial review procedure and the different stages at which cases are concluded have significant implications in a number of areas. Legal commentary on judicial review tends to focus on reported decisions of cases at full hearing stage or subsequently in the Court of Appeal or House of Lords. Yet our research shows that it is only a minority of applications that reach a full hearing and that, in numerical terms at least, decisions taken at the leave stage and the processes of settlement and/or withdrawal either side of leave are far more important. This points to the need to pay much closer attention to the nature of the decision-making process at the leave stage and to the quality of such decisions, as well as for the reasons for withdrawal. We address some of these issues in the following two chapters [see *Judicial Review in Perspective*].

Our findings also have implications for debates on the growing judicial review caseload and the issue of lengthening waiting times. Most discussion of these points assume a direct relationship between the number and type of applications initially made for judicial review and the delay that occurs at later stages of the process.

However, the levels of settlement of applications either side of the leave decision, when combined with refusals of leave, mean that the majority of applications are not directly affected by these delays. Moreover, because refusals of leave or withdrawals occur much more frequently in immigration and homeless person applications, both areas where judicial review has been used as a 'mass' individual remedy, it is arguable that cases in those fields do not have the impact on overall delays in the procedure that is often assumed from their numbers in the initial caseload. By the same token, steps taken to divert immigration or homelessness cases from judicial review might not lead to a proportionate reduction in delays in bringing on the remainder of cases for hearing

* * * * *

Our purpose in this chapter has been to explore the various factors influencing the nature and consistency of decision-making at the leave stage. It is hardly surprising that so much of the current discussion on possible further reform of judicial review has focused on the leave requirement. The leave filter is increasingly being seen, at least implicitly, as a means of limiting caseload pressures on the system, and some judges now appear to consider administrative efficiency of the judicial review process as a legitimate factor to be taken into account in deciding whether or not to grant leave to particular applicants.

The leave requirement is likely to be perceived by applicants as a confusing and difficult barrier to be surmounted in order to gain access to the court (or to a reasonable settlement of their disputes with public authorities). Certainly, lawyers are confronted with growing difficulties in advising prospective applicants as to the precise criteria for being granted leave to apply for judicial review. In this situation there is some irony in the suggestion that these same lawyers are at least partly to blame for increased rates of refusal of leave, because of their supposed lack of expertise or poor preparation and vetting of applications. Although representation of applicants for judicial review is becoming more fragmented among solicitors, there is nothing in our research to indicate that poor quality of legal advice and preparation is a major contributory factor to higher leave refusal rates.

On the other hand, the difficulties in obtaining leave may be compounded by the actions of government and other respondents. We have noted that the current proliferation of 'alternative remedies' is likely to have a significant impact on the future scope and use of judicial review. The present high levels of post-leave settlement of judicial review actions must also be a matter of concern. It suggests, in particular, that some respondents may be using the leave filter for their own administrative purposes, as a means of minimising the occasions on which they will be required to undertake internal reviews and engage in serious negotiations to settle disputes with aggrieved citizens. There is a risk that any further tightening up of the leave criteria or procedural reforms to give respondents more opportunities to intervene in court proceedings at the leave stage will only increase the tendency for them to delay settlement in appropriate cases.

Other factors, inherent in the administration of judicial review, lead to variations in decisions and add to the uncertainties surrounding the leave stage. The establishment of a relatively small cadre of nominated Crown Office list judges was intended to encourage expertise and consistency in decision-making. The growth in the use of the procedure has led, however, to a need to draw on an increasingly wide band of judges, from both an expanded list of nominated judges and elsewhere in the High Court, to hear applications. There are also very considerable differences in the frequency with which the nominated judges handle judicial review cases, with some individual judges appearing very regularly and thereby having a disproportionate influence on the process.

This in turn has led us to consider the patterns of individual judges' decisions on leave applications. Our data show that there are very wide variations between the grant/refusal rates of individual judges. These differences cannot as yet be explained by such factors as the type of procedure (table or oral application) or subject matter of the application. In fact, although we need to know more about how cases within the Crown Office list are allocated to particular judges, the fact that there is a high degree of consistency in terms of each individual judge's decisions, both over time and across subject matters, suggests that attitudinal factors have an important bearing on the results of leave applications.

It has certainly not been our aim to label particular judges as 'liberal' or 'conservative' but rather to indicate how apparent inconsistency in judicial approaches to leave decisions contributes to the general unpredictability of the process. It would appear that obtaining leave to apply for judicial review is something of a lottery, and our data point to the need to improve levels of consistency in leave decisions and to confine and structure the discretion of the judges in this respect.[23] If steps were to be taken to 'tighten up' on leave and to shift it more toward being a contested, *inter partes* stage in the process, without corresponding measures to clarify the rationale for decisions on leave, there would be a substantial risk that access to judicial review would become so unpredictable as to discredit the whole procedure in the eyes of the public.

Our findings point instead to a different strategy for reform. In addition to the need to make the criteria for granting leave more transparent, steps are required to ensure that fuller information on the reasons for decisions by public authorities are made available to applicants, their legal advisers, and judges at a much earlier stage. This may require both that more time should be allowed before formal applications for leave are required to be lodged and that specific pre-leave procedural mechanisms be introduced to provide applicants with access to information held by prospective respondents. In addition to making leave decisions more open and better informed, there is also the possibility that these reforms would serve to encourage more pre-leave settlements and in that way reduce the caseload pressures which are now becoming increasingly apparent at the leave stage.

Notes

1. Growth in the number of applications has continued into 1993. In the twelve months up to the end of March 1993, 2,682 applications for judicial review were lodged at the Crown Office.
2. Sunkin, 1987.
3. The need to obtain leave even in order to have an initial judicial determination of a case is a relatively unique feature of the judicial review procedure in England and Wales. A requirement of leave is much more common in appellate jurisdictions. Judicial review does not involve an appeal as such, although as its name implies, it does entail a review of decisions made by inferior courts and tribunals or other public bodies. It is notable that the parallel procedure for judicial review in Scotland does not include a preliminary application for leave, while in Northern Ireland leave is required but may in most circumstances be granted by a Master rather than a High Court judge.
4. [1986] 1 All E.R., 467.
5. Ibid., at 469, 474.
6. Sunkin, 1987, p. 441.
7. Ibid.
8. Time has not allowed a detailed examination of criminal cases, as the other main category of judicial review, at this stage.
9. It will include refusals to grant asylum to those who possess a visa.
10. It will also include refusals to grant asylum to those not in possession of the necessary entry documents.
11. Per Lord Bridge in respect of appeals under s. 16 of the Immigration Act 1971 (illegal entrants). *Khawaja* v. *Secretary of State for Home Affairs* [1983] 1 All E.R. 765, at 786.
12. Refusal to allow entry on the grounds that the immigration officer was not satisfied that the individual was seeking entry as a genuine visitor under Rule 17 of the Statement of Changes in Immigration Rules (H.C. 169, 9 February 1983).
13. Sunkin, 1987, at p. 446.
14. [1986] 1 All E.R. 717. See further, Sunkin, 'Trends in the Use of Judicial Review Before and After Swati and Puhlhoffer' (1987) *New Law Journal* 731.
15. Or *habeas corpus*. See, for example, *R* v. *Secretary of State ex parte Cheblak* [1991] 1 WLR 890.

16. See further Sunkin, 1991. As noted in the Introduction the figures shown here vary slightly from those given in the earlier paper.
17. Ibid.
18. For the purposes of the diagram 'entry cases' includes both 'entry clearance' and 'leave to enter'.
19. The Court of Appeal continues to impose a tight rein on the leave criteria to ensure compliance by the nominated judges with *ex parte Swati*. See, for example, Lord Justice Woolf's criticism of Paul Kennedy J's approach in R v. *Secretary of State for the Home Department ex parte Doorga* [1990] C.O.D. 109.
20. As will be seen subsequently, the House of Lord's decision also appears to have had little effect in influencing Crown Office List judges toward refusing leave in more homelessness cases (see Chapters 4 & 5 [see *Judicial Review in Perspective*, Chs. 4, 5]).
21. D. Forbes and S. Wright, *Housing Cases in Nine CABx*, unpublished, 1990.
22. See R. Sainsbury and T. Eardley, 'Housing Benefit Review Boards: A Case for Slum Clearance?' *Public Law*, Winter 1992, p. 551.
23. See LeSueur and Sunkin, 1992, pp. 127–129 for particular proposals in this respect.

References

LeSueur, A. P. and Sunkin, M. (1992) 'Applications for Judicial Review: The Requirement of Leave' in *Public Law* 102.

Sunkin, M. (1987) 'What is Happening to Applications for Judicial Review?' in 50 *Modern Law Review* 432.

—— (1991) 'The Judicial Review Case-load 1987–89' in *Public Law* 490.